T0269184

CAMBRIDGE LIBRARY COLLECTION

Books of enduring scholarly value

Earth Sciences

In the nineteenth century, geology emerged as a distinct academic discipline. It pointed the way towards the theory of evolution, as scientists including Gideon Mantell, Adam Sedgwick, Charles Lyell and Roderick Murchison began to use the evidence of minerals, rock formations and fossils to demonstrate that the earth was older by millions of years than the conventional, Bible-based wisdom had supposed. They argued convincingly that the climate, flora and fauna of the distant past could be deduced from geological evidence. Volcanic activity, the formation of mountains, and the action of glaciers and rivers, tides and ocean currents also became better understood. This series includes landmark publications by pioneers of the modern earth sciences, who advanced the scientific understanding of our planet and the processes by which it is constantly re-shaped.

Volcanos

While a student, George Poulett Scrope (1797–1876) visited Vesuvius and Etna and developed a passionate enthusiasm for volcanos. He did pioneering fieldwork in France in 1821, witnessed the eruption of Vesuvius in 1822, and was elected a Fellow of the Royal Society in 1826. Scrope became increasingly involved in economics and politics, but later in his career published revised versions of two pioneering books on volcanism he had originally published in the 1820s. *Volcanos* (1862), reissued here, was based on his *Considerations on Volcanos* (1825, also reissued in this series) and dedicated to his life-long friend and colleague Charles Lyell. This influential work on volcanic phenomena includes a substantial catalogue of 'all known volcanos and volcanic formations' as well as a dramatic illustration of Vesuvius. It was translated into French and German, went into a second English edition in 1872, and was one of the foundational texts of volcanology.

Cambridge University Press has long been a pioneer in the reissuing of out-of-print titles from its own backlist, producing digital reprints of books that are still sought after by scholars and students but could not be reprinted economically using traditional technology. The Cambridge Library Collection extends this activity to a wider range of books which are still of importance to researchers and professionals, either for the source material they contain, or as landmarks in the history of their academic discipline.

Drawing from the world-renowned collections in the Cambridge University Library, and guided by the advice of experts in each subject area, Cambridge University Press is using state-of-the-art scanning machines in its own Printing House to capture the content of each book selected for inclusion. The files are processed to give a consistently clear, crisp image, and the books finished to the high quality standard for which the Press is recognised around the world. The latest print-on-demand technology ensures that the books will remain available indefinitely, and that orders for single or multiple copies can quickly be supplied.

The Cambridge Library Collection will bring back to life books of enduring scholarly value (including out-of-copyright works originally issued by other publishers) across a wide range of disciplines in the humanities and social sciences and in science and technology.

Volcanos

The Character of their Phenomena, their Share in the Structure and Composition of the Surface of the Globe, and their Relation to its Internal Forces

GEORGE POULETT SCROPE

CAMBRIDGE
UNIVERSITY PRESS

CAMBRIDGE UNIVERSITY PRESS

Cambridge, New York, Melbourne, Madrid, Cape Town,
Singapore, São Paolo, Delhi, Tokyo, Mexico City

Published in the United States of America by Cambridge University Press, New York

www.cambridge.org
Information on this title: www.cambridge.org/9781108072533

© in this compilation Cambridge University Press 2011

This edition first published 1862
This digitally printed version 2011

ISBN 978-1-108-07253-3 Paperback

ERUPTION OF VESUVIUS,
as seen from Naples.
October, 1822

ERUPTION DU VESUVE,
Vue de Naples,
Octobre, 1822.

VOLCANOS.

THE

CHARACTER OF THEIR PHENOMENA,

THEIR

SHARE IN THE STRUCTURE AND COMPOSITION
OF THE SURFACE OF THE GLOBE,

AND

THEIR RELATION TO ITS INTERNAL FORCES.

WITH

A DESCRIPTIVE CATALOGUE OF ALL KNOWN
VOLCANOS AND VOLCANIC FORMATIONS.

BY

G. POULETT SCROPE, M.P., F.R.S., F.G.S.,

MEMB. ROY. ACAD. NAPLES, ETC.

SECOND EDITION, REVISED AND ENLARGED.

WITH A MAP OF THE VOLCANIC AREAS OF THE GLOBE, FRONTISPIECE,
WOODCUTS, ETC.

LONDON:
LONGMAN, GREEN, LONGMANS, AND ROBERTS.
1862.

PRINTED BY TAYLOR AND FRANCIS, RED LION COURT, FLEET STREET.

Dedication.

SIR CHARLES LYELL, F.R.S., V.P.G.S.,
ETC.

My dear Lyell,

WHEN the First Edition of this Work saw the light, now seven-and-thirty years ago, I had the honour of being your Colleague in the Secretaryship of the Geological Society.

At that time you expressed a warmer interest in, and more agreement with, the views it contained than they met with from the bulk of our Associates. It was an attempt to investigate one important class of the Agencies of Change now in operation on the earth's surface, and to trace their analogy, or rather identity, with those which have apparently prevailed through earlier geological periods—a portion, in fact, of the great task at which you have so long laboured, as respects the entire range of terrestrial phenomena, with an originality, persistence, and success that have placed you, by common consent, at the head of the followers of the Science.

I was led into other paths for a time ; and was, perhaps, only induced to recur to the studies which we commenced together so long ago, by a wish to aid you in dispelling that signal delusion, as to the mode of action of the subtelluric forces, with which the Elevation-crater theory had, during the intervening period, mystified the Geological world. Once re-embarked in these speculations, I found (or thought I found) a want existing of some general treatise on Volcanic action of the character of my former work—long since out of print, and in some respects out of date. And thus originated the present (I hope improved) Edition.

In asking permission to inscribe your name upon its first leaf, I by no means propose to claim you as sponsor for the opinions it contains (which must rest, of course, upon their own merits), but only to be allowed to record my admiration of the perseverance with which you have so long devoted your high powers to investigating, and giving to the public in a popular form, the great truths of our favourite branch of Natural Philosophy.

I remain, dear Sir Charles,

Yours very sincerely,

G. POULETT SCROPE.

1 Prince's Gate, London,
March 10th, 1862.

CONTENTS.

CHAPTER I.

INTRODUCTORY.

Page

Importance of study of subterrestrial forces.—Plutonic action distinguished from Volcanic.—The last the subject of the first edition of this work, published in 1825.—Its chief object to show the laws of volcanic action to be simple, uniform, and consistent 1

CHAPTER II.

GENERAL VIEW OF VOLCANIC ACTION.

§ 1. Its definition.—§ 2. Connexion of earthquakes and volcanic eruptions—their interdependence and common origin.—§ 3. Locality of volcanos.—§ 4. Number at present in activity.— § 5. Their geographical position.—Extraordinary band of vents circumscribing the Pacific Ocean.—§ 6. Volcanos sub-aërial or subaqueous.—§ 7. Their activity generally irregular. —Assume different phases 5

CHAPTER III.

PHENOMENA OF ORDINARY SUBAËRIAL ERUPTION.

§ 1. Permanently eruptive phase.—§ 2. Phase of moderate in-termittent activity—examples.—§ 3. Phase of prolonged in-termittences and paroxysmal eruptions.—§ 4. General simi-larity of these phenomena.—§ 5. Description of them.—Earth-quakes.—Explosive bursts of steam.—Column of vapour and ejected fragments.—Rise of lava in vent, and outflow.—Crisis, and discontinuance of eruption.—Examples of such paroxysms. § 6. Succeeding intervals of repose.—Crater in state of solfatara —examples.—§ 7. These phases often pass into one another. 16

CHAPTER IV.

EXAMINATION OF VOLCANIC PHENOMENA.

Page

The main agent the igneous ebullition of subterranean lava
—shown in Stromboli and other permanently active volcanos.—Boiling lava present in every eruption.—Cause of
its ebullition the water it contains.—Heated water or
steam discharged always.—Equilibrium of opposing forces of
expansion and repression.—Rupture of overlying rocks.—At
what depth ebullition takes place.—Changes of volume in
subterranean mineral matter through changes in temperature or pressure.—Law as to formation of fissures.—Rocks
usually crack in straight planes.—Two orders of fissures—
their direction and character.—Injection of downward-opening
fissures.—Dykes.—Rise of lava through fissures—its explosive ebullition occasions the circular or elliptic form of craters.   30

CHAPTER V.

DISPOSITION OF THE FRAGMENTARY EJECTIONS.

§ 1. Mode and force of projection.—Character of ejected matters
—volcanic bombs, scoriæ, pumice, lapillo, puzzolana, ash.—
§ 2. Circular bank or cone formed round an eruptive vent.—
Compound cones, ridges, and strings of cones—examples.—
§ 3. Internal structure of cones.—Dip of inner beds towards
the vent, of outer beds the reverse.—Modifications of form—
their causes   55

CHAPTER VI.

OUTFLOW AND DISPOSITION OF LAVA.

§ 1. Cones breached by expulsion of lava—their durability.—
§ 2. Varying fluidity of lavas.—§ 3. Aspect of flowing lava.
—Rapid consolidation and fissuring of surface.—Mode of progression of lava-streams.—§ 4. Figures assumed by viscous
lavas—some coagulate vertically.—Domes and cupolas of lava.
—§ 5. Structure of minor lava-cones.—Connexion of dykes
with lava-currents.—§ 6. Subsidence of their surfaces.—
Rifts or longitudinal rents—arched gutters, caverns—bubble-
shaped domes on fumaroles.—§ 7. Vast volume of some

Page

lavas.—§ 8. Progress of lava-stream—its high temperature —its small momentum.—Influence of obstacles—trees, &c., enveloped by it.—§ 9. Its metamorphic effects on other rocks. —Lava-filled fissures form dykes.—Lava flowing over wet ground or into water.—§ 10. Consolidation of lava-currents.— Assumed divisionary structure.—§ 11. Columnar, how caused. —Joints, ball and socket, &c.—§ 12. Globiform structure.— Compound prismatic division.—§ 13. Concentric structure.— Pearlstone.—§ 14. Tabular and slaty structure, its cause.— § 15. Varieties of divisionary structure 65

CHAPTER VII.

MINERAL CHARACTERS AND COMPOSITION OF LAVAS.

§ 1. Their general division, according to prevailing mineral characters, into, 1. Trachyte, 2. Greystone, 3. Basalt.—§ 2. Older lavas (traps), their different characters.—§ 3. Varieties of texture.—§ 4. How occasioned.—Lavas not always erupted in perfect fusion—owe their liquidity in part to interstitial water or steam.—The crystals partly formed before emission —completed during consolidation.—Analogy to sugars.—§ 5. Lavas altered within or below the vent.—Separation of different minerals—consequent alternate eruption of lavas of different mineral characters—examples.—Mineral character no test of age.—§ 6. Differences of texture influence their fluidity.—§ 7. Likewise of specific gravity.—§ 8. Trachytic domes and bosses—their formation and structure.—Why clinkstone affects pyramidal forms.—Massive beds of trachyte.—§ 9. Vesicular structure.—Elongation of vesicles.— Laminar and ribboned structure of lavas.—§ 10. Brecciated lavas.—§ 11. Metamorphic influence.—New minerals formed by action of heat and vapour on lavas.—§ 12. Amygdaloidal lava.—§ 13. Solfataras.—§ 14. Hot-springs.—Theory of the Geysers.—§ 15. Mephitic emanations.—§ 16. Slow cooling of lava 110

CHAPTER VIII.

VOLCANIC MOUNTAINS.

§ 1. Mountains built up by accumulated products of repeated eruptions.—§ 2. Single cone enlarged by subsequent ejec-

Page

tions and lava-currents.—§ 3. Rents in the mountain's flank
usually radial.—Lateral or parasitic cones formed in strings
upon such rents—examples.—§ 4. Earthquakes caused by
these ruptures.—§ 5. Mountain sometimes cleft in two.—
Barancos.—§ 6. Dykes, their number and extent—cause
little disturbance.—§ 7. Partial distension of cone.—Growth
and strengthening of volcanic mountains—their height.—
§ 8. Parasitic cones.—§ 9. Disposition of loose ejecta.—
§ 10. Influence of water-floods, melting of snows.—§ 11.
Bursting of lakes.—§ 12. Mud-eruptions—Moya, trass, &c.—
Infusoria.—§ 13. Formation of tuff-conglomerates.—§ 14.
Their varieties.—Mixture with sedimentary calcareous matter, &c.—Altered tuffs, wackes, &c.—Tuff-cones—their origin.
§ 15. Structure of a volcanic mountain—its generally conical
figure 155

CHAPTER IX.

ON CRATERS OF VOLCANIC MOUNTAINS.

§ 1. Craters formed by explosions—their general character.—
§ 2. Examples from Vesuvius and Etna—recent changes in
their figure.—Craters alternately opened and refilled.—§ 3.
Truncated cones.—§ 4. Concentric craters—cone within
cone.—Shifting of vents.—§ 5. Paroxysmal explosive eruptions—examples.—Ancient crater-rings encircling recent
volcanos—how produced—examples: Santorini, Teneriffe,
&c.—§ 6. Vast area of some craters—their necessarily explosive origin.—Amount of fragmentary ejecta.—Not formed
by engulfment, but by expulsion of matter.—§ 7. Often
enlarged by denudation.—Vast degradation of insular volcanos.—§ 8. Origin of Barancos.—§ 9. Crater-lakes—proofs
of their explosive origin.—§ 10. Pit-craters—their probable
origin.—§ 11. Craters of Kilauea and Mauna Loa.—Tapping of
a crater.—§ 12. Many crater-lakes probably tapped.—§ 13.
Occasional cases of subsidence.—§ 14. Shifting of volcanic
axial vents.—Deserted craters become solfataras.—§ 15.
Craters of the Moon—their lava-streams, &c. 184

CHAPTER X.

SUBAQUEOUS VOLCANOS.

Page

§ 1. Rarity of opportunities for their observation.—§ 2. Instances.—The phenomena observed compared with subaërial eruptions.—§ 3. Depth of water must modify them.—§ 4. Saline emanations.—§ 5. Disposition of products.—Fragmentary rocks.—Origin of stratified tuffs, indurated and arenaceous.—§ 6. Conduct of lava issuing from submarine vent—amygdaloidal.—§ 7. Examples.—Early submarine eruptions. —§ 8. Occasional elevation in mass of volcanic rocks.—§ 9. Subaërial and subaqueous volcanic formations contrasted—examples.—Insular volcanic mountains in part upheaved, in part erupted subaërially—examples of such in frequent connexion.—Coral islands often based on lava-rocks 235

CHAPTER XI.

SYSTEMS OF VOLCANOS.

§ 1. Frequent allineation of eruptive vents along lines of fissure. —§ 2. Result of a general law.—Neighbouring fissures usually parallel or transverse.—Contiguous vents relieve one another. —§ 3. Nevertheless are in some degree independent—examples in proof.—Fluid connexion may be suspended between contiguous vents or eruptive fissures, and yet heat pass from one to the other by convection or conduction.—§ 4. Theory of a fluid nucleus to the globe, or even a continuous fluid envelope, doubtful.—Heat may pass by conduction between adjoining foci or fissures.—§ 5. Also vertically as well as horizontally.—§ 6. The intervals between eruptive fissures experience changes of level 258

CHAPTER XII.

RELATION OF PLUTONIC TO VOLCANIC ACTION.

§ 1. Effect of variations of temperature on subterranean mineral matter.—§ 2. Elevation and fissuring of surface-rocks.—Eruptions and subsidence—examples of subsidence near volcanic vents.—Cause of oscillations of superficial levels the varying upward flow of heat.—§ 3. This relation between plutonic and

Page

volcanic action shown in the general parallelism of coast-lines
or axial mountain-ranges with neighbouring volcanic bands
—shown by examples to be a general law.—§ 4. Subject to
irregularities—examples.—§ 5. These caused by locally-
varying resistances.—§ 6. Structure of plutonic axes.—§ 7.
Nature of granitic magma.—§ 8. The upper layers laminated and crumpled by pressure and movement, *i. e.* friction,
during their upheaval.—§ 9. Repeated axial upthrusts.—
§ 10. Their effect on the overlying strata.—Corrugation of
strata in parallel folds to axis.—§ 11. Zigzag foldings of
schists.—Effects of lateral thrust on strata.—Theories of Pro-
fessors Hopkins and Rogers.—§ 12. Extension and crumpling
of strata.—§ 13. Some not bent, but broken.—Mountain-
ranges generally most disturbed.—§ 14. Earthquakes—objections to Mr. Mallet's theory.—§ 15. Progression of plutonic
action.—§ 16. Metamorphism.—§ 17. Foliation of gneiss not
stratification.—§ 18. The determining cause of plutonic effort,
tidal action on an elastic substratum.—§ 19. The diagonal
dislocations of the earth's crust caused by a sudden tidal
wave(?).—§ 20. Question whether the rate of change is uniform or diminishing.—General Conclusions on Telluric Phenomena 270

APPENDIX.

DESCRIPTIVE CATALOGUE OF VOLCANOS AND VOLCANIC FORMATIONS.

Prefatory Remarks 311
Volcanic Formations of Europe :—
Southern Italy.—Vesuvius—Neighbourhood of Naples—Ischia
—Ponza Isles—Mount Vultur—Rocca Monfina—Lipari Isles
—Etna—North Africa—Sardinia—Corsica 315
Central and Northern Italy.—Alban Hills—Monti Cimini—
Monte Amiata—the Euganean and Vicentine Hills—Tyrol . 350
Spain and Portugal—Southern and Central France—Rhine vol-
canos—the Eifel—Volcanic band of Northern and Central
Germany—Hungary—Styria 358

Page

The Levant.—Thrace—Greece—the Troad—the Ægean—Asia
Minor—Syria—the Red Sea and Eastern Africa—Armenia—
the Caucasus and Crimea 390

Asia.—The Ural—Persia—Himalayas—Hindostan—Tartary
—China 403

Volcanos of the Eastern Atlantic.—Iceland—the Ferroe and
British Isles—the Azores—Madeira—Canary and Cape Verde
Isles—Ascension—St. Helena 405

The Southern Ocean.—St. Paul—Madagascar—Bourbon—the
Mauritius. 428

Western Atlantic.—Caribbean Antilles 430

South America, or Eastern Pacific Volcanos.—Tierra del Fuego
—Patagonia—Coast of Chili—of Bolivia—of Peru—Choco
range 434

Central America.—Nicaragua—Guatemala—Mexico 444

North America.—Sierra Madre—Rocky Mountains—California
—Sierra Nevada—British Columbia—Russian America—
Aleutian Isles 452

Western Pacific.—Kamtschatka—the Kuriles—Japanese, Loochoo, and Philippine Isles—Borneo—Celebes—Flores—Timor
—Sumbawa—Java—Sumatra—Andaman Isles 456

Southern and Central Pacific.—Carolines—New Guinea—New
Hebrides—New Zealand—Australia—Friendly and Society
Isles—Marquesas, Marianne, and Sandwich Isles.—Coral
islands—probably occupy subsiding areas, but frequently
based on volcanic rock.—Conclusion, as to simplicity and
uniformity of volcanic action 470

CORRIGENDA.

Page 29, line 2, *for* 1765–70 *read* 1765–75.
  ,, 37 ,, 17, *after* the *omit* sufficient.
  ,, 46 ,, 11, *after* temperature *add* or diminution of pressure.
  ,, 59 ,, 16, *for* have *read* has.
  ,, 69 ,,   1, *for* in *read* of.
  ,, 74 ,,   2, *for* (fig. 14) *read* (fig. 13).
  ,, 74 ,,   4, *after* activity *add* (fig. 14).
  ,, 75 ,, 17, *after* Etna *add* (fig. 15).
  ,, 75, last line but one, *omit* last.
  ., 78, line 5, *for* no doubt *read* probably.
  ,, 107 ,, 21, *for* more *read* less.
  ,, 107 ,, 22, *after* than *insert* in accordance with.
  ,, 112 ,,   9, *for* overlies *read* underlies.
  ,, 112, last line but two, *before* their *insert* to.
  ,, 129, line 16, *for* volcanos *read* volcano.
  ,, 161 ,, 23, *for* 1784 *read* 1760.
  ,, 161 ,, 24, *for* eastern *read* southern.
  ,, 161 ,, 27, *for* Greco *read* Annunziata.
  ,, 170 ,,   7, *for* 1784 *read* 1760.
  ,, 199 ,,   1, *for* Mexico *read* the Andes of Quito.
  ,, 303 ,,   6, *for* Baffin's Bay *read* Hudson's Straits.
  ,, 323 ,,   2, *for* Horace *read* Juvenal.
  ,, 328 ,,   3, *for* 500 *read* 300.
  ,, 335 ,, 32, *for* composing *read* so as to compose.
  ., 336 ,, 28, *omit* all *before* that.
  ., 336 ,, 33, *omit* probably.
  ,, 341 ,, 18, *before* scoriæ *insert* vapour and.
  ,, 346 ., 25, *for* the earliest *read* an early, *and after* period *insert*
                                  down.
  ,, 347 ,, 11, *omit* these.
  ,, 401 ,, 18, *after* and *insert* some.
  ,, 428 ,, 28, *for* Moreau de Jonnés *read* Bory de St. Vincent.
  ,, 428 ,, 29, *for* 1778 *read* 1804.

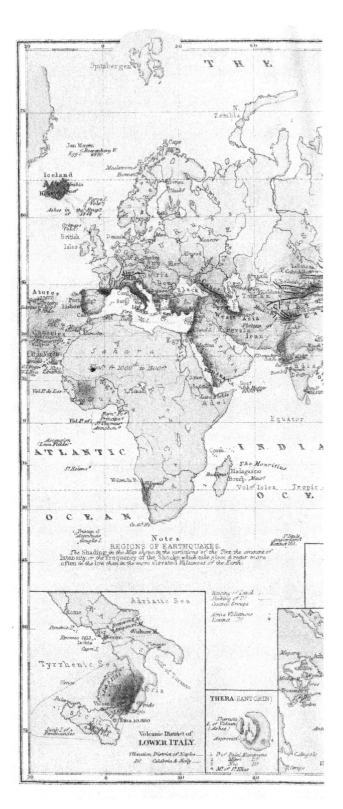

REGIONS OF EARTHQUAKES.

The Shading in the Map shews in the variations of the Tint the amount of
Intensity, or the Frequency of the Shocks; which take place & recur more
often in the low than in the more elevated Volcanoes of the Earth.

Volcanic District of
LOWER ITALY.

THERA (SANTORIN)

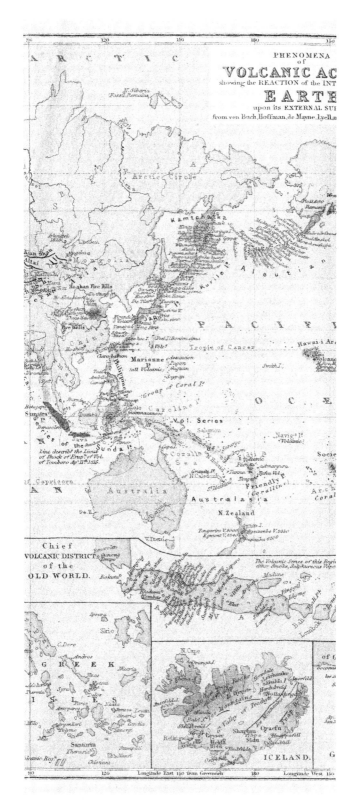

PHENOMENA
of
VOLCANIC AC[
showing the REACTION of the INT[
EARTH
upon its EXTERNAL SU[
from von Buch, Hoffman, de Mayne, Lyell, a[

Chief
VOLCANIC DISTRICT
of the
OLD WORLD.

ICELAND.

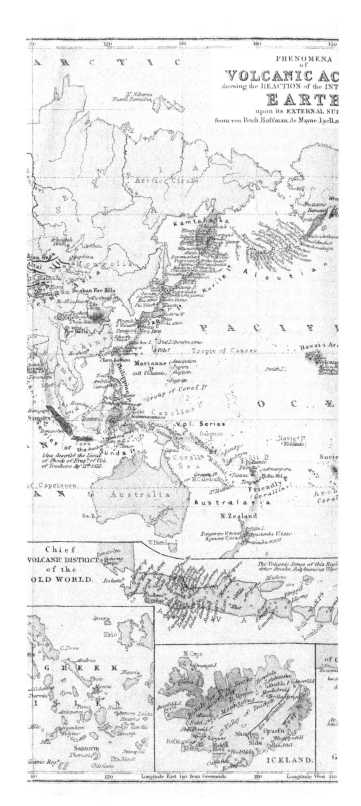

PHENOMENA
of
VOLCANIC AC
showing the REACTION of the INT
EARTH
upon its EXTERNAL SUR
from von Buch, Hoffman, de Mayne, Lyell, an

A R C T I C

N. Siberia
Fossil Remains

Arctic Circle

Wra
RUSSO
Ramano

Kamtchatka

Kurile

Aleutian

P A C I F I

Tropic of Cancer

Hawaii Ar

Group of Coral I

Marianne
Is
(all Volcanic)

Smith I.

Vol. Series

Caroline

Salomon

Navig^s I^s
(Volcanic)

Soci

Corall^a
Sea

Volcanic

Austral-asia

Australia

Friendly
(Coralline)

Arch
Coral

N.Zealand

V.Diem

Tongariro V.8000
Egmont V.8843

White I.
Ruapehu 6200
Taupehu V.9330

Chief
VOLCANIC DISTRICT
of the
OLD WORLD.

Rakata

The Volcanic Zones of this Regi
either Smoke, Sulphureous Vapo

Ipsara

Skio

C. Doro

Andros

Negroia

GREEK

Tino

Seriant

Paros
Naxia

Antiparos

Milo
Argentieri
Polyno

Nio

Santorin
Thirasia

Siang64
Nanti

olcanic Reg^n
Chisiani

N Cape

Orangat I.

ICELAND.

L O N G I T U D E

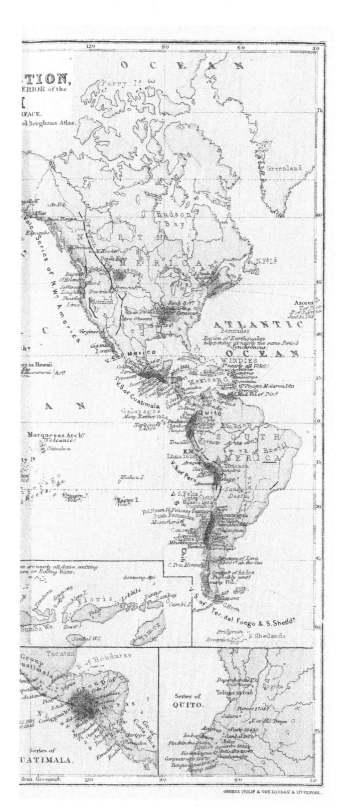

ON VOLCANOS,

&c.

~~~~~~~~~~~~~~~~~~~~~~~~~~~~~~~~~

CHAPTER I.

INTRODUCTORY.

IF the business of Geology is the study of the structure of the accessible portion of the Earth, and of the changes it has undergone, there can be no more important branch of the science than that which examines the nature and mode of operation of the subterranean forces which have everywhere more or less broken up, disturbed, and altered the level of the superficial rocks, modified their internal texture and composition, and brought fresh material upon or towards the exterior of the globe.

The manifestations of these forces, observable in our own days, are usually distinguished as the phenomena of Earthquakes and Volcanos. The first are to outward apprehension purely dynamical, consisting in sudden and transient shocks and wave-like vibrations of extensive superficial areas, attended by fracture, disturbance, and often permanent changes of level in their component rocks. The latter are characterized by 'Eruptions'; *i. e.* the forcible expulsion of heated matters, gaseous, fluid, or solid (usually of all three together), from the interior of the earth upon its surface—these phenomena being very generally accompanied or preceded by earthquakes of a minor and local character.

B

The distinction here drawn is to be recognized likewise in the traces of subterranean action in past times. We see regions, usually more or less superficially broken, composed throughout of scoriæ, ash, tuff, or lava-rock so closely resembling in mineral character as well as in general aspect and disposition the products of recent volcanic eruptions, that we do not hesitate to class them as volcanic formations. On the other hand, we find numerous other rocks, of every variety of character, stratified and unstratified, but especially the former—which we know from unquestionable evidence to have been deposited by water in nearly horizontal beds—exhibiting signs of displacement subsequent to their formation, attended by fracture, folding, elevation, or subsidence,—these changes of position seeming to have been always accompanied by the intrusion of masses of crystalline matter forced up from below through or among the strata, at a high temperature and in a condition of imperfect liquefaction, but yet not so as to reach in that condition the outer atmosphere. And these dynamical results, as corresponding to the observed phenomena of earthquakes unaccompanied by outward eruption, we distinguish from the effects of ' Volcanic,' as those of ' Plutonic ' action, since they seem to proceed from greater depths, and therefore bear some analogy to the powers attributed in the fabulous mythology of antiquity to the infernal Earth-shaking sovereign, ' Ennosigæus.'

Few or no geologists fail to recognize this broad distinction in fact, although all may not employ the same phraseology to denote it. M. de Humboldt bases upon it the entire classification of what he calls the Telluric portion of his great work*. Mr. Darwin expresses it most succinctly, perhaps, when he says, " I believe that a mountain axis differs from a volcano only in plutonic rocks having been injected instead of volcanic rocks having been ejected †." Mr. Mallet looks on an earth-

---

* Kosmos, iv. 1. p. 163 *et seq.* (Sabine's translation).
† Darwin, Volcanic Islands, p. 129.

quake in a non-volcanic region as "an uncompleted effort to establish a volcano."

How far this distinction is justifiable—whether these two kinds of subterranean action are only, as seems most probable, modifications of the same force, but acting from different depths, or under different circumstances—whether the volcanic is only the outward development of internal plutonic action —whether, and to what extent, the two are sometimes, or, it may be, generally, combined—in short, what are the laws, or modes of operation, of each, and their mutual relations—these are evidently questions of the highest geological importance. The study of these laws, and the acquisition of correct views respecting them, are an absolutely indispensable preliminary to any real knowledge of the true history of our planet's surface.

Of the two classes of phenomena, one—that of Seismic, or Earthquake movements—has recently been treated with signal ability by Mr. Robert Mallet, in his four Reports to the British Association for the Advancement of Science,—a work in which all the known facts respecting such phenomena are brought together, reviewed, and reasoned upon in a spirit of sound philosophical inquiry which leaves little more to be done in that direction in the present state of our knowledge and information. Further progress is rather to be looked for in the light that may be reflected upon the subject from a more thorough examination of the other and kindred class of phenomena— those, namely, of volcanic action—than has been hitherto prosecuted, notwithstanding that these phenomena are, from their constancy to particular localities, the most open to observation and scientific examination, and therefore likely to prove the most fertile and reliable source of accurate information as to the true character of Subterrestrial Dynamics. To this end I devoted, many years since, very considerable attention, and published the results in more than one form between the years 1824 and 1829. In the volume on " Volcanos*," I introduced some speculations upon the presumed early condition of the globe, which were then, and might perhaps even at present be considered premature, and unwarranted by our existing information. But, in the strictly volcanic portion of that work, I believe the views taken respecting the normal laws of volcanic action to have been sound and incontrovertible. With regard to them, therefore, I have no alterations of moment to introduce in the present edition. They have the advantage of being simple, and yet perfectly capable, I believe, of accounting for all the recorded phenomena of volcanic formations, whether of ancient or of modern date; for I wholly disagree with the opinion expressed by Humboldt that such phenomena are "isolated, variable, and obscure†." I made, indeed, no pretence of discovery; for I only reproduced in substance the opinions which had been always held upon this subject by the most trustworthy observers, such as Spallanzani, Dolomieu, Sir William Hamilton, and Brieslak, who had made this department of the science their special study, and relied for evidence of their truth on the recorded descriptions of volcanic eruptions and volcanic products in every part of the world, no less than on my own observations.

These views have been, however, disputed by geologists of great repute, chiefly, but not solely, of the Continental schools, and a theory of volcanic action promulgated, with a parade of mathematical formulæ, which almost ignores its eruptive character altogether, attributing the production of volcanic mountains, not to the accumulation of erupted matters, but to the elevation *in mass* of previously horizontal beds in the shape of hollow blisters, each blown up by the sudden expansion of a bubble of aëriform matter beneath. In another publication‡ I have shown—I venture to think con-

---

* Considerations on Volcanos. 1825.   † Kosmos, iv. p. 90.

‡ "Memoir on the Mode of Production of Volcanic Cones and Craters," Quarterly Journal of the Geological Society, Nov. 1859. Translation of the same into French, printed at Paris, 1860.

clusively—that this hypothesis is unwarranted by any facts or
relations of observed volcanic phenomena, and not required
by anything in the composition or structure of any volcanic
mountain,—that it rests, indeed, on wholly untenable and un-
philosophical assumptions.   Beyond this reference to that
theory and its refutation, I do not, therefore, intend to advert to it further, believing that it is now all but universally
discredited.   Its long prevalence, however, through the influence of the great names connected with it, may account for
the general absence among geologists of clear and decided
views as to the true laws of volcanic action, and for the want
(as it appears to me, existing at the present time) of some such
treatise on the subject as I propose now to attempt to supply,
varying somewhat in form, but very little in substance, from
that contained in my earlier volume of 1825-6.

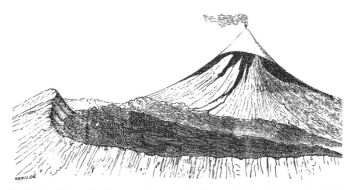

The Peak of Teneriffe (the summit covered with snow), from the margin of the
encircling Great-Crater wall.   The intermediate tract covered with lava.

CHAPTER II.

GENERAL VIEW OF VOLCANIC ACTION.

§ 1. VOLCANIC action, as has been already said, exhibits itself chiefly in the eruption or exhalation of heated matter in a solid, semi-liquid, or gaseous state, from openings in the superficial rocks that compose the crust of the globe. These eruptions usually take place with much violence. Occasionally, however, comparatively tranquil emanations occur of vapour or heated water, containing more or less of mineral matters, as in the well-known instances of hot-springs, and the so-called ' Solfataras ' and ' Suffioni.'   But it is generally to be suspected, as will be shown hereafter, that such placid exhalations are confined to points which have been previously in active and violent eruption, or are in communication through fissures with some subterranean mass of heated volcanic rock.

§ 2. A volcanic eruption—whether from a fresh point of the earth's surface, or from a site previously productive of similar matter, but for some time quiescent—is usually preceded by earthquakes. These, however, are generally of a minor and local character. The great paroxysmal earthquakes, which calamitously affect extensive areas, do not, in the majority of instances, appear to be connected, as to the *time* of their development, with eruptive activity in any neighbouring volcano. Humboldt remarks in proof of this, that " during the great catastrophal earthquake of Riobamba in the year 1797, the neighbouring volcanos of Tunguragua and Cotopaxi remained quite tranquil ; and inversely, volcanos have been subject to great and long-continued eruptions without any earthquake being sensibly felt in the surrounding country*."

Mr. David Forbes, in his recent account of the Geology of Bolivia, says that "the frequent volcanic eruptions breaking out in the district of Titiaca in no way disturb the adjoining main chain of the Silurian Andes," while, on the other hand, "this volcanic district is exempt from the earthquakes which are so prevalent and destructive in the neighbouring regions†."

Mr. Darwin remarks of the great earthquakes of South America of 1835, that "at the instant of time when an immense area was convulsed and a large tract elevated, the districts surrounding several of the great volcanic vents of the Cordillera remained quiescent." After the cessation of the earthquakes the eruptions recommenced with increased violence.

Examples, however, are not wanting of the coincidence of eruptions from neighbouring volcanos with extraordinary seismic convulsions. An earthquake felt on the coast of Caraccas, in 1821, was simultaneous with the outburst of an eruption from the island of St. Vincent‡. "In the great Chilian earthquake of 1826, at the moment the shock was felt at Valdivia, two volcanos near it burst into eruption *for a few seconds,* and then again became quiescent§. At Conception, eruptions are said to have broken out from beneath the sea at the same time. The melting of the chain-cable of the 'Volage' man-of-war, at anchor, during an earthquake off the coast of South America may be adduced as another instance ‖." Still more frequently do we hear of earthquakes having at once ceased on the opening up of volcanic vents more or less distant—or, contrariwise, of regions in which volcanos were once active, but are now dormant, being habitually afflicted by violent earthquakes,—facts strongly con-

* Kosmos, vol. iv. pt. 1. p. 180.   † Geol. Proc. 1860, p. 61.

‡ Humboldt, Kosmos, p. 455.

§ M. Place, Qvart. Journ. G. S. vol. xvii.   ‖ Mallet, Report, 1850, p. 25.

firmatory of the prevalent and very reasonable supposition that volcanos act in a manner as safety-valves for the escape of that excess of subterranean heat, which, by an expansive influence on portions of matter beneath the earth's surface, is probably the chief cause of its spasmodic convulsions*.

In fact, there can be no doubt of the intimate relation between the two classes of phenomena. A glance at the map appended to this volume (and which is founded on that of Mr. Mallet) makes it abundantly clear that both have developed themselves, within recent times at least, *mainly* along the very same linear bands, or about the same seemingly insulated spots upon the surface of the globe. Mr. Mallet sums up the proofs of their local connexion by the assertion that " Seismic energy is greater *or* more frequent as the great lines of volcanic activity are approached." Nevertheless it is certain that the paroxysmal earthquakes—those tremendous shocks and violent undulatory vibrations of the solid surface of the globe, which produce the greatest destruction, and are propagated over wide areas with extraordinary rapidity—are rarely, if ever, coincident as to time, nor, unless viewed on the largest scale, as to the position of the locality they affect, with any volcanic, that is to say, eruptive action. The island of Iceland is nothing else than a vast volcanic mountain rearing itself from the depths of the ocean, and has been very frequently in violent eruption from some of its mouths within the historic period. But, as Mr. Mallet observes, " its seismic commotions are a purely local phenomenon, the shocks being only of moderate intensity ; and the whole island is rarely simultaneously shaken." On the other hand, the Himalayas and the enormous table-land of Central Asia— in which it is doubtful whether any active volcano exists —are very liable to earthquakes of fearful intensity. The Alps also and the Pyrenees, wholly devoid of eruptive vents,

---

* " Volcanos obviate the occurrence of earthquakes, acting as safety-valves."—Humboldt, Kosmos, iv. p. 172 (Sabine's translation).

are frequently shaken; so also the non-volcanic basin of the
Baltic. The valley of the Mississippi is very liable to earthquakes, though far removed from any volcano. The chain
of the Andes, it is true, which is almost always in a state
of commotion, is studded with volcanos. But it appears,
from the authority, already quoted, of Humboldt and Mr.
Forbes, that the latter do not ordinarily show any remarkable
activity during the occurrence of the most violent seismic
convulsions of the neighbouring mountains, and *vice versâ*.
On the whole, then, it seems probable that, in the words of
Mr. Mallet, " both volcanic eruptions and earthquakes are
manifestations of a common subterranean force," but acting
under different conditions of the degree of energy developed
and of the resistances met with. I believe myself that the
greater earthquakes proceed from efforts of subterranean
expansive force taking place at too great a depth, and under
too vast a mass of overlying rocky matter, to effect that free
communication with the outer air which is indispensable for
the development of volcanic phenomena; and that the local
and minor earthquakes which usually precede or accompany
these latter, act from less deeply-seated points, and under less
resistances of the upper crust of the earth, where consequently
channels of free communication with the external surface can
be, and are, created by these efforts. In both cases convulsive
throes in the solid surface of the globe are occasioned by the
same common cause—the expansion (probably from increase
of temperature)—in some cases, perhaps, the contraction—
of some subterranean matter. The result in the case of the
great earthquakes is purely dynamical, consisting of undulatory vibrations transmitted to great distances, with fractures
and changes of level in the overlying rocks. In that of the
minor shocks attendant on volcanic phenomena, these rocks
are so split or opened at one or more points—generally on
lines of previous disturbance—as to permit an extravasation
of some of the heated and liquefied and gaseous matters, the

expansion of which was the cause of the rupture. Hence, as I have said above, the justice of the distinction between Plutonic and Volcanic action. That Mr. Mallet coincides in this general view may be assumed from his describing an earthquake as " an uncompleted effort to establish a volcano."

It will, however, be desirable to postpone further consideration of this branch of my subject to a later portion of this work, and until the character of the volcanic phenomena has been described and thoroughly investigated.

It will be convenient to begin by a brief review of the distribution of the sites of volcanic energy upon the globe, and the description of their ordinary phenomena.

§ 3. Although an eruption has rarely, if ever, been observed to break out from points of the earth's surface upon which no traces existed of previous volcanic development, every volcano, or habitually eruptive vent, must, of course, have commenced its activity at such a point. Moreover, in many regions of seemingly extinct volcanic action, for example, in Central France, the Eifel, the island of Sardinia, Asia Minor and New Zealand, Olot in Catalonia, &c., numerous eruptions have evidently taken place at various points of a surface composed of non-volcanic rocks—granite, gneiss, schists, or secondary and tertiary strata.

The great majority of recorded volcanic eruptions, however, have burst from the summit or side of some mountain composed already of rocks of volcanic origin, attesting the former habitual or intermittent activity of the same vent.

The sites of these volcanic eruptions are not limited to a few particular geographical districts. They occur, on the contrary, either continually, or at intervals of greater or less duration, upon very numerous points, situated in all quarters of the globe,—a fact leading to the important inference that the subterranean force (whatever it be) of which they are the external development, is, if not uniformly, yet very generally distributed beneath the entire surface of our planet.

§ 4. The known active volcanos, or habitual vents of volcanic energy, have been reckoned at various numbers, from 200 to upwards of 400. Humboldt* gives a list of 407 ; but of these only 225 are " known to have been in activity in recent times." Keith Johnstone † gives a catalogue of 270 active volcanos, of which no less than 190 occur in the islands or upon the shores of the Pacific Ocean.

All such enumerations must, however, be considered as approximative merely, and much below the truth. First, because, while a very large proportion of the earth's *subaërial* surface is still unknown or imperfectly explored, it is highly probable that through that far larger area which is *subaqueous* —being indeed fully three-fourths of the whole—many volcanic vents exist which have not yet raised an eruptive orifice visibly above the surface of the ocean. Secondly, because the intervals of quiescence which occur between the eruptions of a volcano are occasionally of such long duration that all accounts of any occurrence of the kind are liable to be lost or forgotten, and the eruptive character of its site consequently ignored, until a fresh outbreak proclaims the continued activity of the subterranean focus ‡.

Indeed, if we were to reckon as volcanos all localities at which eruptions have at one time or other occurred within a comparatively recent geological period—for example, since the close of the deposition of the secondary strata—as attested by the mineral character and arrangement of the superficial rocks (and in such spots the volcanic force may be merely dormant, and liable to break out anew at any moment, for aught we know to the contrary), the number above-mentioned must be multiplied in a very high ratio.

* Kosmos, iv. p. 406 (Sabine's translation, 1859).
† Atlas of Physical Geography. 1859.
‡ Seventeen centuries intervened between two consecutive eruptions from the volcanic focus that underlies the island of Ischia, viz. from 380 B.C. to 1302 A.D.

§ 5. In the Appendix will be found a Catalogue of known active volcanos, framed from the accounts of such phenomena reported by eye-witnesses in different parts of the globe within the historical period. It is illustrated by a Map, exhibiting their geographical position. We find them irregularly scattered over the whole surface of our planet, occurring indifferently in all latitudes, and under every meridian; sometimes detached singly, and at a considerable distance from each other, but generally either concentrated into close groups, or forming a connected linear chain or series; in some rare instances seated in the interior of a continent; usually, however, rising as insular mountains from the depths of the ocean, or at very little distance from its borders, upon a maritime coast.

A reference to the Map accompanying the Catalogue in the Appendix will show the existence of one most remarkable linear range of volcanos which traverses both hemispheres in a great arched curve, commencing in ' the land of fire' (Tierra del Fuego) at the southern extremity of America, running up the entire western fringe of that continent almost to Behring's Straits, crossing the North Pacific through the Aleutian chain of isles, and descending thence southwardly along the peninsula of Kamtschatka, Japan, and the Philippine Isles, to the Moluccas, from which two lines branch off,—one enclosing Borneo in a sudden semicircular sweep to the west and north, and continuing through Java and Sumatra to the Andaman Isles and into Burmah, in which last ' wreath of islands' Dr. Junghuhn reckons no less than 109 lofty fire-emitting mountains—the other threading Papua, and the Saloman and New Hebrides Isles, to New Zealand, whence it seems continued in Victoria Land almost to the South Pole.

It is presumable that this extraordinary lengthened range of active volcanic vents, which circumscribes the entire Pacific Ocean, and almost bisects the globe, indicates some underlying great and prolonged fissure through its crust. But we will reserve to another opportunity the speculations to which this idea may lead, and all other considerations in reference to the general geographical distribution of volcanic orifices.

§ 6. Among the circumstances which must most materially affect the character of any individual volcanic eruption are those which relate to the situation of the point on which it breaks through the solid crust of the earth; and particularly whether this point is placed above or below the surface of any of the great bodies of water which cover so large a portion of our planet.

In the first of these cases, the eruption takes place in open air; in the second, in water; and the different density and other circumstances of these media must considerably modify the nature of the phenomena, and the conduct and disposition of the substances, whether gaseous or solid, produced from the interior of the earth.   These different kinds of eruptions may be called, the first *subaërial*, the second *subaqueous*.

The latter class of eruptions, though we have reason to believe them by no means of rare occurrence, can at least be but very rarely observed; and their phenomena, even when the attention of those who chance to be passing near the spot is called to them, can be only partially and imperfectly perceivable.   Shocks, or sudden impulses, resembling the striking on a rock, have been felt in vessels; and extraordinary, otherwise unaccountable, waves transmitted to the shores, which are probably attributable to such occurrences.   The sea-water in the vicinity has been observed to be more or less agitated, heated, and discoloured, and to be traversed by rising volumes of gaseous fluids, and sometimes even by jets of fragmentary matters; but it is only when the accumulation of the solid substances protruded from below has at length elevated the vent above the water-level, and the phenomena consequently take place in open air, that they become more immediately liable to direct observation, and they then enter into the first

class of eruptions—the subaërial.   It is therefore to this class
that we must first direct our attention.

§ 7. A rapid survey of the information we possess on the
phenomena of the different known volcanos, or *habitual sources
of volcanic products* on the globe, leads to the conclusion that
there exists the most complete irregularity with respect to the
periods and intensity of their activity.

Some volcanos are in a state of incessant eruption; some,
on the contrary, remain for centuries in a condition of total
outward inertness, and return again to the same state of
apparent extinction after a single vivid eruption of short duration; while others exhibit an infinite variety of phases intermediate between the extremes of vivacity and sluggishness.

Every attempt that has been hitherto made to connect this
greater or less activity with the comparative height or bulk of
the volcanic mountain, or its position relatively to sea or land,
or the mineral character or geological relations of the rocks
from which it breaks forth—in short, to discover any general
law by which the frequency or violence of its eruptions may
be determined—has hitherto been unavailing *.   No doubt
some such law exists; but it is probable that, before its character can be thoroughly ascertained, we must acquire a clearer
insight than we at present possess, into the nature of those
subterranean forces of which volcanic eruptions are only one
of the external manifestations.

The first step towards such an inquiry will necessarily be a
correct comprehension of the *modus operandi* or true character
of those manifestations.   With this end in view—and postponing for the present any examination of the several theories
that have been, or may be, suggested upon the cause or origin

* M. Sainte-Claire Deville, I observe, considers that he has discovered
such a law; but, as expressed by him, it merely amounts to this, that
when, or so long as, lava in a liquid and ebullient state is in contact with
the atmosphere, it enters into eruption: which is only saying, that when
a volcano is in eruption, it is eruptive.—See 'Comptes Rendus,' 1856,
p. 435.

of the subterranean force—it will be convenient to distinguish the various conditions exhibited by the different known active volcanos into three general phases, viz. :—

1. That in which the volcano exists incessantly in outward eruption—phase of permanent eruption.

2. That in which eruptions, rarely of any excessive violence, continue in a comparatively tranquil manner for a considerable time, and alternate with brief intervals of repose—phase of moderate activity.

3. That in which eruptive paroxysms of intense energy alternate with lengthened periods of complete external inertness—phase of prolonged intermittences.

View of Monte Nuovo, in the Bay of Pozzuoli: a volcanic cone, 450 feet high, thrown up in three days, A.D. 1538.—See APPENDIX.

CHAPTER III.

PHENOMENA OF ORDINARY SUBAËRIAL ERUPTIONS.

§ 1. *Phase of permanent eruption.*—The instances are rare in which a volcano remains permanently in eruption. Those best known are Stromboli, one of the Lipari Isles, which has been incessantly active since the time of Homer at least; Masaya and Amatitlan, in the province of Nicaragua; Izalco, on the west coast of Central America, which has been in eruption since 1728; Sangay, in South America, which remains permanently in eruption; the Ilha da Fuego, one of the islands off the Cape de Verde, constantly active since 1770. Probably several of the as yet little-known volcanos of Sumatra, Java, and the other portions of that great linear range of vents already mentioned, which, traversing the Pacific, embraces the whole eastern coast of Asia, are also permanently eruptive.

§ 2. *Phase of moderate activity.*—But in general the eruptions of a volcano are intermittent, the intervals between each varying from months to centuries.

When the eruptions are constant, or very frequent, they are usually of a comparatively placid character, as if a permanent issue had established itself, and acted as a kind of safety-valve, for the outward discharge of the redundant caloric or heated matter below, the accumulation of which would otherwise bring about occasional eruptions of far greater violence. In conformity with this supposition, when an eruption breaks out after a long interval of tranquillity, it is usually violent in proportion. Such extraordinary developments of volcanic energy may be styled paroxysmal.

Eruptions of extraordinary violence are, however, not unfrequently preceded by a prolonged phase of moderate activity,
which apparently did not suffice to discharge the redundant
heat as fast as it was transmitted from the inner depths of its
subterranean reservoir or source.   As an example may be
quoted the condition of Vesuvius since the commencement of
the present century; indeed, perhaps, from the year 1631,
when its activity was renewed after a quiescence of a century
and a half.   Certainly throughout the last fifty or sixty years
this volcano has often continued in eruption for periods of
several months, discharging moderate jets of scoriæ, lapilli,
and sand from temporary orifices at the summit or flank of
the cone, or at the bottom of its crater *when there was a
crater*; while streams of lava welled out, sometimes almost
with the tranquillity of a water-spring, from the same or contiguous openings.   These periods of placid activity were generally succeeded by intervals of rest, lasting also perhaps a few
months, at the end of which time eruptions recommenced
from new orifices—the weakest points, it is presumable, of
some fissures broken through the upper framework of the
cone by the efforts of the internal lava to find a vent,—the
former issues being sealed by the consolidation of the residuary
lava which found its way up through them.   This phase of
moderate activity has, however, been occasionally interrupted
by the outbreak of a violent eruption, such as occurred in
1794, and again in 1822, at both which times explosions of
vast force, prolonged for weeks, broke up and blew into the
air the whole upper part of the cone, the product of the preceding minor eruptions, leaving a central caldron of great
dimensions excavated through the heart of the mountain.   To
this latter phenomenon I shall recur presently.

The actual state of Etna affords a similar example of almost
continual moderate activity, with occasionally more or less
paroxysmal outbursts.   During the last half century at least
six principal eruptions have occurred, viz. in 1805, 1809, 1811-

12, 1819, 1831, and 1852, each of which produced a considerable quantity of lava. But the intervals between these epochs of remarkable excitation did not pass without various minor phenomena attesting the continued activity of the focus. Frequent earthquakes were felt, not only by the inhabitants of the mountain's flanks, but often through the whole island; one of these shocks (16th February, 1810) is said even to have extended its influence as far as Cyprus. Smoke was almost continually emitted by the crater, accompanied occasionally by detonations and jets of red-hot scoriæ; the appearance of flames, often mentioned by observers, being in all likelihood caused by the light of these jets, which took place at the bottom of the deep crater, reflected from the impending cloud of smoke and vapour.

Many of the volcanos studding the Pacific, as well as some of that great train of active volcanic vents which stretches in a sinuous line from the northern extremity of the peninsula of Kamtschatka, through Japan, Loo-choo, and the Philippine Isles, the Moluccas, and Java, into Sumatra, and the island chain of the Andamans, &c., seem, from the meagre information we possess concerning them, to exist in this state of prolonged but moderate activity, since the same volcanos are seen to be constantly in eruption by the crews of the different vessels that navigate these Archipelagos.

This is probably the condition of the volcanos of Barren Island, of Arjuna in Java, and of the small island between Timor and Ceram; of that of New Britain, seen successively by Dampier, D'Entrecasteaux, Lemaire, and Schouten, in eruption; of that of Tanna in the Archipelago of the New Hebrides, seen in activity by Cook, D'Entrecasteaux, and Forster; of the Peak of Ternate in the Moluccas; of those of Mutova and Tharma in the Kurile Islands, and many others in Japan, Kamtschatka, and the Aleutian Isles.

Among the American volcanos, we may instance as existing in this phase those of Pichinca near Quito, and of Popocatepetl in Mexico, which have been almost continually active since the period of the occupation of the Americas by Europeans.

The volcano of the Ile de Bourbon offers another remarkable example of the phase under consideration. From accounts given by M. Hubert, who is stated by Bory de St. Vincent to have directed his attention to its phenomena ever since the year 1760, we know it to have existed during the last century in a continual state of moderate activity, vomiting lava at least twice in every year: eight of its currents produced in this space of time have reached the sea, and, with the others, cover a wide slope called *Le pays brûlé*, of a horrible aspect, almost entirely destitute of vegetation, uninhabited, and, from the peculiar glassy asperity of the scoriform surfaces of the currents, nearly impassable. Here, however, as in other instances already cited, this moderate activity is liable to be interrupted by paroxysmal action. From recent accounts we learn that an eruption of prodigious violence occurred there in the course of last year (1859), giving birth to an immense current of lava, which, flowing from the summit of the volcano (5000 feet high) into the sea, interrupted all communication by land round the east coast of the island.

§ 3. *Phase of prolonged intermittences and paroxysmal eruptions.*—But of all the phenomena of volcanos, those sudden and violent eruptions which I have called paroxysmal, and which usually occur after a lengthened period of external quiescence, naturally attract the most attention. And of these the most copious narratives have been recorded.

The stupendous and terrific character of these catastrophes, the rarity of their display, their sudden and unexpected occurrence, and the dreadful extent of injury often resulting from them to the lives and property of the inhabitants of the surrounding country, make them a subject of general remark and relation during and long after the period of their development. Hence accounts of such volcanic eruptions find a place in the earliest annals of history, play their part occasionally in the fabulous mythology of still remoter ages, and form a not unfrequent source of sublime imagery to the poets of antiquity.

§ 4. When we compare together all the received statements of such occurrences, observed in every quarter of the globe and at distant periods, we cannot but be struck by the exceeding similarity of the facts and appearances recounted; nay, more, when due allowance is made for the effects of terror upon the minds of ignorant and perhaps superstitious observers, for the universal proneness to exaggeration of the marvellous, the want of scientific language, and the errors necessarily incidental to the relations of inexperienced persons, as well as for the great distances from which they have been generally viewed, it is not too much to say that a complete identity may be recognized in the main phenomena; no further discrepancies existing than what are fairly referable to modifications produced by local circumstances, or by differences in the intensity of volcanic force developed, and in the mineral quality and chemical condition of the substances elevated.

§ 5. The following is a brief sketch of the phenomena which universally appear to characterize these great eruptions or volcanic paroxysms.

They are preceded by earthquakes more or less violent, frequent, and prolonged, but affecting principally the volcanic mountain itself, which seems to be convulsed by internal throes, comparable to those of parturition in an animal. These shocks are probably owing to the expansive efforts of a subterranean mass of lava in a state of extreme tension, or of the elastic fluids which it contains, to force a passage through the superincumbent rocks. Repeated loud subterranean detonations are heard, resembling, so as frequently to be mistaken at a distance for, the firing of heavy artillery, or the rolling of musketry, according to its intensity.

These sounds are proved, by the immense distance to which

they are propagated, and with a rapidity wholly out of proportion to their loudness near the spot from which they proceed, to be conveyed, not by the air alone, but chiefly by the solid strata of the earth.

Often, it is said, the state of the atmosphere assumes a peculiar character, offering an unusual closeness and stillness, causing a sensation of oppression.

These threatening indications of an approaching crisis last for a greater or less time, and are accompanied by the disturbance or total disappearance of springs, the drying up of wells, and such accidents as the cracking, splitting, and heaving of the substructure of the mountain must naturally occasion. During this time the lava is probably forcing its way upwards like a wedge through one or more fissures broken by these violent throes. When it thus at length obtains a communication with the outward air, the eruption begins, generally with one tremendous burst, which appears to shake the mountain from its foundations. Explosions of aëriform fluids, each producing a loud detonation, and gradually increasing in violence, succeed one another with great rapidity from the orifice of eruption, which is in most instances the central vent or crater of the mountain. This vent has usually been obstructed, during a long preceding period of repose, by the ruins of its sides, brought down by the wasting influence of the weather and the shocks of earthquakes, or by the products of previous minor eruptions.

The elastic fluids therefore, in their rapid escape, project vertically upwards these loose accumulated matters and the fragments of the more solid rocks through which they have forced a passage.

The mutual friction to which these fragments are subjected during their rapid and repeated projection, as they fall again towards the orifice, reduces them to such tenuity that a large portion is carried upwards by, and remains for a time suspended in, the heated clouds of aqueous vapour which are discharged at the same time in prodigious volumes from the volcanic aperture.

The rise of this vapour produces the appearance of a column several thousand feet high, based on the edges of the crater, and appearing from a distance to consist of a mass of innumerable globular clouds of extreme whiteness, resembling vast balls of cotton rolling one over the other as they ascend, impelled by the pressure of fresh supplies incessantly urged upwards by the continued explosions. At a certain height (determined, of course, by its relations of density with the atmosphere) this column dilates horizontally, and (unless driven in any particular direction by aërial currents) spreads on all sides into a dark and turbid circular cloud. In very favourable atmospheric circumstances, the cloud with the supporting column has the figure of an immense umbrella, or of the Italian pine, to which Pliny the younger compared that of the eruption of Vesuvius in A.D. 79, and which was accurately reproduced in October 1822. Strongly contrasting with this pillar of white vapour-puffs, is seen a continued jet of black cinders, stones, and ashes, the larger and heavier fragments falling back visibly, after describing a parabolic curve. This jet of solid fragmentary matter often reaches to a height of several thousand feet, while the vapour-pillar rises still higher. Forked lightnings of great vividness and beauty are continually darted from different parts of the cloud, but principally its borders. The continual increase of the overhanging cloud soon hides the light of day from the districts situated below it, and the gradual precipitation of the sand and ashes it contains contributes to envelope the atmosphere in gloom, and adds to the consternation of the inhabitants of the vicinity*.

These phenomena result from the lava boiling up the chimney of the volcano. The elastic fluids by which it is traversed rend and carry upwards portions of its surface as they explode from it, and form a continual fiery fountain of still liquid and

* See the Frontispiece.

incandescent drops or fragments, which, from the velocity of their motion, present a luminous appearance that has frequently been mistaken at a distance for flame. The internal column of lava, continuing to rise, forces an issue at length, either over the lowest lip of the crater, or from some crevice broken through the side, perhaps even at the foot of the mountain, from whence it flows in torrents over the lowest surfaces it can reach, according to the usual laws of fluid bodies, and often spreads itself to a distance of many miles, and over wide areas, burning, overwhelming, and destroying everything it meets in its course, vegetation, woods, buildings, &c. By night, the running lava appears at a white heat wherever the liquid interior of the current is visible; but as upon contact with the air its surface is instantaneously congealed into a thick scoriform crust, the general tint of the outside is a glowing red, gradually darkening as the solidified coating increases in thickness.

During day the lava is almost concealed from view by the torrents of aqueous vapour which rise from its whole surface in immense volumes, and unite themselves to the clouds of similar nature that hang over the mountain.

In some cases no absolute escape of lava in streams takes place, scoriæ alone being projected.

In all cases where lava is emitted, its protrusion marks the crisis of the eruption, which usually attains the maximum of its violence a day or two after its commencement. The stopping of the lava in the same manner indicates the termination of the crisis, but not of the eruption, for the gaseous explosions continue often for some time with immense and scarcely diminished energy.

At length, however, they cease to throw up liquid or red-hot lava, the surface of the lava-column having seemingly sunk too far within the vent; the fragments projected are either blocks of older rocks, or consolidated scoriæ. By degrees, these fragments, most of which fall back into the crater,

become more and more comminuted by the immense trituration they sustain in the process of repeated projection and fall,
till at length only clouds of sand and ashes, reduced ultimately
to an extraordinary degree of fineness, are carried upwards.

The explosions then gradually decrease in violence, appearing to be stifled by the accumulations of finely pulverized matter which occupy the volcanic vent and impede
their expansion. The column of ashes projected becomes
gradually shorter, the eructations less frequent, and seemingly
from greater depths and through increasing impediments,
until at last all struggle seems to cease—no further explosions
are heard—the eruption has terminated ; usually, however,
not until many days, or even weeks, after attaining its maximum of violence.

After the occurrence of such a paroxysm, the crumbling
in of the crater's sides soon chokes up still further the
volcanic orifice and conceals it from view. Observers are
then enabled to approach, and examine the effect produced
on the figure of the mountain. This is in general the truncation of its cone—the upper part having been blown off, and in
its place a vast chasm formed, of a caldron-like appearance,
and of a size proportioned to the violence of the eruption, its
duration, and, above all, to the mass of fragmentary matter it
has ejected and dispersed over the neighbouring surfaces of
land or sea. The volume of lava, however, poured out by an
eruption does not preserve any constant proportion to the force
or continuance of its explosions. Sometimes the one class of
phenomena predominates, sometimes the other.

These tremendous demonstrations of volcanic energy are
usually accompanied or followed by more or less violent meteoric phenomena, sometimes equally terrific and destructive
with the former, the atmosphere appearing to share in the
convulsion which agitates the earth. The summit of the
mountain attracts, or the coldness of the atmosphere con-
denses, the volumes of aqueous vapours which have risen from

the volcanic orifice and the emitted lava, and hence a fall of rain takes place in prodigious quantity on its sides and base, producing torrents, which, carrying with them the ashes, sand, scoriæ, and fragments with which the slopes are strewed, rush, as deluges of liquid mud, towards the plains or valleys below, and cover them with vast deposits of volcanic alluvium *.

As instances of volcanic paroxysms, amongst many on record, may be mentioned those of

Vesuvius, in the years A.D. 79, 203, 472, 512, 685, 993, 1036, 1139, 1306, 1631, 1760, 1794, and 1822.

Ætna, in 1169, 1329, 1535 (this eruption lasted two years with terrific violence, and occurred after a quiescence of nearly a century), 1669, 1693–4, 1780, 1800, and 1852.

Teneriffe, in 1704 and 1797-8.

San Georgio, one of the Azores, in 1808.

Palma, one of the Canary Isles, in 1558, 1646, and 1677.

Lancerote, belonging to the same group, in 1730.

All the recorded eruptions of Iceland, but especially those of Kattlagaia Jokul, A.D. 1755, which lasted a year, and Skaptar Jokul, in 1783.

§ 6. Such are the phenomena which characterize the display of the volcanic forces from an habitual vent, at its moments of paroxysm. These efforts are generally preceded, and more constantly *followed*, by long periods of complete tranquillity, the energies of the volcano seeming to be exhausted for a time by the violence of their development.

The duration of this quiescent interval is of very unequal

---

* The above description corresponds closely with what I myself observed during the eruption of Vesuvius in October 1822, the greatest, unquestionably, that has been witnessed in Europe since the beginning of the century. I believe it to be equally applicable to the paroxysmal eruptions of volcanic mountains in general, because, after a careful study of the reliable reports of their occurrence at other times and other places, and making allowance for the sources of error natural to such occasions already adverted to, these last appear to me perfectly reconcileable to the belief, in itself, and, *à priori*, most probable, of their general identity.

continuance, extending even occasionally to centuries; and thus it frequently happens that the scoriæ and ashes forming the surface of the cone, and its internal cavity or crater, become so far decomposed as to afford a soil in which various vegetables find sustenance. All appearances of igneous action are effaced; forests grow up and decay; and cultivation is carried on upon a surface destined perhaps to be blown to atoms and scattered to the winds when the crisis arrives for the renewal of the volcanic phenomena. Thus during the quiescent interval between the eruptions of 1139 and 1306, the whole surface of Vesuvius was in cultivation, and pools of water and chestnut-groves occupied the sides and bottom of the crater, as is at present the case with so many of the extinct craters of Ætna, Auvergne, the Vivarais, &c.

In general, after the cessation of a paroxysm, many *fumaroles*, or emanations of vapour, evolve themselves from the lava-currents which were then produced, as well as from the bottom of the crater. These vapours, which, at first, are for the most part aqueous, usually contain a variety of mineral acids, and, as the lava cools down, deposit saline incrustations at the mouths of the fumarole. They are hydrochloric acid, sulphurous acid, sulphur, sulphuretted hydrogen, and their compounds, especially the chlorides of sodium, potassium, ammonium, iron, and copper. Occasionally boracic acid occurs. Specular iron-ore is not unfrequently deposited in the fissures of lavas opening into the air.

When the acidity of the vapours is in excess, and their production continued for a long period, they effect a very considerable degree of decomposition on the exposed parts of the rocks against which they act, and the crater of a volcano in this condition is said to pass into the state of a Solfatara or Soufrière. Such is the actual condition of the Solfatara near Pozzuoli; of the Soufrières of St. Vincent, Guadaloupe, and St. Lucia, in the Leeward Isles; of the great central crater of the Peak of Teneriffe; of the craters of Milo in the Archipelago; of Volcano, one of the Lipari Islands; of Crabla in Iceland; and of Tanna, one of the New Hebrides, according to Dr. Forster; of many volcanos in Java, in the Cordilleras of S. America, and in other localities.

This condition is by no means a proof of the complete extinction of the volcano, as was proved by the eruptions from the Soufrière of St. Vincent in 1812, which had been completely tranquil since 1119, that is to say, for seven centuries!; and by a similar renewal of activity from those of Volcano in 1786, and of Guadaloupe in 1778, 1797, and 1812. Hot-water springs containing sulphur, and mephitic emanations of carbonic acid gas (called *mofette* in Italy), are the most enduring, perhaps, of all the exhalations from a spent volcano. To these phenomena reference will be made hereafter.

All the volcanos of the Atlantic, whether in Iceland, the Azores, Canaries, Cape Verde, or Caribbee Isles, appear at present to exist in this phase of prolonged quiescence, interrupted from time to time, and at distant intervals, by paroxysmal eruptions. A great proportion of those which stud the Cordilleras of the two American continents, and of those which occur in Sumatra, Java, the Moluccas, Japan, Kamtschatka, and the numerous Archipelagos of the Pacific, belong to the same class. Throughout these two great volcanic trains (which in reality form but one, ranging from S. to N., and back again to S.W.) we hear of terrific eruptions occasionally breaking out from mountains which were not previously suspected to be of volcanic nature, or as to which the accounts of former catastrophes of this sort existed but as vague traditionary fables.

§ 7. It has been thought advisable thus to distinguish these three phases, in one or other of which the phenomena of a volcanic vent usually present themselves, in order to simplify the study of their nature and mode of operation. But it must be recollected that the distinction is purely artificial; many volcanos exist in intermediate conditions, partaking of the characters of more than one phase, or occasionally passing from one into another. Thus Vesuvius, whose earlier eruptions appear to have generally been paroxysmal, with prolonged intervals of repose, as far as can be deduced from the imperfect accounts preserved of its phenomena (although, from the reasons mentioned above, only the paroxysms will, perhaps, have been noted), continued during a great part of the seventeenth century in the second phase, frequent eruptions having been mentioned between the years 1660 and 1694, when a paroxysmal outbreak occurred, succeeded by a quiescence of ten years, since which the volcano has been almost constantly active. Ætna in the same manner appears to have passed into the phase of moderate activity towards the beginning of the seventeenth century; since which epoch more than forty eruptions are recorded, with but one quiescent interval of any considerable duration, viz. from 1702 to 1755. Some of these eruptions, however, have been decided paroxysms, particularly those of 1669 and 1787.

Similar changes have doubtless often taken place in other instances, many volcanic mountains bearing marks of having experienced such varieties of condition. It is indeed, as will be hereafter shown, to this alternation of phases of moderate activity with paroxysmal eruptions that is owing a frequent and characteristic feature of volcanic mountains, namely their exhibition of one or more circular cliff-ranges, or segments of such, surrounding, or closely adjoining to, the more recently-formed cones which are at present, or have lately been, in activity,—the outer rings, or annular ranges, being in each case the basal wreck of an earlier volcanic cone which had been blown up by some eruptive paroxysm. Occasionally even a series of concentric crater-cliffs of this kind occur, one within or a little on one side of the other, as in the instance of Vesuvius, which rises from the old semi-crater of Somma (the Atrio), produced by the paroxysmal eruption of the year

79, that overwhelmed Herculaneum and Pompeii; while in 1765–70, and again in 1822–35, it contained within its own crater (formed by a paroxysmal eruption) smaller cones with their craters, one within the other like a nest of boxes. This common but remarkable feature of volcanic mountains will be further discussed in a later page.

How long an interval of repose may argue the complete extinction of a volcano, it would not be easy to say; we know that seventeen centuries occurred between two consecutive eruptions from Ischia. But Vesuvius was during this period in frequent activity, and probably relieved the tension of the subterranean focus to which both belong. In a future chapter on Systems of Volcanos this question will be further discussed.

Summit of Vesuvius in 1774, showing cone within cone.
(From Sir W. Hamilton's ' Campi Phlegræi.')

CHAPTER IV.

EXAMINATION OF THE VOLCANIC PHENOMENA.

THE main agent in all these stupendous phenomena—the power that breaks up the solid strata of the earth's surface, raises, through one of the fissures thus occasioned, a ponderous column of liquid mineral matter to the summit of a lofty mountain, and launches thence into the air, some thousand feet higher, with repeated explosions, jets of this matter and fragments of the rocks that obstruct its efforts—consists unquestionably in the expansive force of some elastic aëriform fluid struggling to escape from the interior of a subterranean body of 'lava,' *i.e.* of mineral matter in a state either of fusion, or at least of liquefaction at an intense temperature. This body of lava is evidently, at such times, in igneous *ebullition.*

That it conducts itself, on obtaining access to the outer air through some crevice sufficiently widened for this purpose, precisely after the manner of a boiling liquid or paste, has, indeed, been ascertained by repeated observations.

It is true that the unexpected occurrence and fearfully destructive character of a paroxysmal eruption, such as has been just described, rarely allow the bystanders to preserve the necessary calmness of judgment for accurate appreciation of its details; still less can they venture upon these occasions sufficiently near to the vent for a careful examination of what takes place *there.* But during the minor eruptions of volcanos whose activity is prolonged through months or years, far better opportunities have been afforded for this purpose. Those especially which, like Stromboli, are in perpetual eruption offer perhaps the most favourable of all. We are there, indeed, admitted almost into the recesses of Nature's laboratory, comparatively open to near inspection at all seasons, without risk, since the explosions which characterize this phase rarely exceed an average ratio, and the crater can consequently be approached and its interior viewed at leisure with complete impunity.

The observations made by Spallanzani upon Stromboli, in 1788, first exhibited the nature of volcanic action in its true light.

This most remarkable island is elliptical in plan and conical

Fig. 1.

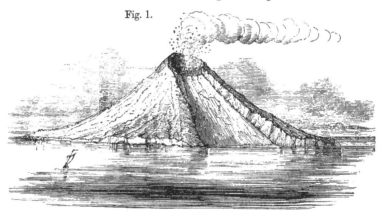

View of Stromboli, from the North.

Fig. 2.

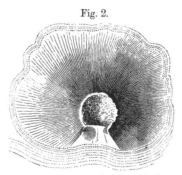

Plan of the Island of Stromboli.

in figure, as may be seen in the accompanying outline-sketches, rising, at an angle of from 30° to 50°, to a height of nearly

3000 feet. It has a crater on the summit, breached towards
the north, on which side a smooth inclined plane slopes, at an
angle of about 50°, almost immediately from the *bottom* of the
crater down to the sea. The steepness of this *talus* prevents any
of the scoriæ which are continually thrown up out of the vent
from resting upon it. Those which fall on that side, therefore,
roll down into the sea, where, after trituration by the waves,
they are probably carried off into deep water by currents.

On reaching the summit edge of the crater by a path that
leads from the inhabited side of the island, the observer looks
directly down into the mouth of the volcano, some 300 feet
beneath him. On my visit to it in 1820, I was able to verify
the accuracy of Spallanzani's account, and ascertain that the
phenomena were at that time still precisely such as he described them in 1788. Two rude openings show themselves
among the black chaotic rocks of scoriform lava which form
the floor of the crater. One is to appearance empty; but from
it there proceeds, at intervals of a few minutes, a rush of
vapour with a roaring sound, like that of a smelting-furnace
when the door is opened, but infinitely louder. It lasts about
a minute. Within the other aperture, which is perhaps 20 feet
in diameter, and but a few yards distant, may be distinctly perceived a body of molten matter, having a vivid glow even by
day, approaching to that of white heat, which rises and falls at
intervals of from ten to fifteen minutes. Each time that it
reaches in its rise the lip of the orifice, it opens at the centre,
like a great bubble bursting, and discharges upwards an explosive volume of dense vapour with a shower of fragments of
incandescent lava and ragged scoriæ, which rise to a height of
several hundred feet above the lip of the crater. Many of the
fragments do not reach so high. Part of them fall back within
its circuit, to be again rejected. A considerable proportion,
however, falling on the steep talus already described on the
north side of the vent, roll or slide down into the sea; and it
is evident, from the crater continuing to retain its depth and

form, that sooner or later, after perhaps repeated ejection, all must find their way there, to be distributed over the bottom of the Mediterranean*.

The volcano of Masaya, near the lake of the same name in Nicaragua, called by our sailors "the Devil's Mouth," exists also, like Stromboli, in a state of permanent eruption, and its phenomena are equally instructive. Incandescent scoriæ are constantly shot up from the bottom of its crater, where large bubbles of liquid lava are seen to rise and fall within a glowing abyss with extreme regularity at intervals of fifteen minutes. The phenomena of its eruptive activity reach at least as far back as the year 1529, when it was visited and described by the Spanish historian Fernando Gonzalez de Oviedo, who, being well acquainted with Vesuvius, was competent to make trustworthy observations on volcanic phenomena. "In its ordinary state," he says, "the surface of the lava, on which black scoriæ appear floating, remains several hundred feet below the margin of the crater; but, at times, by a sudden vehement boiling up, it almost reaches the upper rim, and then discharges a gerb of red-hot stones." The perpetual light seen from a distance is justly attributed by Oviedo, not to any real flame, but to the reflexion from the clouds of vapour that hang over the open abyss, of the glow proceeding from the incandescent lava within it, as also to the brightness of the jet of lava-drops from that surface. The light is, notwith-standing, so strong, that at a distance of more than three leagues on the route to Granada, the illumination of the country around almost equalled that of the full moon. Both

* These observations have been subsequently repeated by M. Hoffman. I may remark, that the scoriæ now ejected by Stromboli are very full of augite, the crystals being very perfect—many of them mackled. Much of the volcanic sand thrown up consists entirely of them. It is evident that these crystals must have been formed in the lava previous to its ejection. The trachyte observed by G. Rose and Humboldt in Stromboli was taken from a rock which forms the base of the island, and is of an early date.

this volcano and that of Stromboli serve, indeed, the purpose of lighthouses to the navigators of the neighbouring waters. These observations on Masaya were confirmed by Mr. Squier on his recent visit, as describing accurately its phenomena at the present time*.

In each of these instances there unquestionably exists, within and below the volcanic vents, a body of lava of unknown dimensions, permanently liquid, at an intense temperature, and continually traversed by successive volumes of some aëriform fluid, which escape from its surface—thus presenting all the appearances of a liquid in constant ebullition.

The phenomena of other volcanos whose eruptions are more or less intermittent, do not differ in character, while they last, from those of Stromboli and Masaya. Thus during the eruption of Vesuvius in 1753, those who ventured to the summit of the cone observed jets of liquid and incandescent lava-drops thrown up in succession from the surface of a boiling mass of lava at a white heat, which occupied the bottom of the crater. A precisely similar appearance is described by MM. Deville, Roth, Abich, and other observers, in the case of the small craters of Vesuvius, which, since the year 1822, have been formed and filled up, re-formed and again filled by more or less placid eruptions, within the great central crater produced by the paroxysmal explosions of that year †. Spallanzani remarked a similar appearance within the crater of Etna in 1788. The volcano of the Ile de Bourbon presents another parallel example. Bory de St. Vincent, who twice visited its active crater and passed a whole night upon its borders, describes it as filled with a body of liquid lava, apparently at an intense heat, but covered with a thin and cracked crust of a dark colour, except at the centre, where the matter was completely incandescent and continued alternately rising upwards and falling again, after having given issue to a jet of vapour

* Squier's 'Nicaragua,' p. 374.
† See Forbes, Edinb. Journ. 1850, and Comptes Rendus, 1856, p. 208.

and drops of lava. The oscillatory motion thus occasioned produced a series of concentric waves or wrinkles on the surface of the pool of lava, which were broken by a network of cracks through which the lava was visible at a white heat, gradually dulling to a faint red as the waves receded from the central fount (see fig. 3).

Fig. 3.

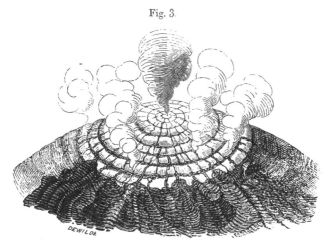

Source of lava on the summit of the volcano of Bourbon.
(From Bory de St. Vincent.)

The crater of Kilauea in Hawaii is described by Professor Dana and other observers as exhibiting the same appearance, on a much larger scale, of a lake of liquid lava more or less crusted over on the exposed surface, through which, at several points, volumes of vapour burst upwards with jets of highly viscous matter, which on cooling takes the form of vitreous filaments or scoriæ.

There exists, indeed, no account of the occurrence of any considerable subaërial eruption wholly unaccompanied by the vertical projection of scoriæ or pumice—that is, of fragments torn by exploding vapours from a surface of more or less liquid lava, which when cooled are found to contain innumerable pores or vapour-vesicles. And this fact is alone sufficient, even though no currents of lava be produced, to show that the elastic fluids explode from the interior of a mass of that nature.

When minor eruptions occur from a deep central crater on the summit of a lofty volcanic mountain, the incandescent scoriæ thrown up from the bottom of the cavity may be invisible to spectators at a distance. In such cases, however, the shock of the repeated detonations is usually felt and their report heard by the inhabitants of the mountain's flanks; while the brilliant light proceeding from the jets of scoriæ, or the glowing lava from which they burst, reflected by the cloud of aqueous vapour impending over the crater, produces that luminous appearance which is erroneously described as *flames* in the accounts given of volcanic eruptions by unscientific witnesses. Whether any real flames are ever evolved from a volcano in eruption, through the inflammation of hydrogen or other inflammable gases, is a point perhaps as yet undetermined. If they do occur, it is only under peculiarly favourable circumstances that they would be visible, their weak light being liable to be effaced by the stronger glow of the incandescent lava. Abich considered that he saw faint but real inflammations of hydrogen gas in the interior of the crater of Vesuvius.

It must then be admitted as indisputable that there exists within and below every volcanic vent, at the time of its eruptions, a subterranean mass of liquefied mineral matter, or ' *lava*,' of indeterminate extent, at an intense temperature, traversed more or less freely by ascending volumes of an elastic fluid, which burst from its surface and rise rapidly into the air. Such a mass of lava conducts itself, as has been said, precisely after the manner of a liquid in ebullition.

If any doubt should suggest itself whether this fluid is actually generated within the lava, or only rises through it, having its origin in some other substance, or in some other manner, beneath, it must be dispelled by the evidence afforded in the extremely vesicular or cellular structure of very many erupted lavas, not merely near the surface, but throughout their mass, showing that the aëriform fluid in these cases certainly developed itself interstitially in every part. And although such vesicles or cells appear at first sight to be wanting in other lavas, at least in the lower portions of the lava-current after its consolidation, the microscope invariably, or almost invariably, discovers them. In those exceptional cases where the rock is to appearance perfectly compact, it is allowable to suppose that the vapour it once contained escaped in ascending bubbles, or by exudation through extremely minute pores, or was condensed by pressure and refrigeration previously to the solidification of the matter.

A question, however, may arise, whether ebullition commences only in the lava when it obtains an approximatively free communication with the atmosphere, or takes place previously in the depths of the volcanic focus the moment any sufficient space is obtained for the sufficient expansion of some portion of the vapour by the yielding and vertical elevation of the overlying rocks. The first hypothesis appears most probable, as the normal law, not only from the evidence of the phenomena themselves, but likewise from a consideration of the character of the vapour, which as it issues from an eruptive vent is that of water—in other words, steam,—in the proportion, according to the experiments of M. Deville and others, of $\frac{999}{1000}$, the fractional portion being chiefly sulphuretted hydrogen and other gases already enumerated.

The fact that the great mass of elastic fluid discharged from a volcanic vent in eruption consists of aqueous vapour, has not only been recognized by all who have experimented upon it directly*, but is evident moreover, as Sir Humphry Davy observed, in the mere aspect of the column of white vapour which every eruption propels upwards, and which, unless dispersed by winds, collects into dense and heavy clouds above the mountain, and is soon seen to fall in torrents of rain upon

* Breislak, Monticelli, Davy, Daubeny, Deville, &c.

its flanks or the surrounding country. The vapour evolved by the permanently active vents, such as Stromboli, forms a constant cloud of a white or greyish colour, which remains stationary above the peak when the air is calm, or encircles it with a wreath of mist, or falls in light showers, or streams down wind, according to the density and temperature of the atmosphere, precisely like an ordinary vapour-cloud. That of Mount Erebus in Victoria Land was observed by Sir James Ross to fall in showers of snow to leeward of that volcano.

The results of the microscopical examination of the cells and cavities disseminated through lavas, by Mr. Sorby, moreover support the conclusion that the vapour which they contained at the time of their formation was that of water; which, indeed, M. Delesse and others have detected in all the crystalline plutonic rocks, whence the lavas of volcanos have probably been derived. If in some compact lava-rocks water is not appreciably present, it has doubtless escaped, as was observed above, and as M. Delesse suggests, through 'fumaroles' at the period of their solidification*.

Now water we know is converted into vapour only at temperatures increased in proportion to the increased pressure to

---

* Bulletin de la Soc. Géol. de France, 2ᵉ sér. vol. xv. p. 750. M. Daubrée, the latest and most persevering experimental investigator of the phenomena of metamorphism, thus expresses himself :—" In the exhalations of volcanos there is a substance which did not at first fix the attention of men of science, because, under the influence of early theories, it seemed altogether inactive towards the formation of minerals, but which appears in truth to have played the foremost part in the phenomena of metamorphism as well as in the eruptions of volcanos. This substance is water, which is found in these exhalations, not in minute quantity only in the vapours, but, on the contrary, as the most abundant and the most constant product of eruptions, in all regions of the globe. . . . We know nothing of the matter situated at a certain depth within our globe, except from what volcanos bring up from thence. Now, these ejected matters, all of them without exception, contain water, either in mixture or in combination. We are then justified in thinking that water plays a most important part indeed in the principal phenomena that proceed from the interior of the earth."—(Etudes, &c. p. 102.)

which it may be subjected*; and when altogether hindered
from communication with the atmosphere, as in a Papin's
digester, or other closed vessel, may be made red-hot without
expanding into vapour.   The moment, however, that an opening is made in the enclosing vessel, reducing the pressure to
that of the atmosphere only, it flashes instantly into steam
with explosive violence.   The same effect, of course, must
take place in an imperfect liquid or paste composed of water
and any solid matter in mechanical suspension or mixture,
such as flour, clay, sand, or any other granular substance.   In
a future page, when we come to treat of the mineral character
of lava rocks, it will be shown that there is reason to believe
nearly all lavas, at the time they are erupted, to be a compound of this mixed character—to consist, that is, of more or
less granular or crystalline matter, containing minute quantities of either red-hot water, or steam in a state of extreme
condensation and consequent tension, disseminated interstitially among the crystals or granules, so as to communicate a
certain mobility to them and an imperfect liquidity to the
compound itself; and that, although at an intense temperature, lava is but rarely emitted in absolute molecular fusion,
such as it would be reduced to by an equal or perhaps less
degree of heat, under the pressure of the atmosphere alone.

Postponing, however, the discussion of this question for the
moment, it is shown beyond dispute, by the evidence of the
phenomena described above, that the rise of lava in a volcanic
vent is occasioned by the expansion of volumes of high-
pressure steam, generated in the interior of a mass of lique-

---

* By the experiments of Regnault, it appears that, under a pressure
equal to twice that of the atmosphere, water does not boil below the temperature of 249°.   Under a pressure of 20 atmospheres it will not boil
below that of 415°.   The observations of M. Bunsen on the pipe of the
Great Geyser, the water in which increases rapidly in temperature from
the surface, where it is below 212°, to the depth of 60 feet, where it is
260°, make it probable that within a narrow and irregular fissure a proportionately less amount of pressure than this will prevent ebullition.

fied and heated mineral matter within or beneath the eruptive orifice.

The great bubbles of vapour that burst from its surface on exposure to the open air, in explosions which form the chief feature of every volcanic eruption, evidently ascend by the force of their differential specific gravity, from a certain depth within the boiling mass. But at what depth these volumes of vapour were generated may be a question. Probably this is determined by the conditions of the lava, in respect to its liquidity and specific gravity, and the temperature and pressure to which it is for the time subjected. The narrowness and intricacy of the fissure through which it rises must also exercise a great influence in this respect. And from what we know of the character of fissures through solid rocks, especially such as appear from their contents to have served as channels of discharge for erupted lavas (the dykes of volcanic districts), it seems reasonable to suppose that if any amount of vapour is generated at considerable depths beneath an active volcanic vent, it only reaches the external surface in a state of extreme condensation, and entangled in the liquid lava which rises with it and escapes outwardly, just as any other thick or viscid matter exposed to heat from beneath, in a narrow-mouthed vessel, *boils up* and *over* the lips of that aperture. The remaining water may be in that globular condition which the experiments of M. Boutigny have shown this fluid to take at intense temperatures, and which would be likely to communicate a high degree of mobility to the particles of mineral matter among whose interstices it was confined. The tendency to vaporization must, however, everywhere occasion an extreme tension or expansive force throughout the mass, only restrained by the enormous weight and cohesion of the superincumbent rocks.

The vast quantity of aqueous vapour discharged during every volcanic eruption must abstract from the interior of the volcano and outwardly disperse a proportionate quantity of caloric. The mass of intensely heated lava usually expelled at the same time carries forth an equal or still greater quantity. These phenomena therefore, if the process of eruption continues for a lengthened period, as it does in the case of the permanently active vents, attest a *continual accession of increments of caloric* to the subterranean mass of similar matter with which that in the vent communicates, proceeding from some source or cause, the nature of which it is not at this moment my intention to discuss. In the case of these permanently active vents, it would seem that this continued accession of heat escapes outwardly, in the way described, as fast as it is received; by which an equilibrium is maintained between the force of subterranean expansion and that of repression, consisting in the weight and cohesion of the overlying rocks, the atmosphere, and, perhaps, the sea.

This equilibrium seems, indeed, in these instances to be so nicely adjusted, that the mere ordinary variations in the weight of the atmosphere are sufficient to disturb it to some extent. The inhabitants of Stromboli positively make use of their volcano as a weather-glass*. The Peak of Ternate in

---

* They are mostly fishermen; and while engaged in their occupations at a short distance from the island, have its orifice constantly in view; and I was assured, by all whom I questioned on the subject, that its phenomena decidedly participate in the atmospheric changes, increasing in turbulence as the weather thickens, and returning to a state of comparative tranquillity with the serenity of the sky.

During the tempestuous weather of the winter season, the eruptions of the volcano no longer preserve the uniformity which characterizes them the greater part of the year. The explosions are then often so terrible that the island seems to shake from its foundations; and these paroxysms, which sometimes last for days, are succeeded by intervals of complete quiescence, of a few hours' duration, which are in turn followed by other eruptions of similar energy. On such occasions the steep flank of the cone, which forms the inclined plane beneath the crater, has been observed to be rent in two by a vertical fissure, which emits a torrent of lava into the sea. The chasm is afterwards sealed again by the consolidation of the lava, and is covered and concealed by the loose fragments ejected from above.

the Moluccas is said to break out with greatest violence during
the equinoxes; and in other habitually eruptive vents a similar
correspondence has been observed between their activity and
any sudden or prolonged fall of the barometer.   Nor is this
other than what might be anticipated.   The boiling-point of
every particle of the water contained within the column of
lava occupying the vent must vary with the weight of the
atmosphere; and even where outward eruption has temporarily ceased, the subterranean expansive force, ever active,
and continually pressing upwards with gradually increasing
energy as it receives fresh increments of heat, must be at
times restrained only by the slightest degree of superiority in
the force of repression, so that the most trifling diminution
in one of the elements of the latter may suffice to give the
predominance to its antagonist and occasion a renewal of
eruption.

For, in the case of an intermittent volcano, it is, no doubt,
the augmentation of internal temperature in the intervals of
external repose, during which heat can only escape from its
focus by the slow process of conduction through the overlying
solid rocks, that brings about at length one of those sudden
and violent ruptures in these, the snap and jar of which are
felt as a sudden shock, propagated in vibratory undulations
through their horizontal extension to a greater or less distance
—the local and minor earthquakes already mentioned as the
usual precursors of a renewal of eruption after an interval
of quiescence.

Every such rupture must be attended by some amount of
vertical displacement or elevation of the overlying beds of
rock, giving room for a proportionate expansion of the lava-
mass beneath.   Sufficient relief will be thus afforded, perhaps,
to admit of the generation of vapour to some extent within
the expanding matter.   Whether this amount of vapour remains disseminated in minute parcels through the mass of the
lava at the points where it is generated (so as to cause a sort

of general intumescence), or combines into globular vesicles which rise to the surface of the mass by virtue of their inferior specific gravity and collect there into larger volumes, will, as I have said, probably depend, *cæteris paribus*, on the greater or less liquidity of the lava—that is to say, on its more or less complete fusion, or disaggregation, and the consequent mobility of its component particles. It is possible that in the case of some of the fine-grained lavas, most nearly reduced to absolute fusion, and especially those whose specific gravity is high, some amount of vapour at a high degree of tension may rise from the inner depths of the subterranean mass towards its surface, and form there blister-like cavities of considerable size; in other words, that in these instances a certain amount of subterranean ebullition may take place before any fissure is opened so far as to afford free communication with the open air, just as a portion of condensed gas or vapour rises to the surface of water in a soda-water bottle or in a high-pressure steam-boiler when the cork of the first or the valve of the last is for a moment relaxed and a partial expansion permitted. The facts favouring this view are the subterranean sounds, resembling the firing of distant artillery, which have been sometimes heard in the vicinity of a volcano before the commencement of an eruption, and which I would compare with the bubbling of the water in a high-pressure engine, and other evidence to be hereafter described, that seem to indicate the existence in some cases of considerable cavities beneath the superficial rocks in the immediate neighbourhood of an eruptive vent. In such cases it will be probably the increased tension of the vapour contained in such a cavity, as it receives further additions from the boiling lava beneath, which brings on, sooner or later, the explosive burst with which an eruption at times commences. But there seems reason to believe that such cases are exceptional, and that, as a general rule, the liquidity, that is, the mobility of the component particles of the lava, is too imperfect to permit

the vapour generated, if any be generated within it by a partial relaxation of the confining surfaces while it is still in
close confinement beneath the overlying rocks, to expand or
agglomerate into volumes of any size.   It remains, probably,
for the most part, where it is first developed, causing a general
intumescence of the mass, or of such portions as local accidents may allow to be thus expanded; but neither uniting
into large vesicles nor rising to the surface until some of the
intumescent matter has forced its way upwards through a
fissure so far as to obtain a free, or nearly free, communication
with the outer surface of the earth, and to be subject consequently to the pressure of the atmosphere only; and there
only enters into positive ebullition, manifesting its presence
externally by explosive discharges of vapour, jets of fragmentary matter, and overflows of lava.

It is, indeed, unquestionable that a very considerable
amount of dilatation *must* take place in every body of mineral
matter so confined, and receiving, as in the case supposed,
augmentations of temperature, *before* the vaporization of any
of its contained interstitial water *can* commence, viz. during
its passage from a state of solidity (supposing it to be originally solid) to one of fusion, or even of imperfect liquefaction.
For although it must be admitted that in the present state
of science little is known of the real nature of heat, and still
less of the changes effected in the internal structure of bodies
by increase or diminution of temperature under varying conditions of pressure, we do know that nearly all mineral matter
expands and contracts with every change of temperature, and
that with intense force.   According to Bischoff, granite, in
passing from a solid to a fluid state, expands from ·7481 to
1·000, trachyte from ·8109, and basalt from ·8960; that is to
say, these rocks occupy from nearly one-fourth to one-seventh
more space in the latter state than in the former.   It is certain that during such a change every molecule of the substance
must be affected in its position by internal movement, which,

irregular as it may be in its local action, cannot but occasion more or less of mutual friction and disintegration among the component solid and angular particles. It would seem that, at some fixed degree of temperature and amount of pressure (according to the nature of the mineral substance), the disintegration caused by the increase of temperature dissolves the cohesion of the ultimate or molecular particles to such an extreme extent as to give them that complete freedom of movement which we call *fusion*; while a proportionate diminution of temperature (or increase of pressure) reaggregates them in a solid mass, though not always in the same crystalline forms. But between perfect solidity, or complete stability of the component particles, and absolute fusion, there must be many intervening degrees of partial resolution and disaggregation, producing more or less of softness or imperfect liquidity in the substance. Such an imperfect fluidity is seen in the gradual softening of sealing-wax and other resinous substances when exposed to moderate heat, short of fusion. A still closer analogy may be found, perhaps, in certain stages of the manufacture of sugar, when the matter consists of a soft mass, or 'magma,' of granules or imperfect crystals enveloped in a liquid (syrup), which, being subsequently dried by evaporation or drainage, consolidates into a hard substance formed of interlaced crystals.

There is reason to believe, as will be shown hereafter, that lava, at the time of its eruption, is generally in this pasty, imperfectly liquid, and loosely granular condition, and, like sugar, acquires solidity and a more crystalline texture by the escape of its fluid vehicle while cooling.

At what depth within the volcano-chimney, or to what extent a subterranean mass of lava exists in this condition of softness, pastiness, or imperfect liquidity—at what depth it still remains solid—or at what points, if any, it is reduced to absolute molecular fusion—must be always uncertain, and, indeed, necessarily dependent on local accidents of composition, temperature, and pressure.   It would, however, be consistent with many physical analogies to suppose that all these several conditions may often—perhaps always—coexist, and graduate into one another, in the reservoir of lava which unquestionably underlies every active volcano.

There is good reason to believe that the matter which has evidently been injected forcibly from below into such fissures as open themselves in the overlying rocks, was at the time in a pasty or semifluid state, and this condition it probably acquired on passing from one of more or less complete solidity through the dilatation resulting from increase of temperature.

The fissures broken through the rigid superincumbent mass by such expansion of a subjacent body of lava will be more or less vertical, because at right angles to the strain—although probably very irregular in their course and figure, owing to varying accidents of composition and structure, and consequently of cohesive force, in the several rocks.   Moreover some of the fissures will no doubt be split in such a manner as to open downwards—that is, to have their wider end in that direction ; others upwards.   The former will, of course, be most readily injected by the intumescent lava.   Upon the subject of the production and direction of such fissures, the general abstract reasoning of Mr. W. Hopkins, in his paper on Theories of Elevation and Earthquakes *, is probably indisputable.   He, however, contemplates chiefly the case of a limited superficial area of solid rock affected by the elevatory force of a subterranean 'lake' of expanding liquefied matter, on the assumption that the elevatory force as well as the resistances are approximatively equal at every point of the lower surface of the solid mass above it.   But this state of things can scarcely, I apprehend, ever occur in nature.   Whatever be the source of that increase in the expansive force which occasions the change from a stationary condition, it must operate sooner, and more powerfully, at some one point than

* Brit. Assoc. Report, 1847.

at any others.   Moreover the resistances cannot be equal throughout the whole area affected ; on the contrary, there is sure to be great local irregularity in them.   The *maximum* of elevation will necessarily take effect at some point determined by the relative amount of these antagonistic influences. This we may call the Centre of Dislocation.

The production of fissures entirely depends on the cohesion and rigidity of the elevated mass.   If that were fluid, or soft and yielding to pressure like mud, or loosely aggregated like sand, there might be much change of place among the particles, but no split or fissure could be formed ; and the fissuring effect upon solid rocks will be determined chiefly by their mechanical structure.   It seems to be a property in the substance of the far greater number of such rocks to yield to pressure in more or less *straight* and prolonged fractures.   Probably this is owing to their component particles being generally flat-sided, and arranged, by some crystalline or concretionary action, in more or less continuous and parallel planes.   In proportion to the degree in which any solid partakes of this peculiar arrangement, it will tend to break in straight and also parallel planes, whether in the case of primary fissures or secondary transverse ones.   Rocks which possess a jointed structure naturally yield most readily in planes coincident with these existing solutions of continuity, which are nearly always straight and parallel.   But even in extended solids of seemingly uniform and uninterrupted structure, it is remarkable how the fractures approximate to rectilinear planes.   In a sheet of ice, for example, fractured by pressure, the cracks in general are prolonged through considerable distances in straight lines, with occasional more or less transverse fissures equally rectilinear. The propagation of such cracks is rapid, though slow enough to be visible, and accompanied by a vibratory movement and sound—reminding us of those that constitute the phenomena of earthquakes, when the crust of the earth seems to be similarly rent by lengthened rectilinear fissures.   Even substances which, like glass, china, resin, &c., have what is called a conchoidal or curved fracture in the solid, yet, when extended in thin planes and exposed to dislocating pressure, break in more or less longitudinal and perpendicular transverse fissures, as is seen when a pane of glass is *starred* by a blow.

Now, if we suppose the rocks above a subterranean mass of expanding mineral matter to have such an internal structure as to give them a general tendency to split in straight longitudinal fissures (and this must almost invariably be the case, as has been just said), the fissures first formed, at the moment when the elevatory force overcomes the resistance caused by the cohesion and weight of the superincumbent mass, will open where the strain is greatest; and therefore, in the upper surface of the overlying rocks at or about the centre of the dislocated area, and in the lower surface towards its lateral limits. A crack formed in the first of these positions will be rapidly propagated downwards to the point, or plane rather, at which all strain ceases and compression begins, which may be called the pivot or 'neutral axis' of fracture. A crack, on the other hand, taking place at the lateral limits of the affected area, adjoining the parts that remain unmoved, will first open at the lower surface of the solid rocks, and be propagated upwards. The first order of fissures will widen towards the upper side as they are prolonged down to the neutral axis, but will be kept forcibly closed below that level by horizontal compression. The second or *lateral* order of fissures will, as they are propagated upwards, widen towards the expanding matter beneath, which will experience a sudden relief from pressure at that spot, and by rapid intumescence inject itself into and up the fissure so opened. The rocks on either side of the upper portion of these last fissures will also be exposed to violent horizontal compression perpendicular to the direction of their planes. But inasmuch as the outer surface presents no appreciable resistance to vertical upward movement, the effect of this squeeze will probably there be to elevate, more or less, portions of the

surface along the line of these lateral fissures, and, by multiplying parallel superficial cracks in the surface-rocks, to weaken
their resistance to the effort of the lava injected into the inferior portion of the fissures to rise through them into communication with the outer atmosphere. On the other hand,
should the superficial strata be soft and pliable, this horizontal
squeeze will tend to compress them into undulating folds.

The accompanying figure is intended to represent, rudely
and approximatively, the probable relative position of these
different classes of fissure, and of the masses of displaced rock
divided by them.

Fig. 4.

Ideal section of a mass of superficial rocks of limited area elevated by the
expansion of subjacent mineral matter.

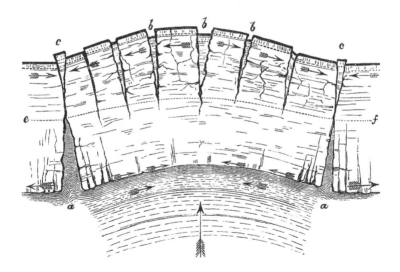

a a, Fissures gaping downwards, and injected by the intumescent lava beneath.
b b, Fissures gaping upwards, and allowing wedges of rock to drop below the
level of the intervening masses. c c, Wedges forced upwards by horizontal compression. e, f, Neutral plane, or pivot axis, above and below which the directions
of the tearing strain and horizontal compression are severally indicated by the
smaller arrows. The large arrow beneath represents the direction of the maximum of expansive force.

I shall have occasion to refer again to these considerations when treating of the elevatory action of the deep-seated Plutonic forces; to which order of facts, indeed, they more properly belong*.

* I am anticipating here, perhaps, what might have been postponed with advantage to a later page; but it is, in truth, impossible to distinguish the action of the subterranean force which pushes up the crystalline axis of a mountain range, elevating, fracturing, and contorting the stratified rocks above it, from that which at the same time, and by the same effort in all probability, opens the crevices through which some portion of the intumescent and heated matter beneath rises into free communication with the outer air in a volcanic eruption.

With respect to the particular direction of the primary fissures that are likely to be formed in a mass of elevated rocks, it is, no doubt, determined chiefly (as I understand Mr. Hopkins to indicate) by the accidental position of two or more points of least resistance yielding simultaneously, and influencing therefore the formation of a connecting fissure, which, once begun, will be rapidly propagated in the same direction to a distance determined by circumstances. And similar circumstances will determine also the position of transverse fissures, which must often be produced by the unequal yielding of the overlying rocks in the direction along which the primary fissures are propagated. Secondary parallel fissures will be formed, I conceive, on either side of the primary ones by the dislocating effect of the passage of waves of vibration caused by the sudden snap of the splitting rocks, and propagated in directions perpendicular to the split. The primary fissures, in this view, impress their position on all the lateral ones, except, of course, so far as deviations may (and often must, no doubt) be occasioned by local irregularities of resistance. I strongly incline to the opinion that it is the jar and vibration propagated through the contiguous rocks by the splitting of each fissure that occasions all earthquake-shocks, whether on the largest or the smallest scale; differing on this point (though with much deference) from the view taken by Mr. Robert Mallet, who ascribes them to a blow caused by the sudden condensation of volumes of intensely heated steam issuing from some fissure beneath the sea. Mr. Hopkins appears inclined to take the same view of this question as myself (p. 90, *loc. cit.*). So also Mr. Darwin (Geol. Trans. 2nd ser. vol. v.). Mr. Hopkins refers to the observation frequently made in regard to the position of displaced strata on either side of *faults*, viz. that the side which has risen highest is that towards which the fault inclines in ascending, and explains it by the suggestion that when wedge-shaped portions of superficial rock intervene between non-parallel fissures, those wedges which have their narrow ends downwards will be less acted on by the elevating force than the neighbouring portions whose

In the meantime it is certain that, in whatever part of the disturbed area they may be formed, those fissures which open first at the lower surface of the overlying rocks, and, as they extend themselves, gape or widen downwards towards the intumescent mass beneath, will be instantly injected with its most

broad ends lie downwards—that is, towards the point whence the movement travels (p. 63, *op. cit.*).   This is perfectly just, so far as relates to the fissures or faults formed about the central portion of the elevated area, and with the understanding that the effect does not depend on the broad-bottomed wedges actually reaching down to the intumescent lava-mass below; since it would be equally the case through whatever intervening thickness of irregularly-dislocated or fragmentary rock the movement might be transmitted—being caused by the greater leverage afforded to the upward pressure by their extended lower surfaces, and also by the outward *gaping* (as I may call it) of the superficial fissures, which gives room for such wedge-shaped slices of rock as have their fine edge downwards, to *drop* by virtue of their own weight, or remain stationary, while the broader-based portions are forced up on either side (fig. 4).   But with respect to such fissures as may reach the surface about the lateral limits of the upheaved area, where the rocks are in a state of compression, such wedge-shaped slices as are there broken off will tend to rise (as already said), rather than fall, above the level of the adjoining beds—being forced outwards by the squeeze resulting from the vertical and horizontal pressures combined (unopposed by any appreciable pressure from above),—just as we see wedge-shaped chips split off and forced outwards from the edges of any crack formed in the same relative position through a rigid mass of stone or metal which is broken across by pressure.

Mr. Hopkins's idea seems to be, that the horizontal compression or lateral crushing of the elevated strata is owing to their subsequent subsidence in portions leaning against each other at irregular angles, " for want of sufficient support to maintain them in their more elevated position" after the outward escape of the elastic vapours to whose generation the elevatory action is attributed.   And Mr. Darwin (*loc. cit.*) agrees with him in this view.   In my opinion, as will be seen from the statement and diagram given above, the horizontal compression, the effects of which are so conspicuous in all greatly elevated strata, proceeds from the same cause as that which is known to be produced in a beam fixed at either end, and broken by upward pressure at its centre,—a compression taking effect in the central part beneath, and in the lateral parts, adjoining the fixed extremities, above a neutral line or 'pivot axis.'   This view will be further insisted on when the elevation of mountain-chains and the convolutions of their lateral strata come under our consideration.

fluid matter, which will be in a manner forcibly pumped or sucked up into the smallest no less than the largest rents there formed, by the tremendous hydrostatic pressure to which any liquid in such a position must be exposed. Such injected fissures, when the lava is subsequently cooled and consolidated, appear as more or less vertical dykes traversing the surface-rocks, but not generally effecting any disturbance in their position.

It is a well-known fact that up-filled dykes of the kind are very numerous in all volcanic districts, especially in the vicinity of the eruptive orifices. For example, in St. Helena, Mr. Darwin describes the surface of some of the plains as reticulated with numberless dykes of basaltic lava, intersecting each other almost like the threads of a spider's web. Multiplied dykes of the kind are seen in the cliffs bordering the ancient crater of Somma (the Atrio del Cavallo), in that of Etna (the Val del Bove), in the central gorges of the Mont Dore and Cantal, and generally in the axial regions of all volcanic mountains,—indeed, wherever lavas have been erupted, and the structure of the inferior rocks has been disclosed by subsequent denudation or other destructive influences. They are naturally most abundant in those compound cones or volcanic mountains of which we shall presently treat, the product of the accumulation of repeated eruptions. The horizontal extension of some vertical dykes is very great—the fissure occupied by the lava having probably been opened by degrees to farther distances. In the great Icelandic eruption of Skaptar Jokul in 1783, lava was emitted consecutively at several points on a linear range of 200 miles. No doubt an underground fissure of this length at least was injected with lava by that eruption, and remains now as a dyke traversing the substrata, similar to those which in the North of England are found cutting through the coal-measures and oolites of Durham and Yorkshire for distances of sixty miles and more in a straight line. Such fissures, in the neighbourhood of active volcanos, are, of course, generally concealed from view by the showers of ejected matters. Nor is it necessary to suppose that they everywhere reach the outer surface. Their existence is often only disclosed by a sinking-in of the loose ground along their course. The earlier dykes of this kind, observable in districts where the surface has suffered much denudation, occasionally swell out suddenly at spots which probably mark the points where lava reached the surface and was outwardly erupted at the date of the injection of the fissure.

For when penetrating such a fissure, the injected matter pressing, with all the force of its own tension and the impulsion it receives from beneath, against the sides of the cleft, must operate like a wedge to widen and extend it. Hence, sooner or later, perhaps, according to the rate of increase in the temperature and consequent tension of the inferior lava-mass, at some weakest point or points of one of the fissures— a point weakened possibly by the superficial rupturing referred to above as arising from horizontal compression in the upper lateral surfaces of an elevated area—an opening will be forced sufficiently high to permit the lava to boil up into contact with the atmosphere, and the volumes of vapour, which must upon this instantly develope themselves within the intensely heated matter at greater or less depth, and struggle up through the vent thus obtained to the surface, in rapidly enlarging bubbles, will burst thence with explosive violence, in the manner already described as characteristic of a volcanic eruption.

The pressure upon such bubbles of steam as they rise through the fissure must be so intense, from the great weight of the column of lava above them, and the pinching squeeze of the rocks on either side—which the swelling wedge of lava expands and keeps apart—that they probably occupy a very small comparative bulk until they approach the open air, when they will flash out at once with an energy proportioned to their tension and previous compression, much in the manner of ignited powder in a gun-barrel. And this flash taking place at some depth within the throat of the vent must tend to break up the sides of the fissure at this point, and discharge the fragments broken from them in a vertical jet, just as the ignition of gunpowder within the breech of a gun-barrel tends to burst it, and drive out any opposing substance with great force from its muzzle in the direction of the barrel's length. In truth, the aëriform explosions of a violent eruption have all the characters of successive continuous discharges of steam from the mouth of a colossal Perkins's steam-cannon in a vertical position. It is, no doubt, the equal pressure of the elastic steam-bubbles on every side, as they explode in the widest part of the fissure of discharge, which gives to that orifice ultimately the circular or nearly circular form so distinctive of volcanic craters. No other cause is, I conceive, capable of accounting for this universal peculiarity of figure.

Nevertheless the pinch sustained by the ascending bubbles from the sides of the fissure does frequently occasion the cavity to take an elliptical form. Indeed, many volcanic craters, when the eruption has broken through solid rock, are not merely elliptical in their horizontal section, but show, at the extremities of their longest axes, evident signs of the further prolongation of the fissure, through the weakest point of which the eruption forced its way. Such a cleft was visible in the cone of Vesuvius after the great eruption of 1822, in a deep depression across the south-eastern lip of the crater; and a similar fissure was remarked during the eruptions of 1852–6, according to MM. Deville and Roth. I shall have occasion to mention many more instances in a future chapter.

CHAPTER V.

DISPOSITION OF FRAGMENTARY EJECTIONS.

§ 1. IT is, then, by the rapid, continuous, and explosive dis-charge of steam-bubbles, ascending from some depth within lava which has forced its way up some fissure to the outer surface, that this orifice is gradually enlarged, the intercepting matter blown into the air, and a ' crater' formed of a magnitude proportioned to the violence and duration of the eruption. But though the matters ejected at first consist principally or wholly of the debris of the solid rocks broken through, no sooner has the surface of the lava, boiling up in the way we have described, risen sufficiently high within the vent, than jets of this liquid and incandescent matter are tossed up likewise, and ragged fragments torn off from the solid crust that instantly tends to form upon its exposed surface. Thus a rising foun-tain and falling shower of such incandescent and fragmentary matter is seen reaching upwards to a greater or less height— in paroxysmal eruptions, to one of several thousand feet.

The force with which these discharges occasionally take place may be conceived from the fact recorded by Ulloa*, that by the eruption of Cotopaxi in 1533, witnessed by the Spaniards under Sebastian de Belelcazar, the plain around the foot of the mountain was strewed through a radius of fifteen miles and more with great fragments of rock, many of which measured as much as nine feet in diameter.

Humboldt, indeed, speaks of one rock, weighing upwards of 200 tons, as having been launched into the air to a height of several hundred feet during an eruption of this same volcano.

* Voyage Historique de l'Amérique Méridionale, t. i. p. 264.

Some of the more liquid portions of the lava shot up assume, by rotation through the air, a globular or pear-shaped figure. These are the volcanic 'bombs' often found in the vicinity of an eruptive vent; and as they can only be produced in the manner here indicated, their occurrence on any spot affords a useful indication of the site of an eruption at some former period, when other signs are perhaps wanting*. Their nucleus is usually compact, and sometimes consists of a solid fragment of some earlier rock caught up and enveloped in the liquid lava; but towards the surface they have a vesicular envelope beneath an outer shell of compact texture. In size they vary from that of the largest block in a man-of-war's rigging, to that of a nut or almond.

The great bulk of the ejected fragments of lava, cooling rapidly in their passage through the air, possess ragged, tattered shapes, and when examined are found to be full of vesicles. Those of the heavier ferruginous lavas are called '*scoriæ*,' from their resemblance to the cinders or slags of iron-furnaces. The scoriæ of the felspathic lavas, which have an inferior specific gravity, are usually still more vesicular or filamentous, and have a vitreous fracture. They are called '*pumice*.' In some cases where the lava is peculiarly tenacious and tough, it is drawn out into filaments, having a silky lustre almost equal to that of asbestos.

As the eruption progresses, the surface of the lava sinks within the vent more or less rapidly, owing either to its outflow from some orifice at a low level in the flank of the volcano, or to the exhaustion of the subterranean eruptive energy by loss of heat through the amount of vapour discharged from it. The eructations, therefore, by degrees lose their power of projecting fragmentary matter beyond the borders of the

---

* The greater number of ancient basaltic currents which cover the slopes of the Monts Dore and Cantal may be traced up to some knoll in the higher parts of either mountain, within and about which such 'bombs' are found, as well as numerous scoriæ, attesting the source in each case of the lava-stream.

crater, and the repeated trituration of the fragments that fall within it, and are continually re-ejected, seems to stifle the explosions beneath an accumulation of fine but heavy dust. They become weaker and weaker, until they terminate altogether. At times, however, the explosions from the crater first formed cease more suddenly, being transferred to some other orifice opened in the vicinity. And this migration of the eruption is occasionally repeated to several vents opened in succession along a straight line, which indicate, no doubt, the course of a subterranean fissure. Sometimes also, after a partial cessation, the eruption recommences with renewed vigour, as if from a fresh effort of the intumescent lava beneath causing the sudden enlargement of the fissure.

Of the smaller fragments ejected vertically from an eruptive vent, those which, by their mutual friction in the air, have been reduced to a sort of gravel of rounded scoriæ, are called *lapillo* by the Italian geologists—a word sometimes corrupted into *rapilli*; when, by still further attrition, they are comminuted into sand, *puzzolana*; and when brought to the condition of fine dust, *ceneri*, or ashes. The *lapillo* is generally of a deep-black colour; the *puzzolana* is red, like burnt brick-dust; the fine ashes are of a whitish grey. But, of course, the three kinds are often found mixed together, as well as with the detritus of older rocks through which the explosions have forced their way. It is probably by the intense mutual friction, in the air, of these ejected solid matters, that the electricity is generated which often manifests itself abundantly in forked lightnings darting from the edges of the dense ascending column. During the great Vesuvian eruption of 1822 they were continually visible, and added much to the grandeur of the spectacle.

§ 2. The finest ash, reduced to an impalpable grey powder, especially towards the close of a paroxysmal eruption, when the crater has been much enlarged, and the lava has sunk deep within it, is borne by the winds that may prevail at the time

to distances which are occasionally prodigious—measured indeed by many hundreds of miles. The coarser fragments heap
themselves necessarily around the orifice of projection, in more
or less abundance according to the violence and duration of
the eruption,—accumulating there in the form of a circular
bank, which increases, as the eruption continues, into a hill
having the figure of a truncated 'cone,' usually with a funnel-
shaped hollow or 'crater' at its summit, marking the mouth
of the eruptive vent. The outer sides of such a circular
mound or cone, the product of a single eruption, slope at an
angle of from 20° to 35°, or even 40°—determined, like the
slope of any ordinary talus of debris, by the average size, form,
and coherence of the component fragments. The scoriæ,
however, ejected from a volcanic vent, being mostly in a heated
state, will often cohere and become immoveable at a steeper
average angle of inclination than ordinary debris. Cinder-
cones of this kind, the product of loose fragmentary ejections
from a single vent by a single eruption, are among the com-
monest and most characteristic features of almost every volcanic district. They vary in size from that of a hay-stack to
a hill more than a thousand feet in height and two or three
miles in circumference. On the flanks of Etna, according to
Sartorius von Walterhausen, more than seven hundred such
are to be seen; in the island of Hawaii, according to Professor
Dana, several thousand!

A favourable field exists for the study of the varieties of
figure assumed by such hills of scoriæ thrown up by single
eruptions, and of the circumstances to which they owe their
numerous modifications, in the French provinces of Auvergne,
the Velay, and the Vivarais. The chain of Puys near Clermont (Dépt. Puy de Dôme) contains above sixty volcanic
cones strung together on nearly the same line, reaching about
twelve miles in length; and again, in its continuation (probably marking a subterranean fissure) through the Velay and
Vivarais, upwards of two hundred similar cones are closely

scattered in a narrow band about twenty miles in length. From among this great number, examples may be observed of *compound* cones, evidently thrown up from two, three, four, or many more points of explosion on the same fissure. Sometimes a string of cones has been produced from vents so near as to mingle their ejections, but still sufficiently removed for each hillock to retain a certain degree of regularity towards its centre, by which its individuality can be recognized ; or a long narrow-topped ridge (such as the Italians appropriately call *schiéna d'asino*, ass's back) has been thrown up by the simultaneous action of numerous vents on the same fissure so close to each other that their products are completely confounded.

The long axis of such a ridge, as well as of the neighbouring isolated cones, most of which will be more or less elliptical, have usually the same direction with the general chain of which they form a part, and which probably attests that of the original fissure in which the overlying rocks first yielded to the force of subterranean expansion.

Vesuvius affords an instance of the production of five small closely connected cones on the same fissure immediately above Torre del Greco, created successively by the eruption which destroyed a part of that town in 1794. In fact, all the parasitical cones thrown up by lateral eruptions on the flank of a volcanic mountain are of this character, and usually arranged more or less in strings radiating from its centre.

In the Prussian province of the Eifel, on the left bank of the Rhine, are a considerable number of independent cones presenting very similar features to those of Auvergne. So, too, in the Australian province of Victoria, it appears, from the report of Captain Smyth*, that several hundred cones of scoriæ are to be seen spread over the country, which have poured forth vast beds of basalt covering an area of some 3500 square miles. Mr. Heaphy describes a large number of

* Geol. Journ. 1858, p. 228.

similar cones in New Zealand, near Auckland, each of which has produced its lava-current.   In the island of Lancerote nearly a hundred such were thrown up by the terrific eruption of 1730, described by Von Buch, in a straight line crossing the whole island.   At Jorullo, in Mexico, six cones were produced upon a fissure of this character by an eruption in 1759.

§ 3. In its internal structure every such cinder-cone will possess, of course, some rude stratification, analogous to that of alluvial gravel,—one layer of fragmentary matters covering another, of a somewhat varied character, according to the changing accidents of their ejection into and fall from the air; and these beds, being superposed more or less conformably, will necessarily have what is called a quaquaversal outward dip, from a vertical plane passing through the curved ridge of the cone.   Moreover, where a crater has been left at the termination of the eruption, there will be another and opposite talus formed in its interior by the accumulation of the fragments that, especially towards the close of the eruption, have fallen into or rolled down the internal slopes of this cavity; and these will, in like manner, form beds having an *internal concentric* dip in the opposite direction, that is, *towards* its centre, from the vertical plane of the rim.   Thus, a section of such a cone will present more or less the appearance given below.

Fig. 5.

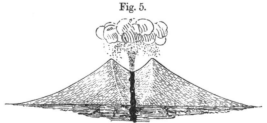

Ideal section of a simple cinder-cone, thrown up by a single eruption.

This is not a mere hypothesis, for such a double or anticlinal annular dip may be seen in natural sections afforded by

numerous volcanic cones ; as, for example, among those of the
Azores, according to Mr. Darwin, and those of the Phlegræan
fields near Naples. The accompanying woodcut (fig. 6) exhibits that of the cone forming the Cape of Misenum, the
northern horn of the Bay of Pozzuoli.

Fig. 6.

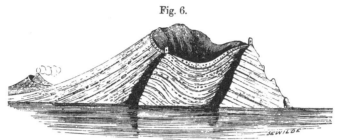

Natural section of an eruptive cone, forming the Cape of Misenum.

Fig. 7 is a view of Graham Island, off the S.W. coast of
Sicily, taken by M. Joinville in September 1831, just before
its final disappearance, in which may be seen the same internal
dip of the beds that composed the nucleus of the cone. These,

Fig. 7.

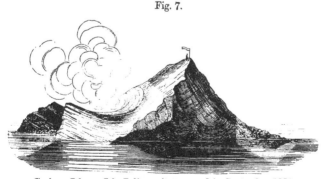

Graham Isle, or Isle Julie, as it appeared in September 1831.

from their proximity to the vent, were no doubt more firmly
compacted by heat than the outer strata, and from this cause
no less than from their central position were likely to resist
longest the destructive action of the waves. Such, too, is

exactly the structure of the small crateriform island on the coast of St. Michael in the Azores, near Villafranca, described by Mr. Darwin *; and other instances will, no doubt, occur to every one who is conversant with volcanic districts.

The strata of the external mass of the hill will be on all sides approximatively parallel to its outer surface (unless this has been subsequently degraded); and their strike will be circular, inclining to elliptical, should the orifice from which the explosions proceed be wider in one direction than in another, or (as must often happen) a fissure which allowed of their escape from more than one contiguous point in the direction of its length. Indeed, the multiplication of such contiguous orifices occasions, as I have said, in many instances, the production, in lieu of a regular cone, of a long ridgy hill, having perhaps several craters in a line, marking, no doubt, the direction of the internal fissure; but sometimes without any crater at all, the irregularity of the ejections having obliterated all such hollows.

The recognition of the dip of the inner beds of a volcanic cone *towards*, not from, the point of eruption is of some importance, since otherwise there would be a difficulty in accounting for their apparent unconformability and irregularity. This, indeed, becomes still more complicated when

Fig. 8.

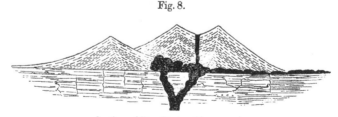

Section of Puy Pariou (Auvergne).

there has likewise been that shifting of the explosive action to fresh points upon the subjacent fissure, already alluded

* ' Volcanic Islands,' p. 108.

to as very frequent, by which a double or compound cone is produced. One example of the simplest kind may be found in the very perfect double cone called the Puy de Pariou, among the Monts Dôme, of which a plan and ideal section are appended (figs. 8 & 9).

Fig. 9.

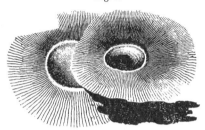

Plan of the Puy Pariou.

Violent winds prevailing in any one direction during the eruption, by causing the ejected matters to fall most plentifully to leeward of the vent, will also modify to a considerable extent the form assumed by them. According to M. Moreau de Jonnès, the fragmentary ejections of the volcanos of the Leeward Isles have uniformly accumulated in much greater quantity to the westward of the vents than on the opposite side; evidently through the influence of the *trades*, or constant east winds, that predominate in those seas. Mr. Darwin makes the same remark of the volcanic cones of the Pacific that occur within the range of the ' trades.'

It is evident that many other modifications of the normal figure and structure of a volcanic cinder-cone must be occasionally brought about by a variety of accidental circumstances —such as the original inequality of the surface upon which the ejected matters accumulated; the degrading effects of the waves or currents of the sea, if it be exposed to them, or of torrents of water, caused by the violent rains which are often observed to accompany or follow eruptions; or the bursting of lakes subsequently occupying the bottom of the crater; or

the sudden melting of snow, should the eruption occur during
the winter; and other causes, to which I shall have occasion
to recur.

But, above all, the form and structure of a simple volcanic
cone, the product of a single eruption (which alone is now
under our consideration), are most often materially modified by
that outburst of lava which usually accompanies the ejection
of fragmentary matters, to the consideration of which I now
proceed.

View of Vesuvius from near Sorrento, showing the modern cone rising out of
the old crater of Somma.

CHAPTER VI.

OUTFLOW AND DISPOSITION OF LAVA.

§ 1. THE emission of lava is usually as characteristic a feature of volcanic eruptions as those aëriform explosions which have been described above. Some volcanos, it is true, send forth vast quantities unaccompanied by any corresponding explosions,—the expanding steam and gases issuing in close intermixture with the intumescent matter, instead of rising through it and escaping in those vast bubbles already described as usually bursting from its surface so soon as a communication is obtained with the atmosphere. This difference is, probably, in great part due to the different specific gravities of the lavas in the two cases.

It generally happens, however, that a quantity of lava is simultaneously expelled in a more or less fluid state from the same orifice as the fragmentary ejections. Should the cone have been formed first, this matter may rise so as to fill the crater up to the lowest lip of its ridge, and thence flow down the outer slope, leaving there a crust or bed of scoriform or solid rock; but more frequently its enormous weight and pressure break down the entire side of the cone, composed, as this always is, of loose incoherent matters, whence it escapes in a flood or ' current' towards some lower levels. Many of the points of eruption on the flanks of Etna are marked by such a cone of scoriæ broken down on one side by an outburst of lava. Breached cones form, indeed, one of the most ordinary features of every volcanic region. In Iceland and Teneriffe they are numerous. Several occur in the extinct volcanic district of Sardinia, according to the report of General La

Marmora ; in that of New Zealand, according to Mr. Heaphy; in Lancerote, in Asia Minor, at Olot in Spain, and elsewhere.   One of the most striking examples, from among the cones of the chain of Puys near Clermont, is given in fig. 10.

Fig. 10.

The Puys Noir, Solas, and La Vache, Monts Dôme (Auvergne).
Breached volcanic cones.

Sometimes the remnant left of the crater's lip will exhibit marks of the height to which its cup had once been filled with lava, in an inner ledge of lava-rock still remaining there, and cementing the adjacent fragmentary matters.

Or if the lava, after filling the crater, escaped, without breaching the cone, by a sort of tapping, through some lower passage forced under its base, a cylindrical shelf or wall of rock may be left entirely round the upper rim.   Darwin observed several such round the craters of the Galapagos Isles. That of Kilauea, in Hawaii, described hereafter, offers a remarkable example of the kind, on the largest scale.   The Peaks of Teneriffe and of Cotopaxi, according to Humboldt, have each a circular parapet of this kind, composed of scoriaceous and vitrified rock, at their summit (looking like a sort of chimney-pot).   The materials were cemented, no doubt, by heat and acid vapours to such a degree of solidity as enables the rock to offer effective resistance to meteoric abrasion.

But it more often happens that, the rapid congelation and hardening of the surface of the lava as it flows out of the vent having formed a sort of covered channel, the cone is built over this by the ejections that accompany the outflow, and remains undisturbed, the lava having continued to the end of the eruption to issue from beneath its base.

Some writers on volcanic phenomena appear to suppose that every stream of lava must necessarily have proceeded from a crater, and ought to be traceable up to one. But though it is true that the gaseous explosions which throw up a cone, and usually leave a hollow, or crater, at its summit, are generally accompanied by an emission of lava in a more or less liquid form, this, as I have already said, is not invariably the case. Moreover it not unfrequently happens that the lava issues by a different orifice from that which is the source of the contemporaneous fragmentary ejections, although in its vicinity. And even when both kinds of products issue from the same vent, the very abundance of the ejections is likely to conceal the immediate source of the lava-current. Other superficial changes, such as denudation for example, will often have the same effect. There is an instance in point near Le Puy, where a wide bed of basaltic lava (the Plateau de Fay) is separated from the well-preserved cinder-cone (Puy de Chaspinhac), which marks the probable point of its emission, by a gorge subsequently worn through granite rocks, by the river Sumène, to the depth of more than 800 feet !

This example affords a remarkable, but by no means solitary, proof of the great durability of such loose cinder-cones, in positions where no diluvial action has operated, and they have only been exposed to the vertical downfall of rains, while the solid rocks in their vicinity have suffered an amount of erosion on the greatest scale. In fact, wherever they exist they present decisive evidence of the non-occurrence of any sweeping aqueous action of a diluvial character since their formation ; while the lava-currents which proceeded from them, in their position as plateaux on the summit of hills now surrounded almost by deep ravines, or even valleys, which could

not have existed when the lava was in a fluid state, tell an
equally convincing tale as to the vast erosive power of the
ordinary meteoric waters to whose action alone the surface
can have been since subjected. We thus acquire a test, and
indeed to some extent a measure, of the time during which
many large areas of the earth's surface have existed in a subaërial condition, uninvaded by any extraordinary 'diluvial'
action, which, properly employed, cannot but prove of high
geological value.

§ 2. From whatever point it issues, the lava thence pursues
its course in the manner of a stream of molten metal or other
imperfect fluid, in obedience to the laws of gravity, down such
slopes as may present themselves, flooding flat surfaces, and
occupying any hollows that may be open to it, with more or
less rapidity, under similar circumstances of the superficial
levels, in proportion to its greater or less fluidity.

This character (its fluidity) varies very widely,—some lavas
being, from causes perhaps somewhat obscure, but which will
be discussed presently, very much more stiff, tenacious, or
viscous than others, and consequently less fluid, although to
appearance equally heated, even to incandescence. Among
the lava-streams produced by the same volcanic vent (for example, from Vesuvius), some have been seen to flow from the
top to the bottom of the slope of the cone with extreme
rapidity *, while others dribbled languidly down its steep
sides, and stiffened there without reaching its base at all, just
as a run of wax or tallow in a 'guttering' candle hardens on
the outside of the upright shaft. This difference is, no doubt,
in some degree owing to the more or less copious discharge
of lava, in given times, from the volcanic vent ; but in most
cases it is unquestionably rather referable to differences either
in the temperature, or specific gravity, or consistency, or

* In October 1822, I myself witnessed, in the company of MM. Monticelli
and Covelli, a lava-stream descend the entire slope of the cone of Vesuvius, from the edge of the crater to the Pedamentina, in fifteen minutes.

mineral characters in the lava at the time of its emission. Indeed, the aspect of different lavas, after their cooling and consolidation, clearly indicates that they possessed at that time very different degrees of liquidity.

§ 3. The surface of flowing lava, on its exposure to the air, cools and hardens, as has been already said, with remarkable rapidity—indeed almost instantaneously, owing, no doubt, chiefly to the immediate expansion and escape of the abundant aqueous vapour developed in and discharged from it, by which a large portion of its caloric must be suddenly absorbed. The slag-like crust thus rapidly formed is so bad a conductor of heat, that it may be stepped upon without risk almost immediately, even while the stream of lava is still to be seen at a white heat through the crevices which in great numbers quickly open in- it. Observers have thus often had the opportunity of approaching and closely inspecting a stream of lava at its very source, as it issues from or flows beneath some crevice.

The heat radiated by it is by no means so intense as would be expected from such a mass of apparently molten matter. Its liquidity, when greatest, does not appear to exceed that of honey, but is generally so imperfect that considerable pressure is needed to cause the point of a stick or of an iron rod to penetrate its surface. Indeed its consistency more usually resembles that of coarse half-dry mortar, or of meal as it issues hot from between the stones of a mill—to which Sir William Hamilton (a frequent and sagacious observer of the eruptions of Vesuvius) compared it—than of a substance in complete fusion. By day, the still liquefied parts appear of a dull red colour, but by night they are nearly white, or flame-coloured; and at places where the current cascades down some steep slope, or over a cliff, its aspect is very brilliant and fiery, every freshly-opened crevice, or portion of the interior uncovered by the fall of the outer slag-like crusts, glowing like the live coals of a furnace.

From these crevices a fresh burst of the yet liquid and

incandescent lava often breaks out, and is in turn superficially congealed and fissured. It is by these innumerable
shrinkage cracks—which, like all such crevices, take a direction generally at right angles to the planes of superficial
refrigeration or desiccation—that much of the vapour contained in the lava escapes; and by the force of this escape,
but still more, perhaps, by the friction and irregular motion of
the matter flowing on underneath, the surface of most lava-
currents is broken and tilted up, as it moves on, into coarse
scoriform crusts or slabs, ragged or angular, which give it a
resemblance to the frozen rivers or seas of northern latitudes,
where thick ice, shattered by the movement of currents or
waves, has been jammed together in a multitude of shapeless
projecting masses. Some of these tilted slag-masses of lava
may be seen to rise 10, 20, or even 50 feet above the average
level of the current, and might easily be mistaken for erupted
dykes, especially when, as happens occasionally, they have
taken a rudely prismatic divisionary structure*.

Hence arises the generally rough and savage surface of the
great lava-streams of Iceland, Teneriffe, Etna, and other volcanic districts, which from their bristling, harsh, *saw*-like protuberances are called *Serres, Cheires,* or *Sciaras* (saws) by the
natives of the country. In Spanish America they usually go
by the name of *Mal-pais,* owing to their desert and almost
impassable character.

In some lavas the fissuring by shrinkage and loss of heat
and vapour on exposure to the air penetrates at once to the
depth of many feet, dividing the rock into loose blocks more
or less cuboidal in form, especially towards the sides or termination of the current—the tendency to break up in this
manner being increased by the impulse of the matter flowing
beneath.

In the Cordilleras of South America, some fields of shattered

* The lava of Graveneire, above the village of Royat, shows some upright slaggy masses of this kind on its surface fully 60 feet high.

lava of this description are called by Humboldt *trainées de blocs*, and are supposed by him, as well as by M. de Boussingault, to have been possibly erupted in a fragmentary form—an improbable hypothesis. No doubt, however, the 'blocks' were often, during their formation by shrinkage, more or less set in motion as the current beneath or behind them rolled on, and this motion will have increased the apparent confusion in which they now lie. Such surfaces bear no appearance of previous liquidity, but resemble rather the rocky debris of a shattered mountain. Professor Dana describes some of the lava-currents of Hawaii, which he calls 'clinker-fields,' as composed of loose angular blocks of all shapes and sizes, from that of a bushel to that of a house—" a surface of horrible roughness." The photographs in the volume of Professor Piazzi Smyth, on Teneriffe, offer a similar idea of the clinker-fields of that volcano.

Owing to this rapid consolidation of its external surfaces, a current of lava advances generally with a rotatory motion, —the slags and crusts that cake upon its front rolling down before it, and forming a sort of broken pavement, over which the central part protrudes and falls, and on which it finally reposes. It is thus seen why a bed of scoriæ usually underlies every lava-current, even where ejected matters had not previously covered the ground. A stream of lava which I had the opportunity of observing on Etna in 1820, and which was advancing at the slow rate of about a yard an hour (the eruption which produced it having ceased for nearly a year), had all the appearance of a huge heap of large cinders rolling over and over upon itself by the effect of slow propulsion from behind. The motion was accompanied by a crackling metallic noise, occasioned by the contraction of the crust as it solidified and the friction and fall of the cindery slag-cakes against one another, and, on the whole, suggested any other idea than that of fluidity. Yet within the crevices of this sluggish mass a dull red heat might still be seen by night, and a considerable quantity of vapour issued from them by day.

The upper and under scoriform portions of such lava-streams are often many feet in thickness; while the middle bed of compact matter, which cooled more slowly, is comparatively shallow and thin, especially where the descent was rapid*.

§ 4. Some lavas, however, differing from the former in liquidity, probably by reason of differences in texture or mineral character, assume ropy and filamentous figures, and present a smooth and even glazed exterior. This is particularly the case with the glassy lavas passing into obsidian or pumice. Such, too, are the more ferruginous but equally viscous lavas of Bourbon, and also of the Sandwich Isles. The latter, indeed, is habitually of so glutinous a consistency at the time of its emission as to be drawn out, as it is tossed up by the gaseous explosions, into vitreous filaments, silky and fine as spun glass, which float for some time in the air before they fall, and are called by the natives 'Pele's hair.'

Lava of this extremely viscous and tenacious consistency coagulates superficially on the instant of its exposure to the air, just like so much melted wax, in glossy or wrinkled crusts, which present singular forms, resembling cables or stalagmites, or the roller-like folds of drapery, or sticks of barley-sugar, or cauliflowers, or the trunks or branches of trees, &c. When such a lava has welled out slowly from an orifice without the accompaniment of much explosive discharge of vapour, the surface of the cooled mass may show concentric folds, like those of which Mr. Heaphy gives a sketch from the lava-fields of New Zealand † (fig. 11).

When the matter is still more viscous, its accumulation over a minor spiracle occasionally produces a hillock of curving and concentric ridges, or even a dome or spire-like protube-

---

* See Sir C. Lyell's remarks on the recent lavas of Etna, Phil. Trans. 1859.   † Quart. Journ. of Geol. Soc. 1860, p. 246.

rance. Professor Dana describes some of those upon the slopes
of Mauna Loa in Hawaii as actually taking the figure of a

Fig. 11.

column or upright bottle (fig. 12), or a petrified fountain.
Some, he says, are as much as 100 feet in height, and have an
aperture at the top or sides, left by the last explosions of the
elastic fluid, whose rise and escape spirted up the jets of liquid
lava that coagulated into these strange shapes as they fell
round the lips of the orifice.

Fig. 12.

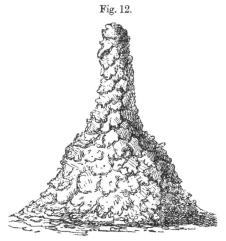

Column of lava in Hawaii.
(After Dana, ' Geology of American Exploring Expedition.')

Bory de St. Vincent describes similar hillocks (likewise of
considerable magnitude—80 or 100 feet high) on the volcano
of Bourbon (fig. 13), whose lavas are highly ductile and viscous,
having a great analogy to those of Hawaii.   He gives a drawing of the appearance of this lava-spring as he saw it by night,

welling-out almost tranquilly, and continuously, from the

Fig. 13.

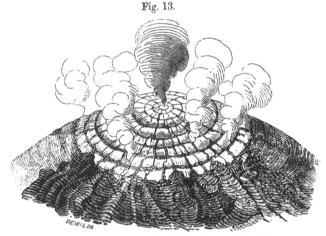

Source of lava on the summit of the volcano of Bourbon.
(From Bory de St. Vincent.)

summit of one of these 'Mamelons' (fig. 14), and forming a
crust of concentric wave-like ridges, as also of another, which
was not at that time in activity.

Fig. 14.

One of the 'Mamelons'—small lava-cones on the summit of the volcano
of Bourbon.   (From Bory de St. Vincent.)

Minor lava-cones of this character are peculiar to the very
ductile and viscous lavas which approach to glass in texture,
and must not be confounded, as has been done by some geologists, with the ordinary eruptive cones composed entirely of

loose fragmentary ejecta. The internal structure of the former class of cones must, however, no less than the latter, exhibit a concentric arrangement of layers of lava, more or less irregular, and having a quaquaversal dip, parallel to the outer slopes of the hill. They will resemble in structure, as they do in the mode of their production, the bell-shaped cones formed by the mud-eruptions of Macaluba in Sicily, or of Beila near the Indus; being, no doubt, like them, solid, except when a small opening has been left on the summit, and composed of more or less concentric and mantling layers, the result of the overflow of one coat of the semi-fluid matter over another.

§ 5. If, however, scoriæ are ejected from the same vent during the formation of such a hillock, its structure will show an irregular interbedding of lava and fragmentary matter. A very instructive natural section of a minor cone of this class, formed within the crater of Vesuvius in 1835, is given by Abich in his ' Views of Vesuvius and Etna.'

Fig. 15.

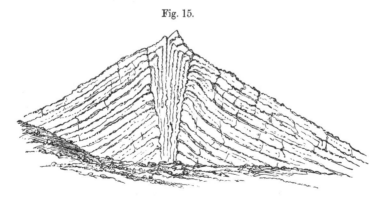

The sloping beds seen in it are of lava, separated by thinner layers of scoriæ and lapillo. The axis of the cone is seen to be an upright dyke, evidently the eruptive chimney, filled with basaltic rock in vertical layers by the consolidation of the last columns of liquefied matter that were successively propelled up this channel. These layers are seen to turn over and flow down outwardly, as they reach the level of the bed which at the time of the emission of each formed the lip of the orifice. This latter feature (which is just what we should expect to find there, from the mode of production of such a cone already described) is seen also in another interesting natural section (fig. 16), given in the same work, of a dyke observed by M. Abich in one of the cliffs of the Val del Bove.

Fig. 16.

Natural section of a basaltic dyke in the Val del Bove.
(From the Quarterly Journal of the Geological Society, 1858, p. 127.)

Here, the scoriæ which originally enclosed the dyke, and through which the lava forced its way, have been washed out by rains; and the dyke stands clear, showing the way in which the lava composing it, when liquid, overflowed the lip of the vent and took its course down the outer slopes*.

§ 6. A stream of lava while flowing down any slope will, owing to its imperfect fluidity, usually be thickest towards its centre, and consequently possess a convex cross section on its upper surface, the sides rising as steep banks from the uncovered ground adjoining. But when the supply of fresh lava from the vent diminishes or entirely ceases (the still liquid interior at the central part of the current continuing for some time to

* Strangely enough, M. Abich imagines the sloping layers of the first small cone to have been formed in a horizontal position and subsequently elevated! while in the latter case he allows that the lava-bed flowed down and consolidated in its present position upon a slope of 25°. But this is not the only instance in which the 'elevation-crater' doctrine has drawn its disciples into the most obvious inconsistencies. 'Views of Vesuvius and Etna,' plate 5. Berlin, 1837.

flow on, urged by its own weight alone, down any slope that offers itself), the upper crust, being unsupported, will have a tendency to subside in proportion.   Hence we often find narrow lava-streams confined within banks, which they seem to have raised themselves, and having a *concave* surface in their cross section (see fig. 17); the sides, which necessarily cooled

Fig. 17.

Transverse section of a lava-current.   The dotted line represents
the original surface.

sooner than the central part, preserving their thickness, and taking the form of high banks or ridges, between which the internal lava-stream flowed on for some time, as in a canal, the level of its surface gradually lowering as the supply from above ceased.

The most remarkable example of this kind, perhaps, is to be seen in the valley of Thingvalla in Iceland, which, it appears to me, was evidently caused by the subsidence, in the manner above-described, of the upper crust of a vast lava-stream that flowed from an adjoining mountain into the Thingvalla lake. The valley is a depression four miles wide and about 800 feet deep, bordered on either side for seven or eight miles before it reaches the lake by a great rift or chasm, perfectly straight, averaging perhaps 100 feet in width and 180 feet or more in depth.   The sides are black precipitous lava-cliffs, which have been evidently torn asunder, since their opposite indentations exactly correspond.   These ' rifts,' the largest of which goes by the name of the ' Almanagia,' are, no doubt, fissures broken at the sides of the lava-stream by the parting of the central mass from the lateral banks as its surface sank, through the escape of the lower and still liquid lava into the unfathomable depths of the lake, after the supply from the volcano had ceased. There are numerous parallel and equally deep, but smaller, fissures or ' crevasses' on the surface of the lava between the two larger lateral ones, which were produced, no doubt, at the same time. Two of them inosculate, and all but enclose a space within which, for the security its moated position afforded, the ancient Althing, or Court of Judgment and Parliament of the island, used to be held. Icelandic travellers usually speak of this subsidence as caused probably by a sinking of the surface into *some great cavity beneath.* I think it more probable that the true cause is that given above, namely, the escape of the still liquid lava from beneath the indurated surface of the stream, on the cessation of the abundant supply from the volcano which had previously filled the valley to the height of the banks that still remain. An exact parallel to this effect may be seen when a swollen river has been frozen over, and the waters run off before the ice is melted. Its surface sinks, leaving great longitudinal rents along and parallel to the sides of the channel. Many fissures of the same character and mode of formation are visible in some of the other great lava-fields of Iceland, and, indeed, may be looked for wherever lava has been poured out abundantly, under circumstances admitting of its continued flow for some time after the supply of fresh matter has ceased. These fissures, being always longitudinal, or in the direction of the current, no doubt have frequently facilitated the formation of a new channel for the escape of the waters of rivers or lakes which the streams of lava have temporarily obstructed, and gave direction to their excavating force *.

Sometimes the lateral longitudinal banks of a lava-stream will be multiplied, several alternate ridges and furrows, on a large scale, appearing in the cross-section of the current's surface. These mark, in fact, so many distinct currents. At

---

* For a plan of the Almanagia and its environs, see the 'Story of Burnt Njal.' 1860.

other times these corrugations form a series of curved or wrinkled folds, having their convexity in the direction of the flow of the lava, and resembling the wrinkles on the surface of a glacier. These variations are evidently due to the varying rates of motion of different parts of the current, or to successive streams—some hardening, while others continued to flow on. Any one who has seen cart-loads of liquid mud emptied upon a flat or slightly sloping surface will have observed, on a small scale, in the forms it assumes when dried by exposure to the air, many of the varieties of superficial wrinkling which characterize lava-streams.

In the case of the very viscous lavas, which tend, as has been already noticed, to assume ropy or stalagmitic forms on the surface, the matter not cooling so rapidly inwards as the more coarse-grained and readily-splitting lavas, its downward escape from beneath the upper crust often leaves behind hollow gutters, arched over by a thin and brittle roof—so thin sometimes as to yield to the weight of a person stepping on them. Such vaulted roofs have pseudo-stalactitic projections left by the subsidence of the liquid, and are coated with a glossy varnish. Sometimes, as among the lavas of Etna, Bourbon, Iceland, St. Michael in the Azores, Teneriffe, and many other localities, caverns of very large dimensions are thus formed beneath the surface of a lava-stream, and even imitate in their extent and windings the well-known caves worn by water in limestone-rocks. Their mode of production by the escape of the highly liquid lava down a sloping surface, from beneath the hardened crust above, is easily understood.

Other cavities of various sizes, but occasionally very large, are produced in some extremely liquid lavas by the union of many bubbles of the vapour generated through or entangled in their mass. These volumes of vapour, influenced by their elasticity and inferior specific gravity, rise towards the surface of the lava as it flows on, and, when sufficiently powerful to break through its crust, burst from it in minor jets, which imitate, on a small scale, those of the principal eruptive vent. Just as the smaller vesicles that, from being unable to escape, remain dispersed through the lava, resemble those of the glutinous varieties of *bread,* so the larger often blister the surface of a lava-stream, like those which, on a smaller scale, are seen on the upper crust of our loaves, in dome-like or conical protuberances.   Sometimes these have opened at the summit or cracked at the sides, and discharge vapour in greater or less abundance.   They are, while in that state, called ' fumaroles.'   Such spiracles remain, no doubt, for a considerable time in communication with the still liquid lava flowing beneath the hardened surface.   This supplies them with their jets of vapour, and sometimes causes them to throw up a few scoriæ and spirts of lava, producing the appearance of minor eruptive vents.   In the Vesuvian eruption of 1855, many such small cones, from 10 to 20 feet high, were formed in succession upon the surface of the main current of lava; they are figured by Schmidt, who (I think erroneously) considers them to com-

Fig. 18.

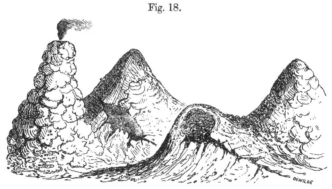

Small cones formed on the surface of the lava of Vesuvius, 1855.
(From Schmidt.)

municate with the interior of the mountain, and not merely with the lava-stream beneath them.   I observed many such, having certainly the latter origin only, on the surface of the lava of 1822, which, after flowing from the summit of the cone, formed a sort of lake on the Pedamentina and in the Atrio del Cavallo. So Mr. Darwin describes the surface of a highly vitreous lava-current in one of the Galapagos Isles as studded with "small mammiform hillocks," and some circular depressions, which seemed to him to have been formed by the falling-in, or blowing-off, of the arched roofs of such great bubbles of lava. These pits are some of them 20 or even 40 feet deep, and circular.

Professor Dana speaks of similar bubble-shaped domes on the surface of some of the lava-streams of Hawaii. They were generally cracked or broken, the thickness of the roof being but from one foot to ten; "and the loosened fragments had sometimes fallen into the *oven-shaped* cavities they covered."

Such too, no doubt, was the origin of the ' *hornitos*,' or *oven*-shaped hillocks, described by Humboldt as studding the high surface of the great lava-pool of Jorullo, called the Malpais, and which appeared so problematical to the great traveller. Coated over to a thickness of a foot or two with fine ashes (the last product of the eruptions from the six vents, now marked by as many cones), and this ash-conglomerate having (as is not unusual under the influence of hot vapours) assumed a globular-concretionary structure, they presented certainly a singular aspect, if the view of them given by Humboldt in his ' Atlas Pittoresque,' from which the accompanying woodcuts (figs. 19, 20, 21) are copied, is not (as may be suspected) somewhat exaggerated. Later travellers report that the *hornitos* have almost entirely disappeared, their loose coating being probably washed away by the tropical rains, or the intervals filled up by the growth of vegetation.

With regard to the disputed question of the origin of the raised plain of the Malpais, M. de Saussure, the last and most trustworthy visitor, entirely confirms the opinion which I ventured to proclaim in 1825, that Humboldt was mistaken in supposing it to have been " blown up from beneath like a bladder," and that it is merely an ordinary current of lava, which, owing to its very imperfect liquidity at the time of its issue from the volcanic vent, as well as to the overflow of one sheet or stream upon another, had acquired great thickness about its source, gradually thinning off towards the outer limits of the elliptical area which it covered *.

§ 7. The quantity of lava emitted by a single eruption, and the extent of surface covered by it, are often very great. The volcano of Skaptar Jokul in 1783 sent forth two prodigious streams—one about fifty miles in length, with a breadth, in places, of fifteen; the other forty in length, and seven miles wide in parts. Its thickness was more than 500 feet in some places, and it has been calculated that the entire mass exceeds in bulk that of Mont Blanc. This is, perhaps, the most copious lava-stream known to have been produced by a single erup- tion. But it is not improbable that it may be paralleled by some which have proceeded from the great South American or Kamtschatkan volcanos.

Among the products of earlier volcanic eruptions—the traps belonging to the Tertiary and Secondary ages, and generally of submarine origin—examples of equal or greater magnitude occur. In India, in the Deccan, an area of 250,000 square miles is said by Colonel Sykes[†] to be covered with a continuous floor of basalt. There is no evidence of its having proceeded from a single source; but the flows of lava must have been prodigious to overwhelm such a vast and nearly level area.

Probably some of these enormous sheets of lava, whether of modern or ancient date, have issued from numerous points, or indeed, perhaps, from the entire length of the fissure of discharge. This seems to have happened repeatedly in the volcanic district of South America. Commodore Forbes de- scribes floods both of trachytic and doleritic (basaltic) lavas,

* See " Cones and Craters," Quart. Journ. Geol. Soc. November 1859, and May 1861.

† Geol. Trans. ser. 2. vol. iv. p. 409.

Fig. 19.

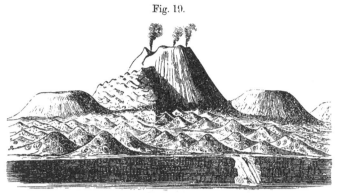

View of Jorullo and its Malpais.　(From Humboldt's ' Mexico.')

Fig. 20.

Section of Jorullo.　(After Humboldt.)　*a*, *b*. Level of original plain.
(The darker part is lava, the lighter scoriæ-cones.)

Fig. 21.

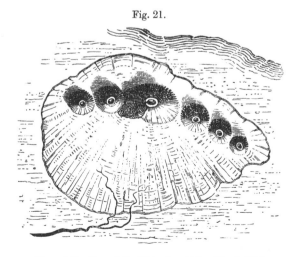

Plan of Jorullo and the Malpais.　(After Humboldt.)

among the Cordilleras of Peru and Bolivia, to have been evidently poured out of lengthened fissures, some of which extend as much as fifty miles. Since, however, the lava, by spreading above and on either side, will necessarily have concealed the fissure through the greater part of its course, it will always be difficult to determine whether it flowed out of several distinct openings in the same line, or from every part of the fissure. It is probable that the lavas of submarine volcanos have spread more widely in proportion to their depth than those which have flowed in open air, owing to the greater difficulty opposed, under such circumstances, to the escape of their contained vapour, and the longer duration in consequence of their fluidity.

§ 8. By reason of the extreme slowness with which its hardened surface gives passage to heat, the interior of a lava-stream often retains a very high temperature for a great length of time after its emission, continuing to send forth vapour from its crevices and fumaroles, and probably remaining liquid, and even more or less in motion, throughout its central or lower portion, for years. I myself saw, in 1819, a lava-stream still slowly moving onwards at its lower extremity, nine or ten months after the eruption of Etna which produced it had ceased.

Owing to this sluggish motion, the slightest obstacle appears to have a considerable effect in retarding the progress of a lava-stream. A bush, a tree, a wall, even a large stone, have often been seen to check the advance of a current to an extent quite unequal to the resistance they can be conceived to oppose to the weight and pressure of the lava advancing upon them, and only to be accounted for by its semi-solid condition and the extremely low degree of mobility possessed by its particles, which the increased compression resulting from the least impediment suffices to destroy to a certain distance back from the opposing surface.

Even the resistance offered by the minor asperities of the

ground over which a stream of lava extends itself, destroys or
lessens the liquidity of this substance to a certain distance
upwards, and hence the rotatory motion, already observed
upon, with which a current appears to advance. The lower
stratum being arrested by the resistance of the ground, the
upper or central part protrudes itself, and, being unsupported
from below, falls to the ground, to be in turn covered by a
mass of more liquid lava which rolls or swells over it from
above.

Hence, too, whenever the advance of a stream of lava is
checked by any material impediment, it accumulates upon
itself, rising in height until it is able either to surmount and
cascade over the obstacle, or turn round it by a lateral deviation. The extreme difficulty with which such a diversion
appears to be effected is remarkable, but by no means surprising, since this circumstance is common to all liquids of
great viscosity or consistency, which, when urged forward
down an inclined plane by their gravitating force, move as it
were in a body,—the component particles, through their
powerful mutual cohesion, retaining almost completely their
relative positions—not rolling over one another in that free
and voluble manner which characterizes the motion of more
perfectly liquid bodies.

Owing to this peculiarity of motion, a lava-current increases
considerably in depth and bulk wherever any impediments
retard its progress. And hence, in Auvergne, the Vivarais,
and other volcanic districts where narrow and winding moun-
tain-gorges have been occupied by descending currents of lava
throughout long distances, every concave elbow is found to
have been filled with a bulky mass of basalt, while the intervening parts of the valley present comparatively narrow and
shallow strips. For the same reason, if the obstacle be of considerable height, and such that it cannot be *turned* by the
lava, but must be surmounted, the accumulation necessary for
this purpose will often attain a much higher level at this point

than the surface of the current for some distance behind it; so that the stream has the appearance of having flowed uphill. The cause of this is, that the hardened crust acts the part of a covered canal, in which the liquid rises towards the level of its supply like water in a pipe.

When lava flowing onward meets any inflammable materials, as dry grass, shrubs, trees, &c., it usually sets them on fire, and the flames produced in this manner are liable to be mistaken at a distance for flames evolved from the lava itself. When trees are rapidly enveloped in lava, the upper parts alone blaze and are reduced to ashes; the trunk is merely carbonized, and, if subsequently removed by aqueous infiltrations, may leave its impression in a hollow cylindrical tube within the solid rock. Such moulds are very common in the lava-currents of the Isle of Bourbon, which have extended their ravages through forests of palm; and one of them is to be seen in the cabinet of the Jardin des Plantes.

Professor Dana relates that when forests have been traversed by some of the glassy lavas of the volcano of Kilauea in Hawaii, the subsidence of the surface of the stream from its earlier level has left numerous stalactites of obsidian hanging from the upper boughs of the trees, like the icicles occasioned by a frost succeeding a heavy snow-storm and thaw; and, strange to say, the twigs and branches to which these pendants adhere, and which were certainly enveloped in the molten matter, show but few signs of having suffered from its heat—even the bark being rarely charred,—a fact to be accounted for, perhaps, by the moisture of their surfaces, when suddenly vaporized, having acted as a sort of sheath to protect them during the brief interval between their immersion in the lava and the cooling of the enveloping coat. It is an effect perhaps analogous to the one I am about to describe.

When a lava-current has met in its course with a flat and extensive surface perpendicularly opposed to its direction, such as the wall of a house, &c., it has been observed to stop, as if by magic, at the distance of a few inches, without coming into actual contact with the obstacle. This and other similar facts may, perhaps, be thus explained.

A quantity of elastic vapour, at a high degree of tension, escapes from every fresh surface of lava successively disclosed as it moves onward ; but as the lava approaches close to a re-sisting surface of considerable extent, the vapour must be pre-vented from escape, and, filling the narrow intervening space or crevice, must create a resistance sufficient sooner or later to prevent absolute contact of the opposing surfaces. The rapid consolidation of the face of the lava-stream then builds up, as it were, another wall of its own in front of that it has met with.

If the momentum of the current behind be considerable, the wall may give way ; but should the motion of the lava be slow, and its fluidity very imperfect, it rises without over-throwing—scarcely even touching—the wall, until it is suffi-ciently elevated to cascade over it, or deviate in a lateral direction. This observation has been repeatedly made during all the destructive eruptions of Vesuvius, and there seems no reason to doubt its correctness. The lava of Etna, which, in 1669, poured down upon Catania, stopped at the wall of that town where it was 60 feet high, and accumulated upon itself until it was able to surmount it and flow over the top in a cascade of fire. The wall was not thrown down, but still exists, and an arch of lava is yet to be seen curling over it like a wave breaking over a beach.

When there has existed a wooden door in the wall, it has been observed that the heat radiated from the lava after some time sets fire to it, and when thoroughly consumed, the lava enters the aperture thus produced, but continues to respect the wall on either side. So, too, it has been actually found possible to give an artificial direction to the course of a lava-stream. While that of 1669 already mentioned was flowing towards Catania, a proprietor of lands which it threatened to overwhelm employed labourers with pickaxes to open a passage through the hardened scoria-crust which embanked and confined the current on one side. By this he succeeded in giving vent to the flow of matter in that direction rather than in the one it threatened to take if left to itself. Just in the same manner, workmen in a smelting-furnace open a course for the molten ore by breaking down the barrier of sand by which it is confined.

§ 9. The fragments of pre-existing rocks accidentally enveloped by lava in its advance are observed to be variously affected by the heat. In general, when quite in the interior of the current, they are partially fused on the surface, and sometimes incorporated so closely with the surrounding substance that it is difficult to trace their outline with clearness. Fragments of limestone still preserve their carbonic acid, but have often acquired a crystalline grain, probably from having been partially fused under pressure, as in the experiments of Sir James Hall. Near Aurillac, in the Cantal, I have found freshwater shells of this character in the lowest stratum of a basaltic current, which appears to have flowed from the heights of the mountain into the tertiary lake that existed at its foot. Fragments of sandstone enveloped in this manner are usually hardened; and clay is frequently converted into a substance resembling jasper (porcelain jasper), and in other instances into *tripoli*.

Under such circumstances, so complete an intermixture has occasionally taken place of the molten lava with the soft sediment, that it becomes difficult to determine which preponderates, and whether the resulting rock should be classed as of igneous or aqueous origin. Thus, in the great basaltic platform of the Deccan (India), the eruption of which in voluminous lava-streams seems to have occurred contemporaneously with the deposition of much calcareous and arenaceous matter in great freshwater lakes, the basalt is described as much impregnated with lime, which fills fissures and seams in it, and as containing freshwater shells, wood, &c., as well as much calcareous spar, jasper, zeolites, quartz-crystals, &c.   Some portions of the arenaceous matter are converted into red or grey chert " deeply imbedded in the basalt," and partly enamelled (quartz-retinite)*.   The lime so entangled was probably often fused.   Mr. Darwin describes, in St. Jago (Cape de Verde Isles), a bed entirely composed of lime, which he believes to have been erupted in a fused and liquefied state, mingled with the molten lava.   " In Quail Island a single sheet of lava, but 12 feet thick, rolling over earthy lime, has changed it into crystalline calcareous spar, or a breccia of glossy black scoriæ cemented by a snow-white, highly crystalline base ; while numerous little balls, composed of spiculæ of calc-spar, occupy the interstices†." This description applies equally to the calcareous peperino of Central France, formed where eruptions have broken through the soft marly sediment of its lakes.   Where the carbonic acid has been driven off in part, and the matter reduced to quicklime, this has often probably been carried away by aqueous vapour penetrating the pores of the rock, and replaced by dissolved silica, which thus either impregnates the whole rock, or fills the casts of shells left by the disappearance of the lime. It is likely that some of the highly siliceous lava-rocks have this metamorphic origin.

The production of magnesia often characterizes the contact of calcareous strata with heated lava, proceeding, no doubt, from the abundant augitic element in the latter.   Hence the frequent dolomitization of limestones in contact with trap, and the production of steatitic and soapy serpentinous rocks in other instances.

So also, lava injected upwards through fissures, and constituting dykes, has occasionally effected changes of a similar

---

* Malcolmson " on the great Basaltic District of Western and Central India," Geol. Trans. 2nd ser. vol. v. p. 537.

† Darwin, ' Volcanic Islands,' p. 12.

character in the rocks which it traversed—especially if the
dyke was very wide, and the lava consequently retained its heat
for a long time. But, owing to the immediate solidification of
lava flowing in the open air as soon as it impinges on any obstacle, it seldom seems to radiate sufficient heat to produce any
important change in the substance of these impediments. The
numerous recent lava-currents in the Auvergne, which have
flowed to a great distance over both granite and limestone,
have rarely effected any alteration except on the mere surfaces
of those solid rocks with which they have come in contact, or
on the fragments they have caught up and enveloped. The
soil or loam which a lava-stream covers is usually reddened
to a brick colour, as if burnt. This change sometimes penetrates to the depth of several feet; at others, it is merely
superficial. Clay is occasionally turned into jasper; marl
into pechstein or retinite. When Torre del Greco was
overflowed by the lava of 1794, some curious circumstances
were observed as to the effect produced on the metallic and
earthy substances which it enveloped. Both were in part volatilized and deposited as crystalline sublimations on neighbouring points. The component metals of some alloys crystallized separately[*]. This took place, it must be remembered,
in the interior of the lava-mass, which retained a high temperature for a long time. In metamorphic changes of this
character, time appears to be an essential element.

When lava flows over an uneven surface, it must fill any
hollows it meets with, and any open fissures or crevices which
may exist in that surface. After the matter has cooled, such
filled-up fissures will have the character of dykes; but the
greater number of dykes are presumably formed by injection
of lava from below rather than by its intrusion from above,—
the heavings of the rocks which overlie or adjoin a volcanic
vent during or previous to its eruptions being, as already
suggested, likely to produce numerous cracks opening inwardly,

* See Brieslak, Voyage dans la Campanie, tom. i. p. 278.

and in immediate proximity to the intumescent lava within the vent or chief channel of discharge.

Dykes of either kind often inosculate, or branch out of one another. They are usually vertical, or nearly so; but some take for a time a nearly horizontal course, the lava being injected into the seams separating contiguous strata of rock, in which direction it may occasionally split more readily than in a transverse one; or, if the lava meet with a stratum of loosely-aggregated fragmentary matter, it may force or melt its way through it. In these cases it is not easy, in sections of moderate area, to distinguish between a dyke and a true bed. Such conformable dykes, however, are rarely found persistent for any distance, but soon branch off in a direction transverse to the planes of the strata they have penetrated. As a general rule, beds of lava-rock, exhibiting a horizontal or nearly horizontal position in a vertical cliff-section, may be assumed to have flowed into that position as a current; while the vertical or nearly vertical beds that cross these are true intrusive dykes,

Fig. 22.

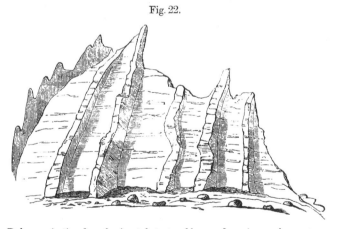

Dykes projecting from horizontal strata of lava and scoriæ-conglomerate, in the Val del Bove. (From Sir C. Lyell's Manual.)

whether filled from above or below. The dykes being composed of hard rock, while the strata traversed by them are often more easily disintegrated, they appear occasionally projecting from an exposed surface in the manner represented in fig. 22.   Hence the name given to them; *dyke* signifying a wall in North-country dialects.

When a current of lava flows over marshy ground, the sudden vaporization of the humidity confined below must occasion explosive discharges, that here and there disturb and tear their way through the superincumbent mass, scattering its fragments into the air, and thus creating rude hollows and scoriform protuberances on the surface.

When a lava-current enters the sea, or any other body of water, a similar effect will be in part produced; but these explosions are of no very extraordinary violence.   It has been supposed that on such occasions a terrific combat must ensue between the two elements, and very poetical images have been employed to heighten the colouring of the picture; but the fact appears to be otherwise.   A certain quantity of the water that approaches nearest to the incandescent lava, as it is protruded from the extremity of the current, is, of course, heated and vaporized; but the superficial consolidation that instantly ensues prevents any further contact.   The fissures that open in this crust, as it solidifies, emit torrents of vapour the instant they are formed, which must impede the entrance of the surrounding water till their sides are consolidated and comparatively cooled.

In other respects we believe that a current of lava advancing below water conducts itself much in the same manner as on dry land.   A stream of lava flowed from Vesuvius in 1794, and, after overwhelming Torre del Greco, poured itself into the sea; and a still more copious current issued from the Monti Rossi, above Catania, in 1669, and flowed into the sea, near that city, forming a promontory which projects more than half a mile beyond the original line of coast; but in neither case was there any extraordinary commotion occasioned by the conflict of the two elements.

The caloric, however, abstracted from the lava, both by actual contact and by the condensation of its heated vapour, communicates a proportionately high temperature to the water in its vicinity, which on such occasions is observed to be heated, discoloured, and rendered turbid to some distance. Fish have been often found killed, in very considerable numbers, by this sudden change of temperature in their native element.   Thus, during the great eruptions of Lancerote, one of the Canaries, which continued unceasingly, and with intense violence, through the years 1730–1736, immense multitudes of fish are said to have been cast ashore dead.   The same occurred in Iceland in 1783.   At Stromboli the fishermen told me that, after any violent eruption of the volcano, during which lava probably issues below the sea-level from the flank of this submarine mountain, shoals of dead fish, parboiled, are often thrown up on the shore.   It is far from improbable that the ichthyolites of Monte Bolca were destroyed by an occurrence of this nature, since the fissile limestone in which they are imbedded is immediately capped by a bed of basalt and calcareous peperino.   The latter rock proves the eruption that produced the basalt to have been submarine; and the remarkable attitudes of some of the fish bear evidence to the extremely sudden nature of the catastrophe that at once destroyed and buried them in the soft calcareous sediment which was at that time in process of deposition at the bottom of the sea.

§ 10. *Consolidation of lava-currents.*—The slag-crust which forms on the surface of a lava-stream on its exposure to the air being, as already stated, an extremely bad conductor of heat, such streams are very slowly cooled and consolidated.   In many instances it has been found that, for years after its emission, the heat, at but a few feet below the surface, is sufficient to set fire to a stick thrust into the deeper crevices.   Hence we may infer that, in places where the current is of great thickness—and some are known, as in Iceland, to have a depth of 500 or 600 feet—a very long period must elapse, perhaps

centuries, before the process of cooling and consolidation can
be completed.   The lower part of the current will necessarily
be the last to consolidate ; and from this cause it happens
that, when laid open by subsequent denudation, the inferior
portion of a lava-stream is always found to be more regularly divided by fissures of retreat than the upper—so regularly in some instances as to exhibit that prismatic or columnar arrangement well-known as basaltic colonnades.   In
the valleys of the Vivarais, the lava-streams which have flowed
into and occupied the beds of their rivers, and have been
subsequently worn down in great part by the running waters,
display sections of the interior of each, often some miles in
length, having a very regular and beautiful columnar arrangement from the base up to about one-third or one-half of their
entire thickness, which is in places upwards of 100 feet.   The
upper part is divided by joints into groups of ruder prisms of
varying size, that assume different directions—generally, however, at right angles to certain main or primary joints, which
are mostly vertical.   The plane of separation between the
upper and lower portion is very distinctly defined, and uniformly parallel to the base; but yet the two portions cohere
strongly, and, indeed, pass into each other.   (See fig. 23.)

Fig. 23.

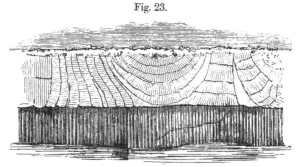

Natural section of part of a basaltic lava-current in the
Valley of the Ardèche (Vivarais).

It has, I believe, been generally considered that these two

portions were distinct currents of lava, one superposed to the other; but the intimate union of the two at the line of junction, without the interposition of a single scoria, completely negatives this hypothesis. The plane of separation probably marks the division between the upper portion, which was first consolidated by the upward escape of its heat and vapour, from the lower, which parted with its heat downwards to the supporting beds, and was perhaps long liquid, and even in slow lateral motion after the consolidation of the upper part had been completed. Indeed this supposition may be said to be demonstrated by the fact that, wherever the lava-bed re-

Fig. 24.

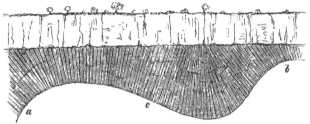

poses upon an irregular surface, the columns are found invariably to have taken a direction perpendicular to its plane, which must therefore have been the surface at which the divisionary process commenced. Thus, when columnar lava rests on a convex surface, as at a, c, fig. 24, the prisms are found to converge downwards; when upon a concave one, as c, b, upwards.

§ 11. It may be worth while to examine the cause of this very regular divisionary structure in some lavas. As any pasty or semi-liquid matter, such as lava, parts with its fluid vehicle, whether this be water or heat, or both combined, its particles tend to come into closer union, and consequently its mass contracts, or *shrinks*. If the process of consolidation begin at the surface, as is always the case when it is effected by contact with either a rarer or a cooler body, the tendency to contract

is exerted on all points of the superficial layer which is undergoing the process. The shrinkage caused by the contractile
force in a direction perpendicular to the surface effects no rent,
because the liquidity of the matter allows of its yielding by
the motion of its particles in *that* direction; but, on the contrary, the contractile force which is exerted at any point in the
plane of the surface is counteracted by the contemporaneous
shrinking of the surrounding parts. From the action of these
opposing forces, the superficial plane or layer must be divided,
by a greater or less number of rents, into distinct portions, in
each of which the aggregate force of contraction overcomes
the opposing forces of the surrounding parts. In each a
centre of contraction establishes itself, the particle which occupies that point remaining stationary, while the others are
drawn from every side towards it. Under favourable circumstances, a concretionary arrangement of the particles accompanies this change of their position, by which they are enabled
to dispose themselves more or less according to their mutual
affinities, and that tendency which all bodies have to collect

Fig. 25.

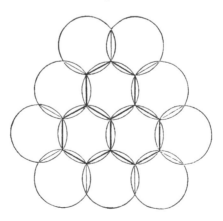

round any nucleus that may present itself at the right time
and in the right place, by a sort of molecular attraction acting

within sensible distances. The rents (or fissures of retreat) dividing the separate portions form themselves along those lines where the contractile and concretionary forces of the proximate centres balance one another. If the process took place while the matter was perfectly tranquil, and if its substance were completely homogeneous, the centres would be equidistant, and all the circles within which their influence operates equal. The rents would therefore divide the consolidated layer into regular hexagons; each fissure being tangential to the two proximate circles between which it is formed (see fig. 25).

But so far as the liquidity of the matter beneath this superficial layer allows it to yield to the contractile force in a direction perpendicular to the surface, no rent will be produced parallel to that plane. Therefore, by the continued inward propagation of the process of consolidation, the fissures will be prolonged towards the interior in planes perpendicular to the surface at which consolidation commenced, by which the mass will be divided into hexagonal prisms; the line described by the centre of contraction forming the axis of each prism.

In all lava-streams, however, the consolidation of the outer surface exposed to the air is so rapid, and disturbed by the continued and irregular movement of the current beneath, as well as by the violent escape of the confined vapours, that the process of contraction must be irregular and tumultuous, and the polygonal blocks into which the upper portion of the mass divides itself are consequently rude, unequal, and often shapeless, though in many cases a tendency to a hexagonal figure is perceivable. Moreover, the prolongation of the prisms inwardly being dependent on the lava of the interior continuing to yield to the contractile force in the direction of their axes, in proportion as its liquidity is too imperfect for this, transverse fissures, or *joints*, will be formed, separating the prisms into so many portions or articulations.

This double system of divisions it is which, where the grain

of the lava was coarse and its texture open, and consolidation therefore hasty, from the rapid escape of the contained
vapour, has broken it up superficially into those cuboidal or
prismatic blocks already mentioned as characterizing many
lava-fields. And in very few cases does the divisional structure, caused by the inward progress of consolidation from the
external exposed surface, exhibit any remarkable regularity.
In shallow and therefore rapidly-cooled sheets of lava, rude
shrinkage-joints often penetrate the *whole* mass. But where
the thickness of the bed is sufficient to cause the inferior
portion to retain its liquidity for a long time, and the process
of consolidation, as respects that part, proceeds from the lower
surface tranquilly and slowly, the prisms often become there
very regular and symmetrical, assuming the well-known columnar figure, in which the section of each column approximates to a true hexagon.

This regularity of divisional structure takes place also in
many dykes (which, no doubt, generally cooled more slowly
than lava exposed to the air), the axes of the columns being
perpendicular to the respective sides of the dyke, that is, the
cooling surfaces. But as the process commences usually on
both sides, the columns will not be likely to meet regularly
and continuously at its middle. Hence the greater number of
columnar dykes have a central seam of amorphous lava, or an
irregular plane, separating the two halves. Sometimes a fresh
injection of lava has penetrated this central seam, producing
the appearance of a dyke within a dyke.

In the same way, in a horizontal bed of lava, a seam or
marked plane of separation generally divides the upper portion,
which was consolidated rapidly and tumultuously by parting
with its heat upwards—and, from the causes already mentioned,
is but rudely prismatic, if at all—from the lower portion, which
was consolidated more slowly and tranquilly by loss of heat at
the lower surface, and is consequently divided with more regularity, often into columns of almost architectural uniformity

and beauty. The contrast between the upper and often quite amorphous portion of the mass, and the lower columnar one, is sometimes striking, and so great as to have led many geologists to suppose them to be separate beds formed by distinct flows of lava (see fig. 23, *supra*); but a close examination will always, I believe, detect the unquestionable connexion and interpenetration of the two portions. No scoriæ anywhere interfere, as would certainly have been the case had the two portions been separate currents. The celebrated columnar range of Fingal's Cave and those of the Antrim coast sections present the same fact in a remarkable manner.

Fig. 26.

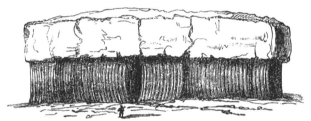

Part of the basaltic colonnade of Portrush (co. Antrim).

As might be expected, the more liquid the lava, the finer its grain, and the more homogeneous its substance, the more uniformly in general, *cæteris paribus*, has the concretionary process operated, and the more regular and numerous are the resulting columns. This also would be naturally favoured by the thickness and consequent slow cooling of the bed. As has been already said, the axes of the columns being always perpendicular to the cooling surface, where this was concave they converge from it, where convex they diverge in the opposite direction. Hence the fan-shaped groups of prisms often observable in columnar ranges*. Sometimes the columns

* Since one of the main causes of the rapid degradation of volcanic rocks, particularly basalt, lies in the facility with which rain-water penetrates between its prismatic divisions, and the consequent effect of frost in rending them asunder, it is obvious that this disaggregation will be

have taken a gradual and graceful curve; probably in consequence of some slow movement impressed upon the lava-
matter during the process of consolidation (fig. 27).

Occasionally (as for example, at La Tour d'Auvergne, in the
Mont Dore), the columns show a cylinder of compact black
basalt within a prismatic case of lighter colour and looser
greatly facilitated by the prisms converging downwards, owing to which
their own weight assists in disuniting them; while, on the contrary, it is
most effectively opposed by the contrary arrangement.

This consideration explains the origin of those insulated conical peaks
of frequent occurrence in basaltic districts, and which have been so often
appropriated, in Central France, Germany, and Italy, to the sites of
feudal fortresses. They will, for the most part, if not universally, be found
to owe their existence to the extreme stability occasioned by a pyramidal
grouping of columns, and the freedom from joints, which has enabled them
to outlive the destruction of the remainder of the beds to which they
belonged.

One of the most magnificent columnar peaks of the kind is to be seen
at Murat (Dépt. Cantal) in France.   (See Plate X. of my ' Volcanos of
Central France,' 2nd edit. Murray, 1858.)

It may be remarked, that the destruction of the basaltic beds in which
such conical eminences were once included has evidently been the result
of the slow operation of the causes now in action. Any violent rush of
waters would sweep away a pyramidal cluster of columns almost as easily
as a vertical one; it is the long-continued influence of vertical rain and
frost that will alone explain the uniform destruction of the one and preservation of the other.

Another frequent cause of dilapidation, peculiar to volcanic rocks,
but which must not be enlarged on in this place, is their superposi-
tion to beds of clay, tuff, scoriæ, and other loose matters, which are
easily washed away, and the overlying rock consequently undermined.
The prismatic structure, by allowing the percolation of water to these
lower beds, obviously accelerates the process. Where columnar dykes
have been denuded by the destruction of their lateral enclosures (which
often consist of a more friable rock, probably a scoriæ-conglomerate), they
project like walls of Cyclopean architecture; the extremities of the
prisms, laid horizontally, showing their polygonal surfaces on either side.
Sir C. Lyell, in his ' Manual' (ed. 1855, p. 487), gives the sketch of such
an insulated mass, called ' the Chimney,' in St. Helena, projecting 65 feet
from its base. In Auvergne, the walls of some of the old feudal fortresses
referred to above, being built of prisms laid horizontally, the resemblance
between these and the natural dykes is such, that the one may be almost
mistaken for the other.

texture, a segregation of dissimilar matter having accompanied the concretionary action.

Fig. 27.

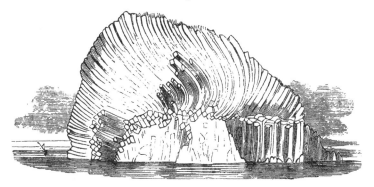

Group of curved columns of trachyte, near the Isle of Ponza.

The error long prevailed of supposing the columnar configuration to be solely confined to the oldest traps. In reality this structure is common to the *interior* of the lava-masses of every age, in which the composition and circumstances of liquidity and disposition were favourable to its production. Such masses, when of an early date, have usually been worn through, or denuded, by long exposure to aqueous erosion and other destructive processes, owing to which the prismatic configuration of their interior has been disclosed to view, while in the currents of modern volcanos opportunities of observing their internal structure are more rarely afforded. But where this *is* the case, as in the sections afforded by the deep ravines or vertical cliffs of Etna, Iceland, Bourbon, and even Vesuvius, a very regular columnar arrangement is often observable, even in lavas of which the date is known*.

Under peculiarly favourable circumstances, basaltic columns

* The basaltic rock of La Motte, near Catania, is beautifully columnar in the interior, and exhibits this structure on one side, where it has suffered great degradation from the waves of the sea; the other sides, as well as the surface, are amorphous.—Ferrara, Campi Phlegrei, p. 322. See also the woodcut given by Sir C. Lyell, Manual, p. 523, ed. 1857.

have been formed in lengths of 100 or even 150 feet, sometimes attaining 50 feet or more without a joint, although only
8 or 9 inches in diameter. The pyramidal hill of Murat in the
Cantal has furnished some of great length and beauty to the
Paris Museum. Those of Fingal's Cave are, I believe, almost
as long, but larger. In some cases, especially where the grain
of the lava is coarse, the prisms are colossal in bulk, measuring six or eight feet across*.

* It appears to me that the explanation of this columnar structure
given in most text-books is scarcely complete or satisfactory. M. Delesse,
indeed, holds it, as I do, to be the combined result of " crystallization "
(rather, I would say, concretionary attraction) "*and* contraction." Dr.
Daubeny denies that contraction has any influence in its production
('Volcanos,' p. 660). Sir C. Lyell scarcely enters upon the question at
all ('Manual,' p. 489). Professor Jukes ('Manual of Geology,' p. 199)
attributes the prismatic arrangement to "contraction on consolidation";
but in the very next page he ascribes it to the "*squeezing together* of contiguous spheroids," following in this opinion, and quoting, Gregory Watt,
who, like Dr. Daubeny and some other recent writers, seems to have
considered the prisms to be necessarily composed of *spheroids pressing
against each other*. But it is evident that the contiguous spheroids must
have been *first separated* before they could (if ever they did) press against
each other. The question to be solved is, what caused the separation of
the prisms, *i. e.* the seam, or solution of continuity, between them, which
is so complete that they fall apart the moment their centre of gravity
is unsupported ? The cause surely must have been contraction. Moreover, the spheroidal-concretionary theory only accounts for the formation of an irregular aggregation of spheroids. The prolongation of the
column in a direction perpendicular to the cooling surface, *without any
joint*, and sometimes to distances 100 times the diameter of the column,
can only, to my mind, be accounted for by a contractile force drawing the
particles towards the already consolidated and fissured surface, while the
fluidity of the interior allowed of their movement in *that* direction.
    There is no reason to suppose these long columns to contain any distinct
spheroids. Where such have been formed, it has been by the multiplication of transverse joints, each articulation containing a single spheroid.
And since the spheroids are always enclosed within the prisms, it seems
clear that the latter were formed first ; in other words, that the shrinkage-
fissures were completed before the concretionary arrangement took effect.
Had the reverse been the case, which is the view taken by some geologists (*e. g.* Dr. Daubeny, *loc. cit.*), the spheroidal concretions would
have been irregularly dispersed through the mass, not confined within the

In very regular basaltic columns, the transverse joints are not flat, but curved into convex and concave surfaces, almost resembling a ball-and-socket joint. In all the cases in which I have observed this, the convexity has been downwards, that is, in a direction towards the surface of refrigeration (which I believe to be uniformly the lower, or base of the lava-bed). And I would account for it by the earlier consolidation of the exterior than of the interior of the column, which would have

bounding surfaces of the prisms. Indeed this difficulty is felt by the writers I refer to, who acknowledge their inability to account for this peculiar arrangement of the spheroids. (See Jukes, *loc. cit.*)

It is well known that a similar divisionary structure is produced in bodies of mud, clay, moist tuff, starch, or other pasty matter, on desiccation. This identity of structure is sometimes curiously observable in close juxtaposition where lava has flowed over a bed of moist clay or shale, which, being dried by the heat so communicated to it, has split into hexagonal columns perpendicular to the drying surfaces, and as regular as those of the lava above. An example is given by Professor Phillips in his ' Manual.' I have seen many such in volcanic countries. So, too, the soft friable sandstone of Rotherham and other places, when used for the lining of kilns or furnaces, is found to harden and split into very regular columnar concretions. Surely it is irrational to suppose that in *these* cases it is contraction, but in that of basalt, &c. the very opposite, viz. compression, that has caused their identical structure.

It has been denied that in the case of basaltic columns the prismatic configuration can have been produced by contraction, because the prisms are often in such close contact that it is impossible to insert a knife or a sheet of paper between them. It must be recollected, however, that the force by which the contraction is produced acts only within the most minute and imperceptible distances, and that a space consisting of a great number of these distances may yet be almost insensible to our powers of observation. It is also certain that the fissures of retreat are subsequently narrowed, and often filled up by infiltrations, usually of iron oxide, which are deposited on the surfaces of the columns. The reality of the contraction, and the extraordinary force it exerts, are attested by a circumstance which I had an opportunity of observing at Burzet in the Vivarais. The basalt of this locality contains numerous large knots of olivin, often of the size of a fist; it is at the same time very regularly columnar, and the columns are in very close contact. Yet it has happened in many instances that the fissures of retreat have divided one of the large olivin nodules into two; half being enclosed in one column, and the other half in another proximate column. Though the division is complete, and the

the effect of producing the joint further from the base, whence
the refrigeration is propagated, at the sides than at the centre.
From the same cause, namely, the angles consolidating sooner than the flat sides, these are
occasionally found to project some way above
the latter in angular processes, which give to
the articulations the figure of so many thick
cups fitting into one another; and these processes are sometimes themselves separated from
the lower portion of the cup by a distinct joint
(see fig. 28).

Fig. 28.

§ 12. By an extreme multiplication of joints, the columnar
tends to pass into a spheroidal or globiform structure. Indeed,
the prisms of some basaltic colonnades appear to be composed
of spheroids piled one upon the other. This structure, however,
is usually disclosed only by a commencement of decomposition; the angles in excess of the spheroids first becoming discoloured, and ultimately falling off in concentric folia. The
columnar lava of the volcano of Bertrich in the Eifel looks
as if formed of piles of Dutch cheeses, whence the name of
" the Cheese-Cellar*" given to a grotto in it. In the Isle of
Ponza, prisms of a green vitreous pitchstone separate readily
into very perfect spheroids or ovoidal masses from a foot to
a few inches in diameter, which when struck desquamate into
concentric laminæ exactly like the coats of an onion (see fig. 29).

More than one separation of parts appears sometimes to

fracture-surfaces smoothed, probably by time and aqueous filtration, they
correspond so completely that it is impossible to doubt their former union
in the same nodule. Their separation can only be accounted for by a
powerful contractile force exerted during the formation of the columns—
a force partaking of the double character of a *grip* and a *wrench*. No
mutual pressure of contiguous spheroids could have had this effect. Just
so, in well-consolidated conglomerates, the hardest pebbles are often
found cut clean through by joints, formed by the same concretionary-
contractile process. In many granites, likewise, the large felspar-crystals
are similarly bisected, one half facing the other on either side of a joint.

* See a woodcut of this in Sir C. Lyell's Manual, fig. 637, ed. 1858.

take place successively in the same mass.   The upper part

Fig. 29.

Prismatic obsidian passing into globiform.   (Chiaja di Luna, Isle of Ponza.)

having been rapidly fissured through the diminution of bulk
which ensues from the escape of its vehicle of fluidity by percolation through the pores of the hardening rock, a second
series of fissures may subsequently be formed, perpendicular,
or nearly so, to the first, when the temperature of the lava is
lowered still further by slow transmission of vapour and heat
through the first series.   In this manner the rude blocks or
layers first formed are divided into numerous small prisms
perpendicular to their bounding surfaces.   The former will be
large and very irregular, owing to rapidity of formation; the
latter small and more perfect.   Amongst the basaltic colonnades of the Vivarais are some beautiful examples of this
double prismatic structure.   A part of one may be seen in
fig. 23, *supra*.

So, too, it occasionally happens that a mass of lava, after
assuming a globiform structure, has, by subsequent contraction,

been divided either into prisms having their axis perpendicular to the surface of the spheroid, and therefore converging to its centre, or into concentric lamellæ—according as the contractile force, in its progress from the surface to the centre of the mass, experiences least resistance to the movement of the particles in the direction of the radii, or of the tangents, to the sphere.   At St. Sandoux, in Auvergne, is a beautiful example of the former variety of divisionary structure, viz. an enormous spheroid of basalt, more than 50 feet in diameter, composed of very regular and jointed columns radiating from its centre, where they are closely united, to the circumference, where a considerable space is left between them*.   In the Sieben-gebirge a concentrically laminated spheroid of basalt is to be seen, having a diameter of 500 or 600 feet !   The laminæ in this case show a tendency to columnar division likewise, the prisms being at right angles to the curved planes of the laminæ†.

§ 13. The division into concentric spheroids is common and well known.   It occurs on every scale, from spheroids many feet in diameter to the smallest oolitic grains.   Many lava-rocks are thus separated into angulo-globular pieces about the size of a pea, occasioning a pisolitic structure.   These spherulites are sometimes only disclosed by decomposition. In the greystone or semi-augitic lavas, each has a grain or crystal of felspar for a nucleus.   Rocks of this character have been called Variolites.   In the very felspathic, and espe-cially in the vitreous lavas, this structure is frequent; the spherulites assume a pearly lustre, and the resultant rock is called Pearlstone, vast masses of which occur in Hungary and the Andes of Peru.   The spherulites, or crystallites, as they are sometimes called, are more or less crystalline, having at times a radiated, at others a concentric lamellar structure, occasion-

---

* An engraving of this rock is given by Faujas de St. Fond, ' Sur le Vivarais et le Velay.'

† Nöggerath, Rheinland, vol. ii. p. 250.

ally both combined. They are of exactly the same character as those which form themselves in the refuse slag of glass-furnaces; and the multiplication of such spherulitic concretions throughout the mass transforms its vitreous texture into a more or less stony or scaly one, just as happens by slow cooling of the fused Rowley Rag basalt in Messrs. Chance's manufactory of artificial stone at Birmingham. Passages of this kind from a wholly vitreous to a stony texture may be observed in many felspathic lava-rocks. In some—for example, in those of the Ponza Isles and Ischia—the movement of the matter, probably under considerable pressure, has evidently dragged out the felspathic spherulites into flattened disks, and ultimately into thin planes, which give a patched or ribboned and laminated structure to the rock.

§ 14. Some lavas, by the multiplication of joints parallel to the cooling surfaces, assume on consolidation a tabular or lamellar divisionary structure. This occurs most frequently where the component inequiaxed crystals or scales are disposed conformably; owing to which, the fluidity is greater in the direction of their longest plane surfaces than in the transverse direction, occasioning the more frequent production of retreat-fissures across, than with, the *grain* of the rock; and thus the lava is divided into concretionary plates or lamellæ, of greater or less thickness, according to the greater or less difference in the fluidity of the lava in these opposite directions. When the thickness is comparatively great, the structure is called *tabular*; when less so, *lamellar*; and when the plates are very thin, *slaty* or *schistose*. The cause of the parallel disposition of the crystalline particles of such lavas is probably the squeezing or dragging out to which they have been subjected while moving on or under pressure, already referred to as productive of a laminated structure in many trachytes,— the same process, in fact, which is now recognized as the cause of cleavage in clay-slate.

The lamellar structure is often accompanied by the prismatic or columnar, of which, however, it is wholly independent. The prisms, being perpendicular to the exposed surfaces, are generally (though not necessarily) transverse to the planes of the lamellæ; which being always parallel to the direction in which the lava moved, and by that motion led its crystalline particles to arrange themselves in planes, are also, of course, mostly parallel to the outer surfaces of the bed, current, or dyke.

A remarkable example of this united structure occurs in the Roche Tuilière in the Mont Dore. The clinkstone of which this rock is composed (an insulated fragment of the vast current descending from the Puy Gros) is regularly divided into nearly vertical columns. It is also extremely schistose, as the vulgar name of the rock implies, the laminæ being used as slates for roofing. The direction of the lamellæ in the centre of the rock is horizontal, and therefore perpendicular to the axis of the prisms, but gradually declines towards the north, until it becomes parallel with their axis. The fact that the former arrangement communicates great solidity to the prisms, while the latter affords easy access to the agents of division and degradation, perfectly accounts for the isolation of this rock; the part in which the latter arrangement prevailed being wasted away. This curvature of the lamellæ is probably, as has been said, owing to some change in the direction of the movement or pressure to which the matter was subjected during the progress of consolidation. It is seemingly analogous to the curvature of the blue and white veins in a glacier.

Some of the basaltic lavas of the Mont Dore, as at St. Bonnet, are remarkable for their tabulated structure—the tables measuring frequently 10 or 12 feet in length, nearly as much in breadth, and from 2 to 6 inches only in thickness. They ring like a bell. The grain of the rock is close, fine, and minutely crystalline. Clinkstone rings in the same manner, whence its name.

Just as the incomplete yielding or subsidence of the lava in a direction vertical to the cooling surface occasions the transverse fissures, or *joints,* by which the columnar concretions are divided, so, when the fluidity of the slaty lavas in the direction of the scales, or plates, is insufficient entirely to compensate all the tendency to contract in that direction, transverse joints are produced in them at intervals more or less distant; and by the multiplication of these, the tables assume a cubical or rhomboidal form, such as is conspicuous in the earlier traps, in syenite, and even in granite*.

§ 15. It has been seen that the divisionary structure, or regularity of internal form, which a mass of pasty lava may assume during its refrigeration or consolidation, according to circumstances, comprehends the following varieties :—

1. The prismatic or columnar.
2. The tabular, lamellar, and schistose.
3. The rhomboidal or cubiform.
4. The globiform.
5. The angulo-globular.

And that two, or even more, of these varieties of form may be combined in the same mass.

But before we proceed further to consider the varying modifications assumed by erupted lavas after their emission on the surface of the earth, some attention must be directed to the various mineral characters and composition which they present on examination. This subject has been partly anticipated already, but requires further examination.

---

* In fact, the *stratification* and *jointing* of rocks of all kinds, where the strata are separated by *seams,* are produced by this concretionary process, accompanying the final arrangement of the particles of the mass influenced by pressure, and probably the drainage of its fluid vehicle.

CHAPTER VII.

MINERAL CHARACTERS AND CONSTITUTION OF LAVAS.

§ 1. LAVA-ROCKS—that is to say, rocks which have been observed to flow as lavas from volcanic orifices, or which appear, from their position (for example, as forming the substance of dykes, or currents, or massive hummocks, on the flank or at the base of a volcanic mountain), to have been manifestly erupted in a condition of igneous liquefaction—although differing more or less in mineral character, are all found, upon analysis, to be composed of silicates of alumina or magnesia, with some protoxide of iron, potash or soda, and lime; these elements being generally crystallized in some of the various forms of felspar, quartz, augite or hornblende, mica, olivin, or titaniferous iron. When the augitic or ferruginous minerals abound, the rock has a higher specific gravity than when the felspathic elements predominate, in the extreme proportion of 5 to 4*. The latter class of lava-rock is known to mineralogists by the generic term of Trachyte; the former of Basalt (*dolerite*, Brongniart). There are many lava-rocks of a middle class, to which, from their prevalent light-grey tints, intermediate between the usual black or dark slate-colour of basalt and the ashy-grey or brown or yellowish-white shades of trachyte, I long since gave the name of 'greystone'—corresponding to the *tephrine* of Brongniart and the *trachy-dolerite* of Abich; and I still think the name of 'greystone' more con-

---

* Mr. Darwin defines the specific gravity of the usual component minerals of lavas, as ranging, in felspar from 2–2·74; hornblende or augite, 2·4–3·4; olivin, 3·3–3·4; quartz, 2·6–2·8; and, lastly, in oxides of iron from 4·8–5·2. (' Volcanic Islands.')

genial to our language than either of the above. The term 'clinkstone' (*phonolite*) is given to the schistose varieties of greystone. It has usually a large proportion of the aluminous minerals *. Granular or crystallized quartz is not very frequently met with in lava-rocks, but occurs in some trachytes, the crystals being hexagonal prisms with a pyramidal truncation at each end, conveying the impression that they crystallized *before* any of the other component minerals. Generally, however, the quartz is granular, and appears as if externally fused; or it is mixed with the compact base of the rock. Among the trachytes of Hungary, the Ponza Isles, and the Andes, some are highly siliceous—indeed, occasionally, might be mistaken for flint, of which they have the conchoidal fracture and cutting edges, with veins of quartz or crystals lining their cavities. The felspar of trachyte is usually glassy, but not unfrequently opake. Its crystals occur in porphyritic trachytes

---

* Professor Jukes, Sir C. Lyell, and some other geologists adopt the term 'dolerite' for the general class of augitic lava-rocks. I cannot see why the old and well-known name 'basalt' should be discarded for this newer coinage. By some writers it is confined to augitic rocks which happen to contain olivin. But this limited distinction will never, I think, be generally received; and as the term 'basalt' is the oldest, the best-known, and the most generally recognized, it would surely be desirable to continue to employ it for the entire class of lava-rocks in which augite or hornblende and the ferruginous minerals generally preponderate, so as to give a prevailing dark colour and high specific gravity to the rock. If the term 'dolerite' be given up as a synonym for 'basalt' among English geologists, that of 'trachy-dolerite' must follow, and 'greystone' had better be substituted, for the intermediate class. I submit, moreover, that the first word is inappropriate, as this class of rocks has nothing of the roughness of texture implied by the first dissyllable. They are, on the contrary, usually smoother than basalt itself, and sometimes have almost a soapy feel, as in some clinkstones, which are scaly greystones. It is possible to go too far in our acceptance of Continental nomenclature. Some writers, for example, call every current of lava a ' *coulée.*' Where we have a fitting and expressive English word applicable to any subject, it is better to employ it than, with a view to scientific uniformity, introduce unnecessary Gallicisms or Grecisms into our vocabulary. For this reason I also prefer the term 'greenstone' to that of 'diorite.'

sometimes more than an inch or two in length. In some lavas it is entirely replaced by leucite in dodecahedral crystals, also occasionally very large. The augite is at times represented by hornblende, and its crystals also are sometimes large and perfect. The mica occurs generally in hexagonal or rhomboidal tables; olivin in bright olive-green crystals or granules; titaniferous iron in grains or octohedral crystals.

§ 2. The older lavas—*i. e.* the volcanic matter which penetrates and sometimes overlies the Secondary or earlier strata, in such situations as show it to belong to the præ-tertiary age—differ to a certain degree when viewed as a whole, both in aspect and in mineral character, from the lavas of more recent date, just described. They have been usually called Traps, or trappean rocks. Those which represent the basaltic class generally contain hornblende rather than augite, and are called Greenstone (diorite). The augitic rocks of this class have likewise numerous varieties, respectively called Melaphyre, Gabbro or Diallage-rock, Diabase, Kersanton, &c. Even serpentine seems to be justly included among them. These last varieties are usually found in the vicinity of limestone strata, into which they appear to pass; the elements of the sedimentary and eruptive rocks being mixed by joint mechanical and chemical action. The felspathic rocks of an early age are most frequently compact, and then called Felstone, or compact felspar: if the base contain distinct crystals, Felspar-porphyry. Elvanite, or quartziferous porphyry, has the same base, or a granular one composed of the same materials, with disseminated crystals or crystalline grains of quartz. The traps, both felstones and greenstones, are accompanied by their respective conglomerates of ash or tuff, some of which, when laminated, have been called ' trap-shales.' The terms ' claystone' and ' wacke' are usually applied to felstone and greenstone, respectively, or their tuffs, when in a decomposed state.

Viewed as a whole, it may be said that the earlier volcanic rocks are more dense and compact in texture, and less fre-

quently exhibit cellular or scoriform parts, than the recent
lavas and tuffs.   It can hardly be said that they are more
crystalline.   Of course they are likely to have been more
altered by infiltrations, and other decomposing or metamorphic influences; and in this way, to a certain extent, we may
account for such distinctive characters as they exhibit, without
being driven to suppose them produced under different circumstances from the more recent volcanic rocks.   Some, however, among them, such as serpentines, porphyries, and syenites,
occasionally approach so nearly in character and in circumstances of position to granite and its associated crystalline
rocks, as to claim a plutonic rather than a volcanic origin.   We
shall have occasion to revert to this consideration hereafter *.

§ 3. Besides their differences in mineral constitution, as
shown either by mechanical or chemical analysis, lava-rocks
vary much in texture and aspect.   Some are perfectly vitreous ;
others semi-vitreous.   The most glassy and semi-translucent
varieties are known by the name of ' obsidian '; the opake
and resinous or waxy, of ' pitchstone.'   The former do not

* To the above brief description of the mineral composition of volcanic rocks it may be objected, that no notice is taken of the different
species of the genus Felspar, viz. albite, potass-albite, soda-felspar, ortho-
clase, anorthite, labradorite, oligoclase, andesin, ryacolite, adularia, Peri-
cline, &c., as distinguished by Gustav Rose and his disciples.   I have
perhaps been biassed in this determination by an imperfect appreciation of the value of these minute chemical differences, in a geological
view of the subject, and too strong a sense of the complexity introduced
by them into the classification of volcanic rocks.   But it appears to me
that chemists are not yet thoroughly agreed on the specific characters of
these varieties.   On the other hand, there is no difference of opinion on
the generic characters of felspar, as shown in its general aspect, its crystalline forms and cleavage, and its composition, as found on analysis,
almost wholly of silica and alumina, with more or less of lime and potash
or soda, but without any, or merely a trace of magnesia or of iron oxide,
which, on the other hand, compose so large a proportion of the other
minerals, viz. augite, hornblende, mica, olivin, and titaniferous iron, usually
associated with felspar in the various volcanic and hypogene rocks.   See
the analysis of minerals most abundant in such rocks, in Sir C. Lyell's
'Manual of Geology,' p. 479, ed. 1855.

I

occur among the trappean or older lavas.   Both are found most frequently in conjunction with trachyte.   Sometimes, however, as in Iceland, Bourbon, and Hawaii, augitic lavas containing olivin have run into a pure black glass.

The vitreous varieties are occasionally porphyritic, but more frequently, as has been already mentioned, by the formation of crystallites or small globular concretions, chiefly of felspar, within them, pass into 'pearlstone.'   Among the basaltic lavas and traps there occurs a similar small globular structure likewise—most frequently disclosed only by decomposition; but in these the very vitreous or 'pearly' lustre is wanting.   The vitreous lavas are indeed, in both classes, exceptional.   The texture of the great bulk of lava-rocks is stony and granular, often highly crystalline; the component crystals of felspar, augite or hornblende, mica, olivin, and perhaps quartz, being, though interlaced and interpenetrating, occasionally as distinct and large as in granite.   At times the larger crystals are disseminated in a compact-looking felspathic base or paste, as in the older porphyries.   In some lava-rocks the crystalline or granular minerals are loosely aggregated, so as to form a porous, or even an earthy, almost pulverulent, rock; in others closely compacted into a hard, dense, and solid one. The texture is sometimes laminar, slaty or scaly, the inequi-axed crystals lying in parallel planes, as in gneiss or mica-schist; but for the most part they are irregularly aggregated in a confused granitoidal mixture, in which, however compact it may appear to the eye, the microscope seldom fails to discover an abundance of minute pores or cavities.   Sometimes, though rarely, several of these varieties of texture are found passing, on a large scale, into each other in the same rock.   Patches or nodules of one kind are not unfrequently included in a mass of a very different one.

§ 4. It is a question of considerable interest, what are the circumstances which determine these differences of texture in lava-rocks ?   The current notion is—or rather, I should say, *used* to be—that all lavas have issued from the volcanic vent into the open air, and flowed or moved into their present positions in a condition of complete igneous fusion, and subsequently acquired a more or less crystalline texture according as they were cooled and consolidated more or less slowly. Certain experiments of Watt are quoted in support of this doctrine ; which, however, is not confirmed by a juster appreciation of the experiments themselves, or by extended observation and comparison of the texture of different lavas as they occur in nature.

It is true, that if basalt, such as that of the Rowley-rag near Birmingham (the rock operated on by Watt, and now employed by Messrs. Chance in their manufactory of artificial stone), having been completely fused in a furnace, is made to cool very slowly, the process lasting some days, it assumes by degrees a more or less stony texture, through the formation of crystallites or globular concretions, which, as they multiply, interlace and compress each other, until the whole substance has obtained a scaly semicrystalline grain.   But if this same matter in fusion is poured out into the open air, in the same way as molten iron is poured from a crucible or furnace into a mould, or as lava issues from a volcanic orifice during an eruption, it invariably hardens into a complete glass, not differing in appearance from obsidian.   Now, a great number of lava-streams are observable in different volcanic districts composed throughout of a rock of such absolutely vitreous texture.   I may mention the currents of glassy pumice and of obsidian in the Isles of Lipari, Volcano, and Pantellaria, those of Teneriffe, of Mount Ararat, of Hecla, Krabla, and several other volcanos in Iceland, and of many among the Andes. These *glassy* lava-streams we may indeed believe to have flowed in a state of true fusion, as complete as that of metal, or of common glass, or the basalt above mentioned when melted artificially in a furnace.   But in none of these examples has it ever been discovered, nor has any one suggested any

grounds for supposing, that the process of cooling or consolidation was, or could have been, more rapid than in the case
of the numerous similarly disposed currents of the same or
neighbouring volcanos, exhibiting the same mineral character
on analysis, but which in *texture* are entirely stony and granular, often indeed as highly crystalline as granite, and have
not even the most superficial film of true glassy matter. The
difference of texture in the two kinds of rock must therefore
clearly be owing to some other cause than more or less rapid
cooling.

Indeed, the globular 'bombs' and tattered fragments of
lava which are thrown up in a liquid state from the seething
caldron within a volcanic vent, and harden instantaneously
before they fall, as well as the scoriform crust which congeals with extreme rapidity on the surface of a lava-current
when exposed to the air, usually possess the same granular
or crystalline texture, and contain the same proportion of
large or minute crystals, as the innermost parts of the current,
which must have been cooled and consolidated in the most
gradual and tranquil manner. Were the theory correct, that
more or less slow cooling occasioned the more or less complete
crystallization of lava, the inner and lower portions of a deep
current, which has taken years to cool, ought to be infinitely
more crystalline than the upper and rapidly consolidated portion. Indeed, the degree and regularity of crystallization
ought to be found to increase uniformly and progressively from
above downwards. But this is true in no instance that I am
aware of. The most compact fine-grained lava-rocks are usually of uniform texture and grain throughout their entire thickness. So also are those which possess the coarsest texture or
the greatest proportion of large visible crystals.

These considerations led me long since (in 1825) to the
conclusion that, in the greater number of cases, lava, when
issuing from a volcanic vent, is already granulated, or composed of more or less imperfect crystals enveloped in a base or

paste of finer grain, but still minutely granular (not reduced to molecular fusion), and that its liquidity (*i. e.* the mobility of the solid component particles) is owing chiefly to the presence between and among them of an interstitial fluid; which fluid can scarcely be supposed other than that same water, or, rather, vapour of water (at times holding in solution, perhaps, more or less of silex or other mineral matter *), which is observed to issue in abundance from the surface and crevices of incandescent lava at the moment of its exposure to the atmosphere, and in the very act of consolidation.

Dolomieu (a sagacious observer of volcanic phenomena in the last century) had already put forward a similar view upon the nature of the fluidity of lavas, with this important exception, that he supposed the vehicle that gives mobility to their solid particles to be, not water or aqueous vapour, but sulphur —an opinion which the quantity of sulphur to be detected in ordinary lavas is quite inadequate to support†. Since his time, however, chemical analysis has discovered water to be a very abundant ingredient in nearly all the pyrogenous crystalline rocks; and even granite is now acknowledged by all geologists to have owed its liquidity or plasticity to this element as a vehicle of its component crystalline particles. There ought, therefore, to be no difficulty in admitting the same in the case of the felspathic and augitic lavas of a granular or crystalline texture.

* It is well known now that heated water or steam containing a certain proportion of potash, readily takes into solution silex. In this way both the felspar and quartz of volcanic (or plutonic) rocks may have existed in a liquid or viscid condition at temperatures far below their point of fusion.

† Dolomieu, Sur les Iles Ponces, avant-propos, 1788:—" Le feu des volcans ne dénature pas ordinairement les pierres qu'il a mises en état de fusion. . . . . . Le feu agit différemment que le feu de nos fourneaux. Il produit dans les laves une fluidité qui n'a aucun rapport avec la fluidité vitreuse que nous opérons lorsque nous voulons rendre aux laves elles-mêmes leur fluidité. . . . . Il produit la fluidité par simple dilatation, qui permet aux particules de *glisser* les unes sur les autres, et peut-être par le concours d'une autre matière, qui sert de *véhicule* à la fluidité."

It is true that M. Delesse (who stands foremost in advocating the aqueous plasticity of granite and the associated crystalline rocks) declares volcanic (lava-) rocks to be anhydrous, i. e. (*comparatively*) destitute of water. But this is precisely what we should expect in consequence of these having parted with their water in the eruptive discharges or more tranquil exhalations of steam that proceeded from them on obtaining a free communication with the atmosphere; whereas granites and many of the earlier trap-rocks not having, according to all probability, reached the open air in a state of liquefaction, but having while in that state been only forced into, or among, the overlying beds, perhaps at great depths beneath the sea, certainly under enormous pressures, still retain their interstitial water.

Moreover, the crystals of many lava-rocks appear to have suffered considerable attrition, rounding, and disintegration through mutual friction. This is particularly the case with the base of the more fine-grained lavas. Nor can any one look at the large glassy felspar-crystals of some porphyritic trachytes, or the large leucites of some greystones, which show marks not merely of disruption, but also of penetration by the finer paste that envelopes them, and of partial vitrifaction, without being convinced that *they* at least were formed long before the lava in which they occur became stationary. Mr. Darwin describes, in Albemarle Island (one of the Galapagos group), a very fluid stream of lava, black, compact, with angular air-cells, and thickly studded with large fractured crystals of glassy felspar (albite), many of them half an inch in diameter, which he says were evidently enveloped and penetrated by the lava, and rounded by friction as the current flowed on*. MM. Monticelli and Covelli describe the lava of Vesuvius, erupted in 1822, as containing leucite in the proportion of 6 to 1. The crystalline granules of leucite

* See also Bischoff, Chemical Geology, Cavendish Soc. Publ. vol. ii. pp. 223–230.

presented clear indications of having undergone partial fusion within the volcano previous to the emission of the lava. They were melted at the surface, and covered with a bluish-white glaze. Mr. Forbes says of the trachytes of the South American volcanos, "There is reason to suppose them to have frequently been ejected while in a pasty state, after the quartz (at least) had crystallized, and the temperature of the whole had become much lower than the fusing-point of the rock itself *."

Some lavas contain nodules of olivin which have the appearance of boulders, so evidently have they suffered attrition during the flow of the current which contains them. It was remarked by Von Buch of the basaltic lavas of Lancerote (and I have had occasion to make the same observation with respect to those of the Eifel, and also in the Vivarais), that while the nodules of olivin are large, reaching even to the size of a man's head, near the source of the current, they dwindle away towards its extremity, so as to be scarcely visible, the olivin being there broken down and mixed with the other fine crystalline granules of the lava. Had the olivin crystallized from slow cooling after emission of the lava, the reverse must have been the case. So, again, the rock which composes many dykes is observed to have a finer grain towards the sides or selvages of the dyke than at its centre, owing probably to the greater amount of mutual friction to which the crystalline or granular particles must have been exposed, if we suppose them to have been formed previously to the injection of the dyke, in the former than in the latter parts. Many dykes exhibit a laminated structure in planes parallel to their sides, which is probably attributable, as in the case of the laminated lavas already spoken of, to internal friction and consequent differential movements among their component crystalline matter while it was impelled between the walls of the dyke, under severe lateral pressure.

Moreover, the scoriæ of all lavas contain as many and equally

* Quart. Journ. Geol. Soc. 1861, vol. xvii. p. 26.

large and perfect crystals as the central parts of the rock. The explosions of Stromboli throw up quantities of perfect hexahedral crystals of augite, generally double ; much also of the loose ejected sand of Etna and other volcanos is composed of separate well-formed crystals of augite, titaniferous iron, mica, &c., all of which must surely have crystallized before they were thrown out.   It is wholly incredible that any of such crystals could have been formed during the few moments occupied by the consolidation of these scoriæ.   " Geologists," says M. Bischoff, "are now unanimous in the opinion that the formation of crystalline minerals, whatever was their origin, took place very slowly*."   I do not agree with Bischoff's esti- mate of thousands, nay, millions of years being required for the formation of a large crystal of felspar or leucite ; but I think every one will acknowledge that the process could not have been of momentary duration only.

It is not, of course, meant to be denied that lavas have, be- fore their eruption, been melted, nor even that, when erupted, they are not usually in a condition of fusion, if by that term we only mean a certain degree of freedom of motion given by heat to the particles.   This fusion, though imperfect, equals probably in most cases that of the impure earthy and partly metallic matter of the slags and scoriæ that form on the sur- face of melted iron in a furnace, to which, in texture and aspect, the scoriæ of lavas often bear so great a resemblance. Nor does the theory of the already semicrystalline or granular character of lava before it issues from the volcanic vent or settles on the outer surface of the earth, exclude the suppo- sition that many of the crystals observed in the rock into which it afterwards consolidates, may have been formed, or, rather, enlarged and perfected, during that process.   This would indeed seem, à priori, to be very probable, since the escape of the greater part of the contained interstitial water, in the shape of vapour (by exudation through the crevices and

* Chemical Geology, Cavendish Soc. Publ. vol. ii. p. 103.

pores of the rock), and the consequent rapid loss of the heat likewise which kept the crystalline particles more or less separated, must bring them together again, while still possessing a certain freedom of movement, in a manner most favourable to the action of the polarizing force (whatever it is) that occasions crystallization. Some of the finer particles may thus have united, according to their affinities, into more perfect crystals than the rest *. Indeed, it is certain that some action of a character approaching to crystallization, affecting all the particles of the mass, and causing them to cohere more or less tenaciously, accompanies the 'setting' or consolidation of even the most exposed surfaces of the semi-liquid lava.

There is a considerable analogy, as has been already suggested, between the condition of many lavas at the time of their emission from a volcanic vent, and also while undergoing the subsequent process of cooling and consolidation, with that of the syrup of *sugar* during the later stages of the manufacture of that substance. In both cases the matter is not a homogeneous molecular liquid, such as any melted or completely fused substance, but (according to my view of the nature of lava) a '*magma*' or compound of crystalline or granular particles to which a certain mobility is given by an interstitial fluid, which is in both cases heated water or steam ; and in both cases consolidation is effected through the evaporation and escape of this aqueous vehicle, by which the particles are brought together in a manner favourable to their cohesive aggregation in a rigid and more or less crystalline mass. It is worthy of remark, that almost every variety of texture observable in different lava-rocks has its counterpart in some of the various modifications in which sugar is pro-

* Bischoff observes that "imperfectly formed crystals imbedded in lava may be subsequently rendered perfect; for, according to the experiments of Dr. Jordan, the broken edges and angles of crystals are replaced when they are suspended in solutions of the same substance ; and it is only after this replacement of deficient portions that the crystals begin to increase in size." (Chem. Geol., Cav. Soc. Publ. vol. ii. p. 95.)

duced for the market by very slight variations in the process
of manufacture. The extremely viscous, ropy, filamentous, and
vitreous lavas have their analogues in barley-sugar and other
kinds obtained by melting; the ribboned lavas in brandy-
balls or coloured stick-sugars. There are compact sorts of
both rock-sugar and lava-rock in which only the finest granular or scaly texture is perceivable; while the highly crystalline and sparkling grain of loaf-sugar strongly resembles that
of the porous and very crystalline lavas, whether augitic or
felspathic. Finally, sugar-candy may be compared to the
granitoidal and porphyritic trachytes composed of large and
perfect crystals of felspar, augite, hornblende, mica, or quartz.
The parallel here instituted is not a merely fanciful one.
However homely the illustration, it may throw a light, I believe, upon the origin of the various distinctive textures of
lava-rocks, and suggests how small a difference in the circumstances affecting them, of temperature, motion, or exposure
to the atmosphere, may have given rise to these varieties
in the same elementary matter *.

* These views are not in accordance at all points, I am aware, with
those of M. Bischoff, who, though believing (and adducing as grounds for
his belief much the same evidence as I have done) that many of the crystals
composing lava-rocks were formed within the chimney of the volcano
previous to the eruptive outflow (Bischoff, op. cit. p. 231), yet considers
that all the *larger* crystals have subsequently *grown*, as it were, to the
size at which they are now found, by an immensely slow process, requiring
thousands or even millions of years; that process being (as I understand
him) the gradual dissolution of the amorphous base by percolating meteoric water, and its recrystallization upon some earlier-formed crystals as
nuclei (Bischoff, op. cit. vol. ii. p. 95). Hence he thinks the generally
large crystals of granite, syenite, &c., sometimes reaching to a foot or
more in length, to be caused by, and proportionate in size to, the time
that has elapsed since the production or consolidation of these rocks.
Numerous facts oppose themselves to this hypothesis. For example,
many recent lava-streams, of historical date, are full of large crystals, very
much larger than those to be found in the majority of the earliest trap-
rocks. Bischoff himself speaks of leucite-crystals in the lavas of Rocca
Monfina (by no means a very ancient, though perhaps a pre-historic volcano) as large as oranges; and it is notorious that, there and elsewhere,

§ 5. It is in itself very probable that lavas will often have been partially cooled, consolidated, and crystallized while forming the upper portion of a molten mass within the chimney of a volcano. This is likely to occur during every quiescent interval, if sufficiently prolonged. Indeed, a very short time may be sufficient to allow such a process to extend to a certain depth. It is also possible that pressure, no less than temperature, influences the more or less crystalline or compact texture of a lava while confined within the volcanic vent at still lower depths. In this position, indeed, it is conceivable that variations of temperature or of pressure may cause it to be more than once alternately expanded and contracted, perhaps liquefied, even wholly fused, and reconsolidated, before it finally issues in an eruption from the vent, and during such alternations many changes may take place in the characters of the rock—changes not of texture only, but even of the component minerals.

Our knowledge, indeed, as has been already hinted, of the laws which determine the formation of different minerals from their elementary ingredients, is at present too limited for the solution of these questions. Since, however, we have ample proof that the heat and fumes arising from a mass of incan-

when very large crystals occur in a lava-rock, they are also to be found in the contemporary scoriæ and tuffs. Bischoff would probably say (indeed he does express the opinion) that the crystals in the tuff were formed there since its deposition by the aqueous process. But besides that the element of *time*, to which he refers their formation, is often wanting in such cases, it is, I think, impossible to conceive a slow aqueous metamorphic crystallization to have produced the large crystals found in loose scoriæ after their ejection without any change in the crisp and semi-vitreous scoriform matter attached to them. Pilla says that numerous large crystals of leucite and augite were thrown up by the eruption of Vesuvius of 22nd April, 1845. The former were as large as nuts, had a vitreous lustre, and were well formed and very perfect. Small particles of fresh scoriaceous matter were attached to them, and also penetrated their interior. The lava of the same eruption was also full of the same large crystals. These must all have been formed within the volcanic vent previously to the eruption which threw them out.

descent lava can alter a solid rock, with which it comes in contact, from a compact carbonate of lime to a highly crystalline dolomite containing 45·00 of magnesia *, it will certainly not appear incredible that very considerable changes of mineral character may be effected in a subterranean mass of lava itself, during the repetition of the processes of dilatation and recondensation under variations of pressure and temperature, to which we know it must be subjected,—that crystals of augite, or olivin, or leucite, may have been produced from some of the comminuted, fused, perhaps vaporized elements of mica, felspar, and quartz,—even that the same original granitic matter may, under some circumstances, have been altered into trachyte, under others into basalt.

The heated water or aqueous vapour with which the mass was permeated probably played a part in such changes, taking the quartz into solution, to assume subsequently, perhaps, pseudomorphic forms, or to envelope and cement the other minerals in a siliceous base. The experiments of M. Daubrée prove that, by the concurrent influence of heat and pressure, crystals of augite, felspar, quartz, and mica can be produced from water containing alkaline silicates in solution mixed with common clay †. The volatilization of the ferruginous minerals may have occasionally separated their elements from the rest, and given rise to the production of a felspathic lava containing very little iron in one part of the chimney, and of a highly ferruginous lava in another, or lined the cavities of the upper mass with specular or titaniferous iron. Where the felspar, or a part of it, is reduced to an extreme degree of minute division, amounting perhaps to complete molecular fusion, during the intumescence of a mass of lava within a volcanic vent, the reconsolidation of the rock may have occasioned the

* In the Tyrol (see Von Buch, Lettre à Humboldt); in Skye (see Macculloch's Western Islands, vol. i. p. 325, &c.); in the Isle of Zannone (see Memoir on the Ponza Isles, Geol. Trans.); in many of the dykes cutting the chalk of the north of Ireland, &c.

† See Daubrée, Études sur le Métamorphisme, &c. Paris, 1859.

crystallization of its elements in the proportions and forms of new minerals, such as leucite or olivin.

Under some circumstances, it is presumable that the different specific gravity of the component elements of lavas may bring about similar metamorphoses. It is quite conceivable that, when exposed in the focus of a volcano to successive liquefaction and reconsolidation, the heavier minerals may sink by a sort of filtration through the lighter ones, the upper portions of the mass becoming consequently more felspathic, the lower more ferruginous or augitic. Mr. Darwin concurs with me in the opinion that the lighter felspar-crystals in a mass of liquefied lava will tend to rise to the upper parts, and the crystals or granules of the heavier minerals to sink to the lower; the viscidity of the matter preventing, however, any complete separation of elements differing but slightly in specific gravity[*]. Or a somewhat similar effect (perhaps the very reverse as regards the relative positions of the different minerals) may be occasioned by the mechanical squeezing out of the finer particles, or the more readily fusible elementary minerals, from among the coarser or the less fusible, under extraordinary and locally varying pressures [†].

Bischoff says [‡], " When a mixture of lead and tin in any proportions is allowed to cool slowly, one or other of the metals solidifies first, and remains mechanically mixed with the still liquid portion, which is an alloy of the two metals in definite proportions." If the fused metals, while in this state, were poured out and suddenly cooled, the result would be a basis of alloy containing crystals of one of the metals disseminated through it, like those of felspar or augite, in an apparently

[*] Volcanic Islands, pp. 118–124.

[†] Professor Jukes, in the article "Mineralogy and Geology" of the Encyclopædia Britannica, aptly says, " If we could follow any stream of lava to its source in the bowels of the earth, we should probably find it changing, under varying circumstances of depth and pressure, from scoriæ or pumice to granite."

[‡] Vol. ii. p. 93, Chem. Geol., Cavendish Soc. Publ.

homogeneous mixture of the two elements in lava-rocks. In the latter case, we have only to suppose the crystallization of the different constituent minerals of a mass of subvolcanic lava to be disturbed by an eruption (than which nothing can be more likely), to account for the existence of numerous well-formed crystals, as well in its ejected scoriæ as in the more slowly-cooled lava-streams. The basis of both would probably consist of small imperfect crystals or granules of the different constituent minerals, lubricated by a solution of silex in the permeating water-vapour. Felspar, it is known, no less than quartz, is soluble in heated water*.

After any changes of this nature, the subsequent intumescence and protrusion of the lavas might produce alternate currents of trachyte, clinkstone, or compact felspar, and basalt, or leucitic greystone. In this way the fact may be accounted for, that the principal varieties of lava are often, nay, generally, found to have been emitted successively—sometimes alternately, but without any regular order of succession —from the same volcano—at least from the same system of vents. Thus, in France, in the Mont Dore, Cantal, and Mezen (three extinct volcanos), currents of basalt, trachyte, and clinkstone may be seen to alternate. The annexed section of the rocks worn through by the waterfall immediately above Mont Dore les Bains offers one noted example (fig. 30).

In the Cantal similar alternations are visible. In the Mezen a very felspathic clinkstone (scaly trachyte) rests upon basalt, and supports in turn a bed of the same rock; and this fact may be observed on many points. In the Chaîne des Puys, near Clermont, eruptions of basaltic lavas have almost constantly succeeded those of trachyte, or greystone, from the

---

* These views of the changes which probably took place in the subterranean rock whence lavas are derived, were put forward by me so early as 1825, in the first edition of this work (see pp. 145, 146, Volcanos). They are now supported by the higher authority of Mr. Darwin and Professor Jukes, as well as by the subsequent experiments and reasoning of MM. Delesse, Deville, Durocher, and Daubrée.

same or contiguous vents. In Somma, leucitic greystone overlies trachytic tuff.

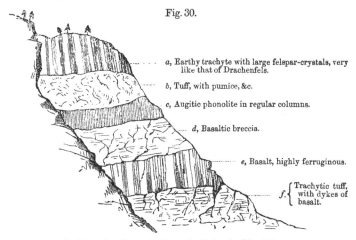

Fig. 30.

a, Earthy trachyte with large felspar-crystals, very like that of Drachenfels.

b, Tuff, with pumice, &c.

c, Augitic phonolite in regular columns.

d, Basaltic breccia.

e, Basalt, highly ferruginous.

f. { Trachytic tuff, with dykes of basalt.

Section of rocks at the Cascade, Bains du Mont Dore.

In the Euganean Hills, in the Monti Cimini, at the Lago
di Bracciano, in the Campi Phlegræi, in the Ponza Isles and
Ischia, amongst the Hungarian groups, the Siebengebirge, in
Teneriffe, in Iceland, and at Guahilagua and at Xalapa in
Mexico, trachyte, greystone, and basalt occur together, and
appear to have been successively erupted from the same or
very neighbouring vents. These, and many other examples
which might be quoted, prove that MM. Humboldt and Beudant, and some other geologists who follow their views, are
in error when they speak of a repulsion or antagonism as
existing between the trachytic and basaltic formations*. It
is natural, and to be expected, that where basalt has been the
latest and most copious production of any volcanic vent, its
currents should conceal entirely or partly from our view those
of trachyte which preceded it, and *vice versâ*. Again, the
generally great difference in the fluidity of these lavas (to which
we shall shortly advert) will have led them to accumulate in

* Humboldt, Essai Géognostique, p. 349 ; Beudant, Hongrie, iii. p. 587.

greatest abundance, the one at a distance from, the other in the immediate vicinity of, the central vent. Thus, in the Mont Dore, the trachytic currents have in no instance flowed more than from four to five miles from the central heights of the volcano; the basaltic currents, on the contrary, have reached a distance of fifteen miles or more. It seems also probable, as Mr. Darwin suggests, that the high specific gravity of the augitic lavas may have caused them to break out more frequently at the foot of a volcanic mountain than from its summit, and thus have given rise to the frequent horizontal sheets of basalt which are found around the base of trachytic volcanos. But, instead of their mutually repelling one another—the production of trachyte preventing that of basalt, and *vice versâ*, as some writers have asserted—the *general* law seems to be, that they occur together, being produced successively from the same or proximate vents, though without any uniform order of succession, and in general at long intervals of time. The complete isolation of basaltic or trachytic formations stands in the light of an exception to the general rule, and may be often accounted for by the concealment of much of the earliest products (perhaps subaqueous) of the vent.

A still greater error was the attempt to limit the production of volcanic rocks of particular mineral composition to separate periods of the history of the globe. According to M. Beudant, all trachytes are of parallel age to the secondary strata, and universally prior to the tertiary. But the trachytes of the Cantal and Mezen rest on tertiary freshwater limestone; those of the Montamiata and Monti Cimini on the tertiary marls and clays of the Sub-Apennine Hills; that of the Euganean Hills cuts through and covers beds of *calcaire grossier*. Moreover the eruptions of many recent volcanos have produced, and to this day habitually produce, trachytic lavas. Most of the active volcanos of America, particularly Popocatepetl, Orizaba, Capac-urcu, Cotopaxi, Sotara, and Rucupichinca, are trachytic, and eject pumice. Those of

Sumatra, Java, and the Moluccas also appear to produce principally felspathic lavas. The seas in the vicinity of these islands are said to be sometimes covered to a considerable extent, after an eruption, with floating pumice.

The Peak of Teneriffe, which has certainly been eruptive at no very distant period, is trachytic, whereas the older dejections of the volcano are basaltic. In Iceland, the older lavas, composing the great bulk of the island, are all basaltic, and chiefly, perhaps, of subaqueous origin. The lavas produced by the recent and still active volcanos are mostly felspathic. Volcano, one of the Lipari Isles, when in eruption in 1786, threw up pumice; and the recent lavas of Lipari itself are of felspathic obsidian or pumice. The Lava del Arso, in Ischia, and that of Olibano, near Pozzuoli (both streams of known dates), are, mineralogically, trachytes. The lavas of the volcanos of Bourbon, still in almost permanent eruption, are highly felspathic. St. Helena is a trachytic volcano rising from a ring of basalt, the basal remnant of an earlier volcano. The trachyte itself, however, is pierced by dykes of augitic lava; so that there has been here (as elsewhere) an alternate production of the two kinds of lava. Passages, indeed, from trachyte into basalt have been observed in the same dyke of volcanic rock.*

In truth, so far from the felspathic lavas being universally of earlier date than those in which the augitic and ferruginous minerals predominate, it is well known that many of the oldest traps are of the latter kind (basalt or greenstone). The opinion that mineral composition is any test of age, as respects the pyrogenous rocks, is fast passing away from among geologists, most of whom now consider that porphyry, serpentine, and even granite may be of recent origin, and in course of formation and upthrust from beneath the superficial strata, at the present day.

§ 6. But, further, it seems highly probable that, under the

---

* See Daubeny, Volcanos, p. 93.

influence of the variations of temperature and pressure to which lavas must be occasionally exposed, within or beneath a volcano, before their eruption, the crystals composing a coarse-grained granitoidal rock may, without experiencing any change in their essential mineral characters, or even complete fusion, be so far disintegrated, or broken up and comminuted, as to give it the *texture* of a fine-grained paste when it issues from the eruptive orifice and is afterwards consolidated in the open air *. At all events, it seems evident that, in proportion as the grain is finer, the greater will be the freedom of motion among its particles, and consequently the liquidity of the matter, and the facility afforded for the aggregation of the parcels of elastic vapour that may be generated in the mass into bubbles, as well as for their upward rise, in virtue of their inferior specific gravity to that of the enveloping mass. On the other hand, where the crystalline particles remain extremely coarse, it is quite conceivable that a general intumescence may take place by the expansion of the interstitial vapour between adjoining crystals or crystalline plates, without occasioning so much mobility as is necessary to admit of its aggregation into visible bubbles. In the one case there will be a porous, intumescent, but still highly crystalline, mass, of very imperfect liquidity, and therefore exerting a greater disturbing and elevating power on the overlying rocks, and a greater wedge-like force against the sides of any fissure into which it may thrust itself. In the other case, a far more liquid matter, through which the expanding vapour-bubbles may force their way upwards, increasing in volume as they rise, and which will more readily inject itself

* I am supported in this view (which I put forward in 1825) by MM. Scheerer and Delesse, both of whom assert that water exists in mechanical combination with all crystalline rocks, " its minute molecules intercalated between the crystals, and when exposed to intense temperature in a state of extreme tension, endeavouring to expand, but prevented from doing so by the pressure to which the enclosing mass is subject " (Bulletin Géol. iv. p. 468) ; and by Bischoff, who says that a crystalline lava " may be disintegrated by the action of strongly heated vapour, without being melted."

into, or through, fissures of moderate width.   So, too, when they have ultimately struggled up to the outer air, lavas of these opposite characters will assume a different disposition; the one flowing away rapidly, almost like water, or any other true liquid, in obedience to the laws which determine the motion of fluids; the other accumulating in a bulky, porous, or semi-solid mass above or around the orifice of protrusion.   Between these two extremes every degree of liquidity may be presumed to exist, and consequently every variety in the external form assumed by the protruded lava may be expected to occur; and this is, in fact, exactly what is observed among volcanic rocks.

Some lavas (and the remark applies especially to the fine-grained varieties) have evidently flowed with such rapidity down the sides of the cone from whose summit they issued, or any other steep slopes to which they obtained access, as to have left but a thin crust of solid rock, or narrow and often tubular channels, through which the extremely fluid matter has escaped downwards, and, when it has reached a flat surface, spread in a thin sheet on all sides, to great horizontal distances; while other lavas, especially the coarse-grained, porous, earthy and crystalline trachytes—and, at times, some basaltic lavas—from their very imperfect liquidity, or extreme viscosity, aided perhaps by a slow rate of issue, are found to have accumulated in massive beds or buttresses at the side of the vent whence they were protruded, or in lumpy dome-shaped excrescences or hummocks, close to, or perhaps even directly above it.

§ 7. It is also probable _à priori_ that the comparative specific gravity of any lava will exercise a considerable influence in determining its fluidity, and consequently the figure it will assume when poured out upon the surface of the earth.   Observation confirms this assumption.   It may be generally asserted that the greater the proportion of the ferruginous minerals (augite, hornblende, mica, and titaniferous iron) in the constitution of a lava, the greater (*cæteris paribus*) in proportion to its bulk are found to be the horizontal dimensions, and the more uniform the thickness of the bed it forms upon a level or nearly level surface.   I may instance the more ancient wide-spreading sheets or plateaux of basalt, such as that of the Deccan, already referred to, or those of the Cantal and of the Haute Loire, which are now seen at a considerable elevation above adjoining valleys of denudation.   But the horizontal dimensions of some ferruginous lava-currents produced by recent volcanic eruptions are also quite astonishing.   The stream which destroyed Catania in 1669 is fourteen miles long, and in some parts six wide.   Recupero measured the length of another upon the northern side of Etna, and found it to be forty miles.   Spallanzani mentions Etnean currents of fifteen, twenty, and thirty miles in length; and two streams which issued from a volcano in Iceland (Skaptar Jokul), in 1783, spread over a surface respectively of forty and fifty miles in length by from ten to fifteen in width.   In the island of Hawaii, a recent eruption in August 1855 gave birth to a stream of lava sixty-five miles long, and from one to ten miles wide *.   These lavas are all either basalts or ferruginous grey-stones, and of a high specific gravity.

§ 8. On the other hand, that a low specific gravity, especially when combined with a coarse crystalline or granular and open or loose texture, will occasion a *minimum* of fluidity, is shown in the disposition general to the large-grained or very porous trachytes, such as compose the massive beds or hummocks and domes that are closely grouped about the volcanic centres of the Monts Dore, Cantal, and Mezen in France, of Hungary, and of the Andes.   Perhaps no better examples could be mentioned of the bulky form resulting from this extremely imperfect fluidity than the three or four domitic Puys of the Monts Dôme, viz. the Puy de Dôme itself, and the neigh-bouring bell-shaped hills of Sarcouy and Cliersou.   These hills

* Coan, Quart. Journ. Geol. Soc. 1856, p. 170.

actually rise each one out of the crater or hollow of a cinder-cone formed of blocks of different lavas, or pumice and ash. The projection of these fragmentary matters into the air evidently accompanied or immediately followed the emission of the trachytic lava, which seems to have risen upwards from the vent in so pasty or imperfectly liquid a state as to have accumulated above it in the form of a dome or bell, just as would a body of melted wax, or one of moistened clay, if forced outwards through an orifice in the cover of any containing vessel. The substance of all these hills is a very earthy, porous, and pumiceous trachyte. An outline-sketch of the Puy de Sarcouy, embraced by its two adjoining scoriæ-cones, is given in fig. 31.

Fig. 31.

Outline of the Grand Puy de Sarcouy (trachyte), between the Puys of La Goutte and Little Sarcouy (cinder-cones), in Auvergne.

It is quite clear in this instance that the central boss was produced by the same eruption which threw up the cinder-cones.

The probable internal structure of such a dome or boss is also shown in fig. 33. One layer of the pasty mass may be supposed to have overlapped another as it welled up from the vent, so as to form either a single lump, or a series of rudely concentric beds dipping outwardly on all sides. The small hummocks or 'mamelons' of glassy felspathic lava upon the summit of the volcano of Bourbon (already noticed, p. 74, *suprà*), which Bory de St. Vincent watched in the act of formation by the welling-up of highly viscous matter at a white heat, and its consolidation as it guttered down the outside of the hill it had itself raised in irregular concentric coatings, may, I think, be looked upon as types of the mode of production of the larger trachytic domes and hummocks likewise (see figs. 32 & 33).

Such, I believe, was the mode of emission of the bosses of trachytic lava which occur in the bottom of the crater of Astroni near Naples, of that of Santa Croce rising in the

Fig. 32.

The "Mamelon Central;" a boss of vitreous lava on the summit of the Volcano of Bourbon. (After Bory de St. Vincent.)

Fig. 33.

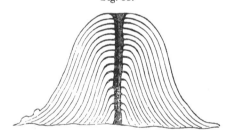

Ideal section of the Mamelon Central.

centre of the crater of Rocca Monfina, that of the Camaldoli, of Montamiata, of Palma, and other examples which have been sometimes adduced (by Dr. Daubeny and others) in support of the theory of 'Elevation Craters'—their protrusion being supposed to have tilted up the sloping strata of the surrounding crater-cones.

It is obvious that if lava of this imperfect liquidity be emitted on many contiguous points of a lengthened fissure, it will, by accumulating on either side or above this, give rise to a chain of bosses or domes, or a lengthened ridgy hummock, with somewhat of an anticlinal structure, which it may sometimes be difficult to distinguish from a bulky current that flowed from a single source.   As examples of this character I may mention the range of round-topped trachytic hills which stretch to the north from the central heights of the Mont Dore, and go by the names of the Puys de l'Angle, Hautechaux, Barbier, Baladou, l'Aiguiller, and Pessade*.

In all probability, many of the massive trachytic formations of the Andes (to some of which M. de Humboldt ascribes the vertical thickness of 6000 metres) were produced in this manner from wide and lengthened fissures disgorging, during a long period of eruption from numerous vents upon their whole extent, an immense quantity of felspathic lava, under circumstances peculiarly unfavourable to its fluidity.

M. de Humboldt expressed the opinion (to which some other geologists appear to assent) that these trachytic domes are hollow blisters, blown up like a bladder, and have or have not, at the summit, a crater, according as the bladder *burst* or not.   There is, however, no reason to believe that such mountains are less solid than any others; and their craters, so far from having been formed by the bursting of a single bubble, were no doubt, like all craters, produced by continuous explosions, often lasting for months, which have indeed been frequently witnessed from their summits†.

Such, too, was probably the mode of production of the great range of clinkstone bosses which has its origin in the Mont Mezen (near Le Puy en Velay) and stretches into the gorge of the Loire, thirty miles off, with an average width of six—covering therefore a surface of about 156 square miles. It may be seen to rest upon basalt in some places, and on

* See my 'Volcanos of Central France,' 2nd ed. 1858.

† See my paper on Cones and Craters, Quart. Journ. Geol. Soc., Nov. 1859, p. 26.

freshwater tertiary marls and clays in others ; more frequently still on granite. This, and its gradually inclined slope, from the heights of the Mezen into the ancient bed of the Loire, where it terminates (the extremity leaning against the foot of the granitic range of La Chaise-dieu, on the opposite bank), make it doubtful, perhaps, whether it was not produced rather as a continuous lava-current than from a lengthened fissure. The atmospheric agents of erosion have, however, effected great changes in this hilly range since its formation, cutting entirely through it in some of the softer parts, and notching it into a number of fantastically-shaped eminences, generally conoidal, from 500 to 600 feet in height.   Local differences of hardness and structure, by affording more or less facilities to the process of degradation, particularly to the action of frost and rain, and the easy destruction of those parts of the rock that were based upon friable strata of marls and clays, to which the fissile structure of the clinkstone suffered the rain-water to percolate from above, were no doubt the causes of this extremely partial weathering*.

* Many geologists have remarked the peculiar tendency of phonolitic rocks to waste into detached masses of a conical form. Nowhere could this remark be better appreciated than along this clinkstone range of the Mezen, which is reduced to a series of rocky eminences presenting every gradation of figure from the rude segment of a bulky bed to the perfect cone.   (See the panoramic view of Le Puy in my 'Volcanos of Central France.')

The cause of this uniformity clearly lies in the much greater facility with which this rock yields to meteoric influence on some points than on others—as well from its frequent differences of texture and consequent aptitude to decomposition, as from its accidental varieties of structure; the columnar and laminar modifications at times combining to hasten a disunion of parts (as was remarked of the rock Tuilière, Mont Dore), at others to afford the utmost power of resistance, as when a sheaf of columns leaning against one another converge into a conical cluster. The same causes continue to influence the aspect of the mass after it has been completely isolated and reduced to a rounded form by the wasting of its angular portions. Where the clinkstone is of a quality that readily decays on exposure, it presents a smooth-sided cone, clothed with a thick layer of white earthy soil, which frequently supports luxuriant forests of oak and

It is, however, I think, not improbable that the clinkstone lavas may have been generally erupted in a condition of extreme consistency, approaching to solidity, and thus enabled to rear a lofty mass in a pyramidal or wall-like figure to considerable heights above the lip of the vent, without lateral supports. Mr. Darwin has expressed the opinion (grounded on this general uprightness of form) that clinkstone rocks are always denuded dykes. The position of the great phonolitic chain of the Mezen, however, is irreconcileable with such a mode of formation, since it ranges through its whole length at heights of 800 feet and more above the granite platform on either side, which we cannot suppose to have been worn away to that extent since the eruption of the clinkstone.

It is likely that masses erupted in such a semi-solid condition should exert a more powerful disturbing influence on the rocks through which they are thrust up than the more liquid lavas. And in accordance with this presumption trachyte is occasionally found to have more or less tilted up the adjoining strata. This effect is, however, exceptional, and confined to the immediate· neighbourhood of the erupted rock. I may instance the remarkable hill called the Puy Chopine among the Monts Dôme. Messrs. Hamilton and Strickland make this remark of the trachytes of Asia Minor, at the same time adding that the general horizontality of the strata remains undisturbed *.

As examples of trachytic or highly felspathic lavas which have flowed as currents over inclined surfaces, but still show, in the great thickness of the beds they now form as compared with their superficial area, how imperfect was their liquidity, I may instance those of Mont Dore (especially the plateaux de l'Angle, de Rigolet, and de Chambon, just above the Baths, generally above 100 feet in thickness†) ; those which slope from

fir. Where the rock is less destructible, its upper outline is cap-shaped, notched, and craggy, and its base encumbered with barren and ruinous piles of slaty fragments.

* Trans. Geol. Soc. 2nd ser.   † See fig. 30, *suprà*.

their high central source to the town of Aurillac in the Cantal, resting there on freshwater marls ; that of Olibano, proceeding from the crater of the Solfatara near Naples; those which compose the substratum of Procida, Lipari, Le Saline, Volcano, and Ustica, in the Lipari Isles ; that of the Montagna della Guardia in Ponza, of Ventotiene, &c. The group of the Monti Cimini, near Viterbo, affords many instances of the same disposition : thick trachytic currents, based on tufa, may be observed to form the slopes of this mountain. There can be little doubt that the volcanic mountains of Iceland and Hungary present many similar facts ; for in both cases massive hills or beds of trachyte are represented as enveloped by, or alternating with, tuffs and conglomerates.

§ 9. *Vesicular structure.*—It has been already stated that some lavas are much more porous or cellular than others ; and this not merely in the upper parts of the current, where the greater number of bubbles of vapour may be expected to gather *, but throughout their mass. Lavas in which the felspathic minerals predominate are, as might be anticipated from their inferior specific gravity, much more frequently vesicular throughout than the heavier or ferruginous lavas. Such are the lava of Volvic, near Clermont, used generally in that neighbourhood for building-stone; that of Piperno, near Naples, there employed for the same purposes; the millstone lava of Nieder-mennig, near the Lake of Laach ; above all, the glassy lavas, or, rather, pumice-currents, of Volcano and Lipari.

In all these and similar cases, the vesicular cavities appear

---

* One of the most remarkable examples of cellular lava that I am acquainted with is the upper portion of the bulky basaltic rock that surmounts the high plateau of Radicofani, on the road from Florence to Rome. The mass of the rock is generally heavy and crystalline, dense, compact, and devoid of pores, while its upper scoriform parts are completely honeycombed with globular vesicles, the intervening partitions being so slight that an axe will make as clean a cut through the mass as through a true honeycomb. Captain Smyth describes some of the basaltic lavas of Victoria (Australia) as " so cellular that it is easy to lift masses several feet square " (Quart. Journ. Geol. Soc. 1857, p. 228).

more or less flattened out and elongated, and present their longest axes uniformly in the direction of the flow of the current ; and this observation becomes serviceable as a guide to its source or point of eruption.

Those parts of the stream in which some accidental condition of movement or pressure has caused the bubbles of vapour to be developed in special abundance, are also drawn out and flattened, in the same direction of the movement, into lenticular masses, or ultimately into zones, alternating with the intermediate more compact parts. This flattening was probably favoured by the hydrostatic pressure and slow subsidence of the mass as it gradually cooled and consolidated. The main cause, however, of the elongation is clearly the differential motion of the vesicular and the non-vesicular portions : the bubbles occasioning a certain amount of resistance to the motion of the enveloping liquid, those parts in which they existed moved less easily than those in which they were absent. The same law evidently has often determined the arrangement of solid crystals, or concretionary nodules of segregated mineral matter, or other heterogeneous substances which the flowing lava may have contained, their longer axes being found in the direction of the internal movements of the mass.

Von Buch describes a stream of lava in Teneriffe containing innumerable thin *plate-like* crystals of felspar, arranged in trains, like white threads, one behind the other, in the direction of the flow of the current. Mr. Darwin observed numerous similar facts in the lavas of Ascension, as I had previously done in those of Ponza, Ischia, &c.

I believe, indeed, I was the first observer who pointed out this peculiar laminated or ribboned structure of many lavas, and suggested an explanation of its origin in a paper read before the Geological Society in 1823 *. To this I was led by observing, in the Ponza Isles, repeated passages from pure glassy obsidian to pearlstone, by the formation of felspathic sphero-

* See Trans. Geol. Soc. Lond. 1824.

lites in the vitreous matter, resembling those which form in the
slag of a glass-furnace, or in the melted basalt in Messrs.
Chance's artificial-stone manufactory near Birmingham. These
felspathic concretions had been evidently, by the unequal
movement of the enveloping veins of more liquid matter under
considerable pressure, broken up and drawn out into planes
or laminæ, sometimes folded and contorted in a striking
manner. Examples of this character abound not only in the
Ponza Isles, but also in Ischia, in the Isle of Ascension, and
among the perlites of Hungary and the Andes. Commodore
Forbes, indeed, speaks of this arrangement of the component
crystals in parallel planes as characterizing a very large proportion of the trachytic lavas of South America.

I have already mentioned, and shall again have occasion to
refer hereafter to this laminar or foliated structure, which I
believe to throw a light on the probable origin of the similar
structure in the so-called metamorphic hypogene rocks, gneiss,
mica-schist, and serpentine.

§ 10. *Brecciated lavas.*—Occasionally lavas assume a brecciated structure, the base of lava enveloping fragments either
of the rocks through which it forced its exit, or of portions of
its own already consolidated substance, broken up and carried
away by some renewal of movement caused by a fresh outburst, or some change in the direction of its course. The
enveloped fragments are sometimes partially fused, and appear
to graduate imperceptibly into the base; sometimes, but not
always, they are metamorphosed by the heat and other chemical
agencies of the enveloping lava. Quartz so enveloped is melted
partially; felspar crushed, and rendered more or less glassy;
mica is bronzed; limestone dolomitized; sandstone partially
fused.

§ 11. *Metamorphic influence of lava.*—Many new minerals
are also found to have been formed in the cracks or cavities of
such enveloped masses. The most remarkable instance of this
is seen in the extraordinary number of rare and highly cry-

stalline minerals that occur in the blocks of altered Apennine limestone and other rocks found among the fragmentary ejections of old Vesuvius (Somma). The volcanic tuffs of Latium are almost equally fertile in this class of exceptional products. So also are those of Auvergne *.

Analogous to these changes are those observable in some rocks when traversed by dykes of lava, as in the well-known example of the dolomitization of chalk by contact with basalt at the Giant's Causeway, and the carbonization of coal under similar circumstances. In many such cases new crystalline combinations have developed themselves. The earlier intruded lava-rocks (traps) appear to have exercised this metamorphic influence upon the strata penetrated by them more frequently than the very recent. Yet perhaps it is only that denudation has more frequently disclosed the deep-seated planes of junction where alone the effect was produced †. I do not dwell upon this point, nor recapitulate the varieties of mineralization or metamorphism so occasioned, because in the works of Sir C. Lyell, Dr. Daubeny, and MM. Delesse and Daubrée, this interesting subject is amply and more authoritatively treated.

These phenomena of metamorphism by contact with volcanic matter are however, it must be remembered, exceptional. The sides of the greater number of dykes present no appearances of the kind. Probably such changes are effected only

* Among these may be mentioned haüyne, ice-spar, sodalite, spinel-lane, melilite, sommite, nepheline, idocrase, zircon, wollastonite, brieslak-ite, garnet, humite, sphene, &c. (See a complete list in Daubeny's ' Volcanos,' p. 236, ed. 1848.)

† Mr. Austin (on the Geology of the South-east of Devon, Trans. Geol. Soc. vol. vi. 2nd ser. p. 470 *et seq.*) describes some good examples of the influence of intruded traps on secondary limestones and slaty shales. Magnesia more or less characterizes the altered limestones (proceeding, no doubt, from the hornblende of the trap). The shales are often fused into jasper and quartz-retinite. The solid limestone is crystallized to some distance from the dyke. The sandstone hardened, the lines of stratification obliterated, and innumerable vertical fissures formed in it, coated with manganese. Such examples are frequent near Dartmoor.

when the injected matter is at a very high temperature, or of great bulk, and consequently very slow in cooling; or when it has been long traversed by heated water or steam containing mineral matter in solution. Similar alterations, it is well known, are often observable where plutonic rocks come into contact with, or into the vicinity of, limestones, sandstones, shales, or other sedimentary strata, especially in the instance of metallic veins, which are of frequent occurrence in such situations.

There is a considerable analogy, as might be expected, and not unfrequently an identity, between the crystalline minerals produced by this metamorphic action and those to be spoken of presently, which occur as zeolites, unquestionably formed in the vesicular cavities of lava-rocks by infiltration of water or vapour containing mineral matter in solution. The vapour, indeed, which escapes by percolation from a cooling lava-mass, through the pores and crevices of the outer and already consolidated part, is always found to be accompanied by a variety of mineral substances, in a state either of solution or sublimation, or, finally, as permanent gases. These substances, it has been already observed, are scarcely discoverable in the vapours that rise in immense volumes from lava immediately upon its protrusion, and which appear to be either purely aqueous, or to contain but a very small proportion of mineral matter. But as the quantity of steam evolved diminishes, and is at length reduced to thin columns, or *fumaroles,* exhaled from the narrow and intricate crevices of the superficial lava, the proportion of other gases or substances that accompany it is found to increase. Upon coming into contact with the external air, a certain proportion of the steam is condensed by refrigeration, and deposits the matters it holds in solution on the sides and edges of the fissure.

The principal of these are sulphuric, muriatic, and carbonic acids, either in a state of purity, or combined with various alkaline, earthy, or metallic bases. Among these combinations, sulphates of lime, magnesia, ammonia, soda, and potass, muriates of soda and ammonia, and carbonate of soda, predominate. It appears probable that the first of these acids is occasionally, if not always, formed on the spot, by the union of a part of the sulphur, derived from sulphuretted hydrogen in the aqueous vapour, with the oxygen of the atmosphere. The remaining sulphur, when it is produced abundantly, is deposited in a state of purity in angular or octahedral crystals, or, when most copious, in stalactitic concretions or crusts, at the edges of the orifices, and upon those neighbouring points where the vapour in which it is suspended undergoes condensation. Some spent volcanos, which have probably long existed in the condition of (so-called) solfataras, abound in massive deposits of pure sulphur, e. g. some of the craters of Lipari, of Iceland, of Java, of Quito, &c. The compounds of boracic acid are rare. A large amount, however, is continually formed at the bottom of the crater of Volcano, one of the Lipari Isles. Much also seems to be evolved by some of the solfataras of Iceland*.

The metallic sublimations are condensed in the same manner on the sides of the fissures of escape. The most frequent of these is specular iron. The muriates of copper and iron, and sulphurets of iron, copper, arsenic, and selenium, are occasionally found in similar situations.

It is not easy to distinguish those of these different substances which were originally contained in the lava, and are merely volatilized by heat, from those which are the product of chemical combinations operated during the consolidation of the lava between its various elements and those of the water or vapour it contains, which may be partially decomposed while percolating the mass at a considerable, though decreasing, temperature. Soda, potass, lime, and iron are constant ingredients in almost every variety of lava. Chemical

* See Mr. Warington's paper in the Quart. Journ. Geol. Soc. 1860; and Forbes's 'Iceland,' 1860.

analysis has occasionally discovered in them ammonia, bitumen, and muriatic acid. Sulphur, in combination with iron
or copper, is also frequently present. These substances, when
brought into contact by the condensation of the lava through
pressure or cooling, but while still at an intense temperature, and permeated by the aqueous vapour which ascends
from the depths of the focus, may give rise to various combinations.

The intense heat of the lava in the interior of the mass is,
perhaps, sufficient of itself to effect the decomposition of some
of its ingredients. The specular iron, so frequently met with
in lava-rocks, is evidently sublimed by this action, and it is
remarkable that it is always found in the upper parts of the
bed or current; while the lower parts of the rock in these
cases do not influence the magnet, having obviously lost all,
or the greater part of their iron by sublimation. Or, as in
many of the currents of Etna, the upper parts, which are at
the same time very porous and cellular, contain much specular
or oligistic iron; while, on the contrary, the lower and com-
pacter division abounds in magnetic iron, in grains or octahedral crystals. Here, the magnetic iron originally contained
in the central and upper parts of the current has been evidently volatilized, and deposited again in the form of specular
iron, while that of the lower part, from which little or no
vapour was enabled to escape by percolation, remains un-
changed.

In the same manner the other mineral ingredients of a lava
are occasionally volatilized, and again deposited, not unfrequently under new combinations and forms, in the cavities of
the lava, as its temperature is gradually lowered. This was
in all probability the origin of those delicate and often capillary crystals of hornblende, augite, melilite, and other minerals,
which occur in the cellular cavities of many lava-rocks (Capo
di Bove, &c.). Crystals of augite are also said by Brieslak to
have been formed by sublimation in the interior of some of the

houses of Torre del Greco, destroyed by a current of lava in 1794; and, indeed, M. Daubrée has lately succeeded in forming them artificially under somewhat similar conditions*.

§ 12. *Amygdaloidal lava.*—The substances taken into solution by the aqueous vapour are often unquestionably deposited in a similar manner, either in crystals, or stalactitic and mammillated concretions, which line the interior of the vesicles, or the interstices of the mass. The minerals found in this situation are principally either siliceous or calcareous, consisting of carbonate of lime, calcedony, fiorite, and some of the numerous varieties of the zeolite family. They all bear the character of having been deposited from aqueous solution. We know, indeed, that water at a high temperature, and particularly with the assistance of an alkali, readily takes both silex and carbonate of lime into solution. It is obvious, therefore, how intensely-heated aqueous vapour, in its passage through the lava, whether by percolation or in the form of bubbles, may become impregnated with either or both of these ingredients; and how, when prevented from any further escape, as it is gradually cooled and condensed, these substances may crystallize on the sides of the containing cavity.

Sometimes the water remains in the centre of this geode, accompanied by a portion of some permanent gaseous fluid, as in the calcedonies of the Monti Berici, and in the cavities of numerous quartz-crystals in granitic rocks. In general the water has escaped, probably by filtration, towards the lower parts of the mass, after these had been entirely consolidated. During this filtration it perhaps continued to deposit the substances it still retained in solution, in the cavities, pores, and minute fissures through which it slowly percolated; and hence it is that many of the cellular cavities of these rocks have been *entirely filled* with successive coatings of the calcareous or siliceous deposit; which could not have been all held in solution by the minute quantity of steam, to whose expansive force the

* See his 'Études et Expériences,' Paris, 1859.

L

original formation of the vesicle was owing. The microscopical examinations of Mr. Sorby have, however, discovered the existence of water in minute cavities even of such zeolites.

In many cases the subsequent filtration of water, carrying in solution mineral particles from *overlying* rocks, may have produced the same effect. This is particularly true of the amygdaloidal basalts which are found underlying calcareous strata, or in such positions as make it probable that they were once covered by them, and which therefore were consolidated beneath water holding calcareous particles in suspension.

§ 13. *Solfataras.*—The sulphuric and muriatic acids which are evolved in a state of freedom, when deposited by condensation on the lava which forms the edges of the crevices of escape, act upon its substance and give rise to fresh changes. The sulphuric acid uniting with the alumine of these rocks produces a sulphate of alumine, which is often carried away by the rains and deposited in vast accumulations in lower situations. The mining operations of the different alum-works of Hungary, Italy, &c. are carried on in deposits of this nature*.

The sulphuric acid uniting with the lime produces a gypseous efflorescence, which is often found coating these decomposed lava-rocks in considerable quantities. The iron of the lava is either attacked by the acid, and aggregated into crystalline or concretionary pyrites disseminated through the disintegrated rock—or, assuming various degrees of oxidation, stains it with stripes of different shades of brown, yellow, red, green, and blue. The silex alone remains untouched; and when all, or a large proportion of the other ingredients have been washed away, the lava appears occasionally changed into a light, harsh, carious, and highly siliceous rock, or into a white powdery or earthy matter like chalk-dust, but consisting of

* There is a crater-lake in Java, called Taschem, of which the water is so strongly impregnated with sulphuric acid that nothing can live in it, nor in the river which discharges its overflow into the sea. The latter has a parallel in the 'Vinegar River,' described by Humboldt as proceeding from the volcano of Puracè.

nearly pure silex, and bearing no resemblance whatever to its original character.

Where exhalations of vapour strongly impregnated with this acid have continued for a great length of time, the changes thus effected in the neighbouring rocks are, from the continued shifting of the fumaroles, extensive and remarkable, and have acquired for such spots the common appellation of Solfataras or Soufrières. These are usually found in the interior of a volcanic crater, as might be expected ; since the evolution of vapours from a current of lava that has flowed away from the volcanic vent upon the surface of the earth must be extremely limited in quantity and duration; whereas after most eruptions a vast body of heated lava is likely to remain beneath the bottom of the crater, in which the process of solidification goes on slowly for an indefinite, and often very long period, affording a continual source of aqueous vapour charged more or less with various mineral substances.

§ 14. *Hot-springs.*—When the superficial parts of such a mass of subterranean lava have cooled down to the temperature of the atmosphere, and its vapours are enabled to escape through fissures in these, or in solid overlying rocks of any heterogeneous nature, they must be more or less condensed by refrigeration before they issue into the air, and make their appearance in the shape of springs of water at elevated temperatures, occasionally even many degrees above the boiling-point.

Such hot-springs are common in all volcanic districts, and, where any more energetic external development of volcanic action has been long wanting, are the only indications of the still elevated temperature of the focus below. There can be no doubt that the quantity of caloric which is enabled to pass off through permanent fissures, in this way, materially contributes to maintain the outward tranquillity of the subterranean focus. It is even conceivable that this regular and placid transmission of caloric may be, for very lengthened periods, exactly proportioned to the supply constantly communicated from below to the volcanic lava-reservoir, and by thus preserving it at a uniform temperature, may entirely prevent that accumulation of heat by which fresh eruptions would be produced. The quiescence, or what is usually called the extinction, of these volcanic foci will in these cases be owing to the fissures through which the excess of caloric is enabled to escape in combination with water. Thus, the permanent hot-springs of Bath, Buxton, Carlsbad, Aix, and other localities may possibly act as safety-valves, letting off the excess of heat from some subterranean focus, which might otherwise, sooner or later, find vent in earthquakes or volcanic eruptions.

The water of thermal springs holds in solution various mineral substances which, in the form of vapour, it carried upwards from the heated lava below, or which it may have abstracted from the rocks with which it has come in contact. Some of these are deposited immediately on its coming into the air, while others remain permanently combined with it. The siliceous sinter deposited by the hot-springs of Iceland is well known, as also the calcareous incrusting springs of numerous volcanic districts (Auvergne, La Tolfa, that of the Bridge of the Incas in the Andes, &c.).

Even when the latter sources no longer retain any elevation of temperature, and are derived from rocks which do not present any indications of recent volcanic action, they may still be supposed in many cases to have the same origin as the hot-springs, but to have parted with more of their heat by a longer passage through cold rocks, and perhaps by their mixture with other superficial springs of atmospheric origin.

Where the traces of volcanic action are recent, and the intensely heated lava very close to the surface whence the springs issue, the curious phenomena of intermittent fountains are sometimes produced, such as afford so magnificent a spectacle in the Geysers of Iceland.

The greater part of the hot water of these springs is probably derived from the filtration of atmospheric water through the crevices or interstices of a mass of lava not yet cooled *.   Indeed,

* The theory of these springs has been the subject of much difference of opinion.   The following considerations will perhaps sufficiently account for their phenomena.   The lavas of Iceland are replete with hollow blisters or caverns.   Let us suppose such a cavity in a vast bed of heated lava (fig. 34), which may be either isolated, and cooling slowly, or con-

Fig. 34.

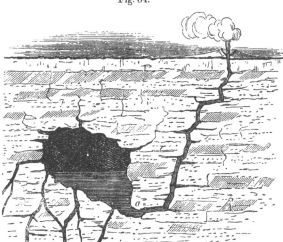

nected with the volcanic focus below, and receiving a constant supply of caloric from thence.   The steam emanating from numerous fissures collects there, and is partly condensed into water by the pressure of the column in the fissure of discharge.   The increasing temperature of the floor and sides of the cavity, and the accumulation of steam evolved by the feeders, augment the expansive force of the steam collected, till at length it overcomes the resistance of this column of water, and discharges it from the orifice above.   As the water issues, the pressure on the steam in the cavity is diminished, more of the water there is vaporized, and the ratio of its expansive force to that of repression in the column increased.   The jets therefore that are thrown up must augment in violence.   As the water decreases still more in quantity, some of the steam escapes together with it, and at length, when all the water has been driven out, the whole remaining body of steam issues in a few powerful bursts.

But by the vaporization of all the water it contained, the temperature of the cavity is reduced to the boiling-point under the pressure of the atmosphere alone, and its sides proportionately refrigerated.   The vapour since all lava-currents (as we have shown) rest upon a bed of loose scoriæ, it will follow, almost as a matter of course, that when a current has occupied the bed of a river, the waters which previously flowed there will still to some extent find their way through the loose matters beneath, and their temperature must be affected by that of the lava above, so long as it remains elevated above the mean temperature of the earth around. An example on a large scale was afforded at the volcano of Jorullo in Mexico. The rivers Cuitemba and San Pedro, which lose themselves on one side, beneath the vast sea of lava that forms the Malpays, issue on the opposite as permanent springs of large bodies of water, which for many years retained an elevated temperature. When Humboldt visited Jorullo in 1804, forty-six years after the eruption by which the lava was produced, their temperature was 52° C.; but travellers who have recently visited the spot report that they have since that time been cooled by the refrigeration of the lava-bed through which they flow, and are now scarcely more than a few degrees above the mean temperature of the air.

At Bertrich-bad, in the territory of Luxemburg, a thermal spring rises immediately below a spot on which three vents of volcanic eruption were formed at no very distant period, and probably owes its qualities to having percolated through part of the focal lava raised at the time of these eruptions, and not yet entirely refrigerated. It is known that the temperature and mineral nature of this spring have progressively dimi-

therefore that rises from the feeders is condensed into water, and collects at the bottom. This water rises in the pipe as the steam accumulates, preserving an equilibrium with its expansive force, until this equilibrium is broken by the latter having acquired sufficient power to discharge some of the water from the orifice of the crevice, when the phenomena of aqueous eruption recommence.

A recent traveller in Iceland, Mr. Forbes, disputes this theory, which was originated by Sir George Mackenzie, and substitutes one which does not appear to me as satisfactory—indeed, which I can with difficulty comprehend (see Forbes's 'Iceland,' 1860).

nished. It is at present below blood-heat, and has as nearly as possible the taste of pure fountain water. In earlier times its thermal and mineral qualities were much more considerable, and acquired for it a great reputation as a bathing-place.

§ 15. The permanent gases that emanate from the crevices of lava during its consolidation are less palpable, and not so easily recognized, mixing of course immediately with the atmospheric air, and leaving no trace of their existence. Those which have been observed are principally carbonic acid, nitrogen, and sulphuretted hydrogen gases.

The first of these was detected by MM. Monticelli and Covelli in the exhalations from the fumarole of the lava of Vesuvius, while still visibly incandescent. The same gas, mingled perhaps with nitrogen, frequently escapes in very considerable quantity from numerous crevices in the sides of a volcanic mountain immediately after the termination of an eruption, that is, when the crater or central aperture of escape is completely closed.

These mephitic emanations are extremely destructive to vegetation, and do great injury to the plants growing near the points from which they proceed. During the terrible eruption which convulsed the island of Lancerote in 1730–1734, exhalations of this kind appear to have been equally deleterious to animal life ; all the cattle of the island are represented to have been suffocated by them. M. Hubert, as quoted by Bory de St. Vincent, mentions that, during an eruption of the volcano of Bourbon, he observed seven or eight birds, flying at gun-shot height above the current, drop suddenly, as if stifled, the moment they entered the cloud of vapour that emanated from it.

In Java there is a crater, called the Guevo Upas, or Poison Valley, half a mile in circumference, always so full of carbonic acid gas, that every living thing that passes its limits is suffocated, and the ground is strewed with the carcases of wild animals, birds, and even of men that have met their fate there. These facts lend colour to the tales related by the classical writers of Lake Avernus, which is certainly a volcanic crater of no very ancient date, from whence carbonic acid and azotic gases may once have been so copiously exhaled (rising through the atmosphere in consequence of their elevated temperature) as really to affect the birds that flew over it. It is much more probable that this fable should be founded on fact, than that the fortuitous invention of a poetic imagination should coincide so well with known phenomena. The banks of the Lago d'Aguano, in the immediate vicinity of the Avernus, still exhale carbonic acid gas, as well as sulphureous vapours, on many points. Any one who has descended into a volcanic crater, either in activity or in the state of a solfatara, and felt the effect on his own lungs of the acid vapours, will readily give credit to the possible truth of the old story.

Exhalations of carbonic acid gas occur abundantly in many other districts which were the former seats of volcanic action, as in Auvergne, the Vivarais, the Eifel, and throughout the whole basaltic range of North Germany, from the Rhine to the Riesengebirge. Bischoff considers this gas to be developed by the decomposition of carbonate of lime by volcanic heat or heated water*.

§ 16. The period that intervenes between the deposition of a bed of lava upon, or its elevation into the vicinity of, the surface of the earth and its complete refrigeration, that is, its attaining the mean temperature of the atmosphere, must be determined by the combination of numerous circumstances; such as,—

1. The figure of the mass. It is obvious that the more equal its dimensions in every direction, the longer will the central part retain its heat; and *vice versâ*, the greater the superficial extent of the bed in proportion to its thickness, the more rapid will be the process of refrigeration, other circumstances remaining the same.

* Chemical Geology, Cav. Soc. Publ. vol. i. p. 237.

2. The situation of the bed, by which it is more or less exposed to the influence of refrigerating media, such as currents of water or air, &c. ; and the conducting powers and thickness of the solid masses with which it is in contact, or through which alone its heat can escape.

3. The tendency of the lava itself to part more or less rapidly with its heat, either in combination with aqueous vapour, or by convection or radiation, which will probably be influenced by its compactness or porosity, the extent to which it is penetrated by shrinkage-fissures, and, perhaps, its mineral composition.

It has been already shown that the coarser the grain of the lava and the more irregular its arrangement, the greater the quantity of vapour that is enabled to escape by percolation, and the more rapid therefore, *cæteris paribus,* will probably be its solidification. The loss of caloric by convection and by radiation follows perhaps the inverse ratio, since the conducting powers of the mass will probably be proportioned to its compactness, and therefore to the fineness of its grain, or the regularity of its arrangement.

Under favourable circumstances, a body of lava will certainly retain an intense heat and liquidity in its interior during a very long period. Sir William Hamilton lighted small strips of wood by inserting them into the fissures of a lava-current from Vesuvius nearly four years after its emission. Currents of Etna are mentioned by Ferrara and Dolomieu as still moving down the sides of the mountain ten years after the eruption by which they were produced, and others as emitting vapour twenty-six years after their escape from the volcano. In the case of Jorullo, already cited above, a massive bed of lava appears to have retained an extreme internal heat, attested by the exhalation of steam in considerable quantity from numerous fissures in almost every part of its surface till within a very few years, though the eruption by which it was emitted dates from 1759.

In respect to subterranean masses of lava once perhaps in eruptive communication with the outer air, of course no limit can be assigned to the duration of their high temperature—all their conditions of position and extent being unknown, and scarcely to be guessed at; as well as whether, or in what degree, they are fed by continual increments of caloric from beneath or laterally. These considerations will be resumed in a later page.

View of Jorullo and its Malpais. (After Humboldt.)

CHAPTER VIII.

VOLCANIC MOUNTAINS.

§ 1. THE preceding chapters having treated generally of the normal phenomena of a single volcanic eruption, the nature of the erupted matters, and the mode in which they tend to dispose themselves outwardly around and about the vent, I propose now to consider the more complex cases, resulting from eruptions continued, or repeated at intervals, on the same point through a long term, perhaps many thousands of years, which we know to be an habitual character of volcanic action.

It is at once evident that such repeated eruptions cannot fail to load the surface of the earth around their source with a mountainous excrescence of a magnitude proportioned to the quantity of matter thrown up. Some of the lateral or parasitic cones, for example, formed upon the slopes of Etna in the course of a few days by a single eruption, measure 800 or even 1000 feet in height, and are bulky in proportion. Each has also given birth to streams of lava which have covered areas averaging very many square miles with beds of solid rock often twenty or thirty feet in thickness, and occasionally much more. And since we know Etna to have been in eruption hundreds of times within the historical era, it is impossible but that the vast amount of matter thus added to the external surface of the mountain during that period must have greatly increased its aggregate bulk. Yet that period most likely forms but a minutely fractional portion of the time through which the volcano has been equally active.

Again, on examining the structure of the mountain, we find

its entire mass*, so far as it is exposed to view by denudation or other causes (and one enormous cavity, the Val del Bove, penetrates deeply into its very heart), to be composed of beds of lava-rock alternating more or less irregularly with layers of scoriæ, lapillo, and ashes, almost precisely identical in mineral character, as well as in general disposition, with those erupted by the volcano at known dates within the historical period. Hence we are fully justified in believing the whole mountain to have been built up in the course of ages in a similar manner by repeated intermittent eruptions†. And the argument applies by the rules of analogy to all other volcanic mountains, though the history of their recent eruptions may not be so well recorded, provided that their structure corresponds with, and can be fairly explained by, this mode of production. It is also further applicable, under the same reservation, to all mountains composed entirely, or for the most part, of volcanic rocks, even though they may not have been in eruption within our time.

In order then to be in a position to determine this question in regard to any particular mountainous aggregate of volcanic rocks, it will be necessary to consider carefully the manner in which the products of repeated eruptions are observed to dispose themselves, or must necessarily do so, in obedience to the laws of volcanic action already ascertained.

§ 2. When, in a former chaper, we discussed the form and composition of the cone or hillock, whether simple or com-

* With the exception of a few comparatively insignificant strata of marine sediments intercalated with the volcanic matters up to a certain level, and by which the argument is not affected sensibly.

† If any geologist still adheres to the opinion of M. Elie de Beaumont, that the lavas forming the cliffs about the Val del Bove and other ancient portions of Etna have not flowed into their present inclined positions, but have been *upheaved* from a nearly horizontal one by some single and sudden operation—because he believes that lava *cannot* consolidate in such thick beds at any angle above 3° or 4°—I refer him to Sir Charles Lyell's ample refutation of this groundless assertion, as respects Etna especially (Phil. Trans. 1859).

pound, produced by a single continuous eruption of fragmentary matter from one vent, it was observed, that the weight
and pressure of the lava ascending within the funnel-shaped
cavity of such a cone often breaks down one of its sides, and
allows the lateral escape of the liquid matter, which disposes
itself at the foot of the hill, according to the circumstances
already enumerated ; but that, should the cone be sufficiently
solid to resist this pressure, the lava will often rise until it is
enabled to escape over the lowest part of the ridge of the crater,
and pour itself down the outward slope of the hill.   In this
case, a part of it congeals in its descent, and remains fixed as a
more or less solid rib or buttress to the fragmentary cone.

The effect of every subsequent eruption from the same
orifice which has already produced a cone of scoriæ and a
lava-current, must be to cover these with fresh products of a
similar character, arranged more or less conformably to the
outer slopes of the cone ; and the frequent repetition of such
phenomena must create an irregular alternation of stony consolidated lavas and conglomerate strata produced by the contemporaneous ejections; these different beds having a quaquaversal dip away from the eruptive centre, and accumulating
together into a mass of more or less magnitude, in proportion
to the violence and number of the eruptions.

The original cone thus by degrees assumes the size and
dignity of a mountain.   Those currents of lava which are able
to force their way through the side of this hill, harden into
massive buttresses at its foot, or upon its skirts; while those
which overflow the edge of the crater add still more efficiently
to the strength of the cone, remaining, as so many solid ribs,
interbedded with, and more or less cementing, its looser materials.   In this manner the mass becomes gradually more and
more fortified; and while the outward pressure of the column
of lava, raised during its eruptions, within the chimney or
central vent of the mountain, increases with its growing elevation, the strength and solidity of its framework, and conse-

quently the resistance opposed to this force, are augmented likewise *.

§ 3. That resistance, however, is frequently overcome by the intensity of the rupturing force, but not in the same manner as in the case of a cone formed solely of fragmentary matter, where the whole side is broken down and carried away at once. Where it consists of alternate fragmentary and solid beds, knit together into a compact framework by heat and pressure, should the force of the lava, ascending in the central vent, and acting like an immense wedge driven from below upwards into the heart of the cone, overcome the cohesion of its sides, one or more vertical fissures will be split through them in an approximatively radial direction; and by these the lava will often make its escape, with a velocity and volume determined by its fluidity, the dimensions of the

* This view of the mode in which volcanic mountains are formed, I need scarcely say, was held by all geologists before the " Elevation-Crater theory " was suggested by MM. Humboldt and Von Buch, especially by those geologists who had made volcanos their peculiar study, such as De Saussure, Spallanzani, Sir W. Hamilton, and Dolomieu. Spallanzani, for instance, in describing the formation of the island of Saline, one of the Lipari group, as composed of repeated beds of lava and scoriæ, one above the other, sloping from the summit-edge of the crater to the sea around, goes on to say, " We must conclude that there were at least as many eruptions from the summit of this mountain as we can count beds of lava. *Thus it is that volcanic mountains are for the most part formed.* In the beginning it is only the *accumulation* of the products of one first eruption ; then a second takes place ; then a third ; and the mass goes on increasing always in bulk in proportion to the number of eruptions. Thus was, no doubt, formed, increased, and extended the colossal bulk of Etna. Such was the origin of Vesuvius, of the Lipari Isles, and of other volcanic mountains,—not, however, forgetting that some minor volcanic hills, like the Monte Nuovo, and the Monte Rosso on the flank of Etna, were produced by a single eruption." (Voyage dans les Deux Siciles, ii. p. 116.)

The opposing theory, which attributes the production of all volcanic mountains to " sudden bladder-like upheaval," has been amply refuted in a separate publication. (See my paper on Volcanic Cones and Craters, Quart. Journ. Geol. Soc. 1859.) It is wholly irreconcileable with the view of volcanic action which it is the purpose of this volume to support and establish.

fissure, and the relative height of the internal column. As it flows out of any lateral opening thus established, the surface of this column must gradually fall, until it reaches the level of the orifice of emission. This then becomes the vent for the jets of elastic fluid likewise, the surface of the column being here brought into communication with the atmosphere. Meantime the pressure and intense heat of the lava, still boiling up the fissure, tend to enlarge and lengthen it, so as probably to force another opening at an inferior level. From this point the same phenomena take place, and are often repeated from fresh orifices successively produced, one below another, in the side of the mountain, until the pressure of the internal column of lava is so far diminished that it can no longer overcome the resistance afforded by the solid substructure of the mass. The internal plethora of the volcano being then relieved, all discharge of lava in currents before long ceases; and the column is still further lowered within the vent by the escape of its vapour alone, projecting scoriæ upwards, either from the last-formed aperture or the central crater, or alternately from both; and the eruption gradually terminates—the sum of resistances to the outward escape, in this manner at that point, of the excess of subterranean heat having regained the predominance over the antagonistic forces.

The process thus described is proved by observation to be the normal action of every habitually eruptive volcano; for the annals of all compound volcanic mountains teem with the records of eruptions characterized by these circumstances. Let us take Etna for example. In the eruption of 1536, twelve different mouths opened successively, one below another, on the same radial line or fissure, each producing lava, while the central crater vomited vapour and scoriæ[*]. In 1669, the south-east flank of Etna is described as having been visibly split open by an enormous rent, reaching from the summit two-thirds of the distance down the mountain. From its

[*] Ferrara, Campi Phlegræi. Borelli, Storia del Eruzione. Hoffman.

lower extremity issued the vast current of lava, which, taking the direction of Catania, destroyed a third of that town, and formed a large promontory projecting half a mile into the sea*. After the emission of the lava-current had ceased (that is, when the internal column of lava had subsided to the level of the lateral vent), aëriform explosions succeeded from the same orifice, and continued to be discharged with violence during fourteen days. The fragmentary matters vomited by them produced the large double cone called Monti Rossi, near Nicolosi, and covered a circuit of about two miles radius with a deep deposit of black sand containing innumerable separate crystals of augite. This district is only now beginning to support a scanty vegetation, in spite of the assiduous efforts made by the inhabitants to fertilize it. Part of the fissure then formed is still visible behind the Monti Rossi.

In 1780 the earth sunk along a straight line from the upper crater to a new lateral vent which produced an eruption, showing the existence of a fissure in that direction. In the later eruption of 1792, Ferrara observed that a fissure was broken through the side of the mountain, whence, during ten days, the lava boiled out very tranquilly, while the aëriform explosions took place only from the principal crater. At the end of this time the explosions ceased from the main crater, and commenced from the extremity of the fissure, at the same moment that the lava ceased to flow out; the liquid column within the vent having evidently lowered itself, by continual emission, to the level of the lateral aperture.

Again, in 1809, numerous orifices emitting lava opened successively upon one line or fissure, reaching downwards from the margin of the great crater†. A similar circumstance occurred during the eruption of Etna in 1811–12, according to the relation given to me by Signor Gemellaro, who was a witness of the fact. It appears that after the great crater had by its violent detonations for some time testified that the ascend-

* Fazelli, p. 212.   † Annales de Chimie et de Physique, 1810.

ing lava had nearly reached the summit of the mountain by its central duct, an unusually violent shock was felt, and a stream of lava broke out from the side of the cone, at no great distance from its apex. Shortly after this had ceased to flow, a second stream burst forth at another opening, considerably below the first; then a third still lower, and so on till seven different issues had been thus successively formed, all lying upon the same straight line, prolonged from the summit nearly to the base of the mountain. This line was evidently in this, as in the former cases, a perpendicular rent split through the internal framework of the mountain; probably not opened through its whole length by one shock, but prolonged gradually downwards by the weight, intense heat, and wedge-like pressure of the internal column of lava, as its surface sank by gradual discharge through each vent. The flowing of lava from each of these orifices was followed by the eructation of scoriæ, creating as many small parasitical cones. In fact, in nearly every lateral eruption of Etna, the production of such a fissure has been observed; the lava issuing from its lower extremity, and successively from different points, as the rent was prolonged downwards.

Other volcanos present the same phenomena. I may particularize the great eruption of Vesuvius in 1784, when five small cones were formed in succession at the eastern base of the mountain, in a line which, if prolonged upwards, would have intersected its apex. They still exist, and mark the points whence the lava-streams issued by which Torre del Greco was overwhelmed.

In all these cases the central craters of the two volcanos continued to discharge torrents of elastic fluids, carrying up scoriæ, lapillo, and ashes, while their lavas escaped in currents at a much lower level. When, however, the aëriform explosions took place from the lateral vents, those of the central crater ceased for a time, and usually recommenced when the former had in turn stopped.

But one of the most remarkable and instructive instances of analogous facts is to be found in the terrific eruption which tormented the west coast of Iceland in 1783, when the lava issued in enormous quantity from several sources opened successively in a plain at the foot of the high volcanic cone (Skaptar Jokul), from which the gaseous explosions had been, and for some time continued to be, discharged.

These sources were about eight miles distant from one another, and were formed along the same straight line, which obviously marked the direction of a fissure broken through the superincumbent strata of the plain, by the upward pressure of the lava below, in communication with that which was forced up the internal chimney of the neighbouring mountain. A fourth source opened itself in the prolongation of the same line, but beneath the sea, and at a distance of thirty miles, producing a rocky island, now reduced to a mere shoal by the erosive action of the waves and submarine currents. The lava poured forth by the three inland sources deluged the plain to a superficial extent of more than 400 square miles; and the extreme distance of the vents, and consequently the length of the fissure then formed, was not less than 100 miles !

In this instance it appears that the framework of the volcanic mountain offered a more solid texture, and a greater degree of resistance to the pressure of the internal column of lava, than the superincumbent beds constituting the plain at its base, which in consequence were the first to give way, and open an issue to the fluid. The fissure so formed was, in fact, only a prolongation or a re-opening of the fundamental one, which ranges south-west across the entire island, and which has given birth to all the recent eruptions of trachytic lava that compose and surround the great central range of jokuls or volcanic mountains, Hecla, Katlugaia, Skaptar, Skalbreide, Sneyfels, &c. The immense quantity of lava produced by the eruption of 1783, and the velocity with which it flowed forth, were obviously in direct proportion to the great height to which the column had ascended in the interior of the mountain, before the fissure was formed.   The distance of the apertures by which the lava was poured out from the crater or central vent of the volcano which discharged the aëriform fluids, proves the vast horizontal extent of the subterranean reservoir of lava, which is the less astonishing from our knowledge that the whole island of Iceland has been produced from the bottom of the sea by the successive eruptions of the same volcanic system— we might indeed say, of the same *volcano*.

The great eruptions of Lancerote in the Canary Isles in 1738 afford a parallel instance.   Some forty vents there opened in succession along the course of a fissure which crossed the whole island, itself the summit of a great submarine volcanic mountain.   Each of these discharged lava-currents and quantities of scoriæ—the latter producing as many cones, the former flooding the surrounding surfaces with basaltic matter spread out in horizontal sheets.   These eruptions lasted through several years.

§ 4. There can be little doubt that most of the minor shocks which agitate the environs of a volcano, during and previous to an eruption, are owing to the rending of some part of the solid framework of the mountain or its supporting strata, by the action of the force we have described as resulting from the pressure in all directions of the liquefied matter which is in communication with that elevated within the volcanic chimney.   The prolongation or widening of a fissure previously formed would have the same jarring or vibratory effect as the creation of a new one. It is, indeed, a remark common to the observations made on almost all volcanic eruptions, that local earthquakes always precede the emission of lava-currents, and cease while the lava is flowing, to recommence when it has stopped.   Those more violent shocks, on the other hand, which are felt throughout considerable distances, are probably caused by new rents produced in the solid subjacent strata supporting or surrounding the mountain ; and some of them belong perhaps rather to that class of plutonic earthquakes which are rarely accompanied by any outward eruption, although we may suppose them to prepare the way for such.

§ 5. The rents thus produced in the frame of a volcanic mountain are sometimes of such width as to cleave its whole mass in two. This occurred to the volcano of Machian, one of the Moluccas, in 1646. The crater of the Soufrière of Montserrat, and the volcanic cone of Guadaloupe, both appear to have been thus split through. So also the Montagne Pélée of Martinique. The eruption of Vesuvius in October 1822, which was peculiarly fertile in interesting phenomena, offered also an example of this rending of the mountain. The crater or, rather, chasm left by that eruption was but a local enlargement of an enormous fissure broken across the cone in a direction N.W.-S.E. The cleft was prolonged through the whole frame of the cone on the S.E. side, and produced a deep notch in its ridge, which, though considerably effaced by the beds of scoriæ and fragments thrown into it, was still 500 feet lower than the neighbouring points of the crater's rim.

These axial rendings of a volcanic mountain have probably been the originating cause of the greater gorges which are found occasionally opening a wide avenue into its central crater, such as the Barranco of Palma, the Val del Bove in Etna, and others. On this subject more will be said presently.

§ 6. The narrower fissures broken through the internal framework of the mountain, and instantly occupied by the liquid lava, become hermetically sealed by its subsequent consolidation, and assume the character of *dykes*. These being usually formed, as has been said, in a vertical direction, and therefore through the mantling beds or currents of lava that compose a large part of its substance, communicate a vast accession of strength to the structure of the mountain, acting as ties to the latter, which may be likened to the main beams of the edifice.

The section of Monte Somma presented by the steep cliffs

above the Atrio del Cavallo, which are the remaining walls of an ancient central crater of this volcano, exhibits an almost infinite number of such dykes traversing the mass of the mountain in various directions, all approaching, however, more or less to the vertical, and crossing each other, as well as the more massive and apparently horizontal beds of alternate lava and scoriæ, so as to give a reticulated appearance to the face of the cliff. They are of very compact leucitic basalt, and frequently divided into prisms lying at right angles to the walls of the dyke. The crater of Vesuvius formed in 1822 exhibited similar features, which are common, in fact, to the central parts of all volcanic mountains of which the internal structure is sufficiently exposed by denudation or otherwise.

Some of these dykes are very narrow, often not exceeding a foot or two in width. The small veins are probably but ramifications of other larger ones, and have never reached the outer surface. It is, however, quite conceivable that a very narrow and intricate fissure may act as a channel for the ascent and outward efflux of large quantities of highly fluid lava, supposing the aëriform explosions, which would necessarily widen the fissure considerably towards its upper extremity, to be discharged for the most part at some other contiguous point—perhaps from the central crater. The dykes visible in the precipitous walls of the crater of Vesuvius left by the explosive eruption of 1822 were vertical, and could be traced at least 400 or 500 feet downwards, penetrating through the horizontal layers of both lava and scoriæ which constituted the bulk of the cone. Many of them terminated upwards in some of the lava-beds, of which they had probably been the feeders from beneath. (See figs. 15 & 16, p. 76 *suprà*.)

Similar dykes, composed of trachyte, are to be seen in the Val de l'Enfer—a chasm towards the centre of the Mont Dore. Three or four of them are only from 5 to 8 feet in width, and rise vertically nearly 1000 feet from the bottom of the ravine to the high projecting peaks called Les Aiguilles,

by which name their upper extremities are known. Mr. Darwin describes one dyke in St. Helena as being 1260 feet high,
and maintaining throughout a uniform width of 9 feet from
top to bottom.

The matter composing such dykes is generally compact
and free from vesicles. It not unfrequently has a finer grain
at the sides than towards the centre. This may be attributed
to friction against the walls disintegrating the crystalline
elements of the lava as it is propelled up the fissure—if we
suppose it to have been already in a semi-crystalline state.
Some dykes, however, have lateral 'selvages' of a vitreous
texture. Is this owing to the disintegration from friction
proceeding so far as to cause the complete fusion of the lava
by the same amount of heat which allowed the central portion
to retain its partial crystallization?—or is it (as Sir C. Lyell *
and others suppose) that the lava, when injected, was altogether in the state of liquid glass, and crystallized subsequently, the sides retaining their glassy texture owing to their
comparatively rapid cooling?   The first of these alternatives
is to some extent supported by the facts, that the lateral matter
of dykes is often laminated, as if by friction;—that dykes of
syenite are occasionally bordered by seams of greenstone, *i. e.*
fine-grained or disintegrated syenite;—and, again, greenstone
dykes, when traversing limestone, by serpentine,—formed, apparently, through a metamorphic process, by mixture of the
magnesia of the augite with lime from the side rocks, and
dragged out into crumpled laminæ by the pressure and friction
of the heated lava forcing its way up the dyke.

The injection of such fissures by liquid lava generally takes
place with very little, if any, derangement of the rocks they
penetrate. " That dykes *shift* or disturb the beds they pierce
is a rare phenomenon," says Sir C. Lyell, speaking especially
of his observations on Etna, Madeira, and the Canary Isles †.

* Manual, p. 532, ed. 1855.
† "On Lavas of Mount Etna," Phil. Trans. for 1858, p. 47.

And this remark agrees with what is generally observed even among the earlier volcanic (trap) dykes, with some exceptions to be noticed presently.

§ 7. It is, however, evident that every fissure so formed and filled from beneath with solid matter must, in proportion to its size, be attended by a certain amount of derangement in the rocks it traverses, and also must occasion in some degree the distension or internal swelling of the mountain, which will thus grow in bulk, not merely by the outward addition of erupted matters, but also to some extent by the internal accretion of injected lavas.   Such inward dilatation has been aptly compared by Sir C. Lyell to the endogenous growth of a tree by the ascent of sap in its veins. The increase of bulk acquired by this process will nevertheless be but trifling in comparison with that occasioned by its outward eruptions,—an inference confirmed by the small proportionate bulk in the aggregate of the dykes which are observable in the interior of a volcanic mountain wherever it is exposed to view, in comparison with that of the repeated beds of lava and conglomerate which, sloping from the central summits to its extreme skirts, make up evidently the far greater part of the mass. It should be remarked, however, that the dykes being more numerous near the central vent, their aggregate effect in elevating these beds will be greatest *there*, and give them a steeper inclination near the summit than lower down the flanks of the mountain. This is one cause (but by no means the principal one) of the angle of slope of the higher beds and of the outer surface likewise, usually ranging from 20° to 35°; while towards the base it diminishes to 10°, and graduates ultimately to horizontality.   The more influential causes of this general result are (as will shortly be shown), the frequency of lateral eruptions on the lower slopes of every volcanic mountain, loading them with parasitic cones and floods of lava, and the abundance of fragmentary matter carried down the heights by rain and floods—all combining to enlarge its base.

Since the expansion of a volcanic mountain by inward accretion (to whatever extent it takes place) must be slow and gradual, accompanying throughout the gradual accumulation in very much larger quantities of sloping beds, both fragmentary and solid, formed by the external dejections of the volcano, it in no degree corresponds with, or can be brought forward to justify the idea involved in the theory of ' Elevation Craters'—that is to say, the formation of volcanic mountains by a " *single, sudden swelling-up of previously formed horizontal beds of lava and scoriæ* into a hollow bladder"—which is the notion so long and perseveringly supported by MM.Von Buch, De Beaumont, and their disciples *.

From this knitting together of its component beds by interlaced dykes and veins of solidified lava, it is clear that as a volcanic mountain grows in height and bulk, the cohesive strength of its fabric must increase in at least an equal proportion, which, combined with its equally augmented weight, enables it to resist the increasing hydrostatic pressure of the column of liquid lava that may be propelled at periods of eruption up its central duct.   Hence it would seem that no absolute limit can be set to the growth of such a mountain in height and bulk.   And, in fact, we know that many volcanos

---

* See " Cones and Craters," Quart. Journ. Geol. Soc. 1859. M. Elie de Beaumont, in one of his later works, speaks of such dykes as forming "the evidence and the *measure* " of the upheaval of the volcanic mountain in which they occur. To that proposition all will assent. But it is just because the dykes form but a fractional portion of the whole bulk of such a mountain, that the amount of its upheaval, of which they are the *measure*, is, in my estimation (and by the rule laid down by M. Elie de Beaumont himself), also but fractional. Moreover the injection of these dykes was certainly not simultaneous, since they interlace and traverse one another. This sentence, however, of M. Elie de Beaumont evidently amounts to a renunciation of the doctrine of a *bladder-like upheaval*; for if any hollow existed beneath the upraised mass, that, and not the solid up-filled dykes, would be the ' measure ' of the upheaval. Moreover it is impossible to understand how fissures in the ' upheaved' crust could be injected with fluid lava, by the same process which left a vacant cavity beneath it (!). These, however, are not the only inconsistencies of the ' upheavalists.'

do reach an extraordinary altitude :—Etna, in round numbers, 11,000 feet; the Peak of Teneriffe, 12,000; Klutschew, one of the volcanos of Kamtschatka, 16,000 ; another in the Aleutian chain, 14,000 ; St. Helen's, north of the Columbia River, on the north-west coast of America, 15,000 ; Popocatepetl and Orizava, in Central America, 18,000 ; Sahama in Bolivia, 22,350; Aconcagua, 23,000 ; Sangay, 17,124; Antuco, 16,000; Chimboraço, 21,000 feet (the last four all in Chili) ; and other volcanic peaks of the Andes, not long since considered the highest mountains of the globe, are examples of the vast elevation occasionally attained by accumulations of erupted matters.

It is true that the circumstance of the Pic de la Teyde, Chimboraço, Aconcagua, and some other of these very lofty volcanic mountains, not having for centuries exhibited eruptions from their summits, may appear at first sight to show that when the mountain has acquired an extreme height it becomes unable to resist the pressure of so enormous a column of lava, which in consequence finds a vent at its side or foot. But it must be recollected that these lateral eruptions tend continually to fortify the flanks of the mountain by the accumulation and consolidation of their products ; it may be foretold, therefore, that should the internal focus of these volcanos continue in activity, the time must come when they will be sufficiently propped on all sides to be able to sustain the pressure of the lava, which in this case will be again, perhaps, poured forth from the principal summit.

§ 8. The lateral eruptions here spoken of, proceeding exactly in the manner of those from fresh vents described above (Chap. V.), like them produce a more or less regular cone of scoriæ, &c., on every point where they find an issue.

The slopes of Etna are loaded with above 700 such *parasitic* cinder-cones, many of them of considerable magnitude. Almost all possess craters, and each marks the source of a current of lava : that from the Monte Rosso, which is 700 feet in height above its base, and two miles in circumference, destroyed Catania in 1669.

Vesuvius has occasionally produced similar hills : one, on which stands the Camaldoli della Torre, is an example; and eastward of this point rise the five other small neighbouring cones already mentioned as thrown up by the eruption of 1784 at the base of the mountain.

Few indeed, if any, of the greater volcanic mountains are unattended by such minor elevations about it, like the satellites of a planet.   All, however, are not scoriæ cones ; for in some cases, especially among the volcanos productive of felspathic lava, the lateral efflux of this has given rise to lumpy excrescences of trachytic rock alone ; the explosions that at the same time threw out fragmentary pumice (felspathic scoria) taking place probably from the central vent at a greater or less distance.

This brings us to the consideration of the mode in which the fragmentary ejecta of an habitually eruptive volcano tend to dispose themselves.

§ 9. Of the fragmentary matters ejected during a violent or paroxysmal eruption from the central or principal vent of a volcanic mountain, the greater part, as they fall from the air, spread themselves mantlewise over its surface and slopes ; but the lighter and more pulverized portion is borne away, often to great distances, by the winds prevailing at the time.   The abundance of these materials, and the extent of country covered by them, are sometimes prodigious.   By the terrific eruption of Coseguina, for example, in the Gulf of Fonseca in Central America, in 1835, all the ground within a radius of twenty-five miles was loaded with scoriæ and ash to the depth of 10 feet and upwards, houses and woods being buried in them at that distance ; while the lightest and finest ash was carried by the winds to places more than 700 miles distant.   The eruption of Sangay, in the Cordillera of South America (1842–43), ejected black scoriæ and ash, which

covered the surrounding district to a distance of twelve miles
with beds 300 and 400 feet thick, according to M. Sebastian
Wise.   On occasion of the eruption of Tomboro, in the island
of Sumbawa, in 1815, the *roofs of houses at the distance of
forty miles were broken in* by the weight of the ashes that fell
upon them ; which were, moreover, carried to a distance of
300 miles in such quantities as to darken the air ; while the
floating pumice in the sea westward of Sumatra formed a mass
several feet thick and many miles in extent, through which
ships with difficulty forced their way.

It may readily be conceived, therefore, to what an extent
the configuration and structure of the mountain itself, and
its immediate neighbourhood, must be affected by the accumulation of such prodigious quantities of fragmentary
ejecta as in each of these and similar instances fell upon and
about it.

§ 10. *Eluvial torrents.*—Their disposition is often largely in
fluenced by the agency of torrents of water, which frequently
pour down the slopes of the mountain at moments of eruption; often owing to violent rains caused by condensation
of the volumes of aqueous vapours evolved—often, when the
elevation or geographical position of the spot, or the period of
the year, has caused the heights to be covered previously
with snow or glaciers, to the sudden melting of these by the
showers of red-hot scoriæ that fall on them, or the contact of
still hotter lava, or the internal heat transmitted through the
sides of the mountain.

In Iceland, as we should expect, phenomena of the latter
kind accompany almost every eruption from its snow-clad
heights, and constitute by far the most destructive feature of
the fearful volcanic catastrophes to which the inhabitants of
that sea-girt caldron are exposed.   During the eruption of
Katlugaia in 1756, prodigious torrents of water, ice, rocks, and
sand, occasioned by the melting of the glaciers, rushed from the
heights and produced three parallel promontories, reaching

several leagues into the sea, which remain above its level in places where the fishermen formerly found forty fathoms of water*.

Sir C. Lyell relates, that on Etna a bed of *solid ice* was lately found under a current of lava. It is very conceivable that a coating of sand and scoriæ, the best possible non-conductors of heat, may enable snow to bear even a red-hot stream of lava over it without being melted. It is probable, then, that in Iceland this circumstance must have been often repeated, and we might expect to find glaciers there alternating with beds of lava and conglomerate. The transmission of heat to such masses by the renewed rise of lava up the chimney of the volcano, or up some fissure traversing these intercalated masses of ice, would be likely to melt them rapidly, and send down torrents of water and rocky débris to the skirts of the mountain. But probably the most powerful of all such causes would be the continual fall of red-hot scoriæ over the upper surface of a snow-covered mountain.

Such an aqueous torrent broke forth from near the summit of Etna in the year 1755 and swept down the Val del Bove, bearing with it a vast amount of detritus. The inhabitants of the slopes of Vesuvius designate the deluges of mud and ashes to which they also are often exposed during eruptions by the name of ' *Lave d'acqua*,' or ' *di fango*,' in contradistinction to ' *lave di fuoco*.'

§ 11. But there is another not unfrequent cause of such ' eluvial' debacles. Whenever the pulverulent ejections of a volcano are of such a nature as to form a clay or paste, on mixture with water—which is generally the case with the ash of the felspathic lavas abounding in aluminous minerals—the finer particles carried down by rain to the bottom of a crater must form an impervious coating to the interior of the basin, and cause the water falling from the clouds to collect into a lake. Under favourable circumstances, this body of water will

---

* Olafscen and Povelsen.   See also Brit. Quart. Rev. for April 1861.

increase in depth and volume till its pressure breaks through
one side of the cup and occasions a debacle, which must carry
down to the lower levels and spread over them vast volumes
of fragmentary matter. Or the bursting of such a lake may
be brought about by an earthquake, or by the occurrence of
an eruption; and in this latter case the aqueous and igneous
ejections must become mingled in extraordinary confusion.

§ 12. Such 'eluvial' eruptions, whether occasioned by the
bursting of lake-basins, or the sudden melting of snow, or
other causes, appear to have always played a great part among
the phenomena of the South American volcanos. In a single
night of 1803, according to Humboldt, all the snow on the
vast volcanic cone of Cotopaxi was melted, and discharged in
torrents of ashes and mud into the Rio Napo and the Rio de
los Alaguos. Such torrents of mud, *i. e.* fine volcanic ash
mixed with water into a paste, have in many other instances
been known to burst from the sides or summits of these stupendous volcanic mountains, and carry destruction into the
valleys and plains at their base. In the province of Quito
this volcanic mud is called '*moya*'; and many fish, especially
a peculiar species, the *Pimelodes Cyclopum*, are occasionally
found enclosed in it. So great a quantity of fish were, it is
said, ejected by the volcano of Imbambaru in 1691, that fevers
were caused in the neighbourhood by their putrefaction. The
carbonaceous matter which also is mingled with the mud
(sometimes even in such quantity as to render it inflammable)
proceeds probably from the algæ and other water-plants that
grew in the crateral lakes, the bursting of which occasioned
the mud-debacle. The abundance of Infusoria detected by
Ehrenberg in rocks of this origin is to be explained in the
same manner, the volcanos among which they are found being
exclusively trachytic. As known instances of such occurrences may be noted the mud-eruptions of Carguirazo in 1698,
of Cotopaxi in 1743 and 1854, of Tunguragua in 1797, of
Nevado de Ruiz in New Granada in 1845, and of Puracè in

1848 *. The volcanos of Java likewise eject mud-lavas for the most part.

Among the extinct trachytic volcanos of Europe such phenomena have not been unfrequent. To them are owing the trass of the Andernach district, of the Eifel, and Siebengebirge †.

The trees and plants growing on the slopes of a volcano—indeed whole forests at times—are rooted up and carried away by these exundations, and buried among the alluvial strata at its foot. This is, no doubt, the origin of the fossil wood frequently met with in such formations. The Surturbrand of Iceland, which is interbedded with its volcanic conglomerates, is by some thought to be derived from drift timber brought across the Atlantic by the Gulf-stream, since the climate is too severe for the growth of trees of any magnitude. It may, however, be the remains of a more vigorous vegetation than the present, belonging to a period when the climate was more favourable. The latter view is supported by the abundant leaf-beds occurring in the Surturbrand. It is to be hoped that some of the recent Icelandic travellers will bring back evidence on this interesting point.

The vegetable fibre in these deposits is occasionally mineralized, but not always. In the Rhine district it is invariably carbonized, and occasionally passes into jet. In Hungary it is sometimes silicified, and even metamorphosed into opal. In the Mont Dore I have seen the trunk of a tree enveloped in tuff, which at one extremity was reduced to perfect jet, and at the other retained the colour and texture of wood, not being in the least degree carbonized. The experiments of M. Daubrée throw considerable light on these metamorphoses.

§ 13. These alluvial conglomerates, when their materials have been forcibly mixed up with water in a muddy mass,

---

* Comptes Rendus, xxii. p. 700.

† See Dr. Hibbert's monograph of the Rhine Volcanos, and my paper on the same (Journ. of Sci. 1826). See also Appendix, on Eifel Volcanos.

acquire on desiccation a great degree of solidity, even without being cemented (as they often are) by ferruginous or other infiltrations. The fine ashes of the eruption of Vesuvius of 1822, which I saw carried down the slopes of the hill by torrents of rain, caked into a hard and tough rock, that required a sharp blow from a hammer to break it : evidently the particles were compacted together by a kind of cohesive attraction like the 'setting' of mortar or cement. The beds of hardened tuff covering Herculaneum to the depth of from 50 to 150 feet were, without doubt, produced in this manner by the eruption of the year 79, which at the same time overwhelmed the more distant towns of Pompeii and Stabiæ with looser ashes, arranged for the most part in layers as they fell from the air. It is probable that the greater part of the similar tuffs which clothe the outer slopes of Somma and encircle its base were thrown out by the same paroxysmal eruption. They are, as Sir W. Hamilton long ago remarked, undistinguishable in character from, and continuous in disposition with, those under which the Greek cities lie buried. Even at the distance of Naples, the ashes ejected by this eruption seem to have fallen to the depth of several feet ; for in more than one spot, for example at the back of the Studii in that town, I observed beds of stratified pumice and lapillo from 8 to 12 feet thick, overlying made ground, which contained numerous Roman and Greek tombs *.

§ 14. From some, or the combination of all these causes, alluvial deposits of fragmentary matter may be expected to form a large portion of the bulk of every volcanic mountain, filling up its hollows, covering its skirts, and extending to some distance from its base, often in alternate beds with those of its lava-currents, which, owing to their superior fluidity, have stretched farthest from the central summits. Such, in fact, is found to be the structure of mountains of this class whenever they have been sufficiently exposed by denuda-

* See plate to Geol. Trans. 2nd ser. vol. ii. pt. 3.

tion to geological examination; for example, those of the Cantal, Mont Dore, and Mezen, in France, of Teneriffe, Madeira, Iceland, Hungary, and the Andes.

The violently tumultuous manner in which many of these volcanic conglomerates were deposited has caused them to accumulate occasionally in massive beds of some hundred feet in thickness; while at other points, where perhaps the waters were embayed in eddies or pools, the finer materials have subsided so slowly as to form foliated rocks, of which the laminæ are thinner than paper. The materials of some are loosely aggregated like drift gravel, and contain large boulders; others are compacted into so tough a rock as to be fit for use as building-stone. The Peperino of the Alban Mount, near Rome, and the hard tuff of which Naples is built, are both examples of this character *. The tuff of the Rocher Corneille and other rocks in the vicinity of Le Puy is a 'Peperino.' Such indurated basaltic tuffs, wherever they occur, are in much use as building-stone, from their cutting well under the axe.

If the sea or an inland lake washes the base of a volcanic mountain, such fragmentary matters as at its periods of eruption are borne down by torrents, or fall there by the mere force of projection, must be distributed over the bottom or shores of the water by currents, and mixed or interstratified with other sedimentary deposits. After the great eruption of Tomboro in Java, in 1818, the sea in the vicinity of the island is described by eye-witnesses as covered by floating pumice and half-burnt trunks of trees, torn of course from the shattered sides of the mountain, and launched with its fragments into the air, by this great paroxysmal eruption. These must have subsided into beds of volcanic ash and pumice-conglomerate at the bottom of the sea, intercalated with others of vegetable matter †. This was, without doubt, the

---

* 'Peperino' is the name given by the Italians to the tuffs, or hardened conglomerate of basaltic or greystone (leucitic) lavas.

† Journal of Science, vol. i. p. 255.

origin of the stratified tuffs of the Terra di Lavoro, which penetrate into the recesses of the Apennines, and of the similar tuffs of the Campagna di Roma, both having been since raised from beneath the sea by an elevatory action affecting the whole western coast of Italy to the extent of more than 200 feet, as shown in the cliffs of Sorrento. In Hungary, pumice-conglomerates alternate with tertiary limestone strata. So, too, in Asia Minor, in Sardinia, in the Deccan, in New Zealand, and many other localities. In the vicinity of Pont du Château and Veyres in Auvergne, a calcareous volcanic peperino alternates with limestone strata full of freshwater shells ; and at Monton, as well as on Gergovia, an adjoining hill, currents of basaltic lava are interbedded with the same strata, eruptions having evidently taken place from the bottom of the Miocene freshwater lake in which much calcareous matter was at the time subsiding. The Euganean Hills, at the southern base of the Vicentine Alps, present a similar result of the intermixture of volcanic ash and calcareous sediment. At or near the points of eruption, in all such cases, the mixture of the two substances seems to have been so forcible and complete, that there are occasional passages from limestone into lava, making it difficult to determine where the aqueous rock terminates and the igneous one commences. And occasionally a sort of segregation of the two elementary matters has taken place, small rounded scoriæ (lapillo) or angular fragments and granular nodules of basalt being cemented with calcareous spar, which sometimes also forms veins in the igneous rock. Similar calcareo-volcanic conglomerates, alternating with trachytic ash, were deposited in the same manner in the præ-carboniferous period.

Much variety in rocks formed chiefly of eluvial volcanic ash is necessarily occasioned by the different mineral characters of the lavas from whose comminution they proceed, and their consequent greater or less tendency to ' set ' or cohere either under water, or as the water gradually drained from, or

N

was squeezed out of them. It is well known that both puzzolana and trass, when mixed up with lime, *sets* readily under
water. They frequently contain crystals of augite, leucite,
felspar, or mica, giving them a porphyritic aspect, which,
together with their tough cohesiveness, makes it sometimes
very possible to mistake them for earthy or lithoidal lava.
Bischoff is of opinion that these crystals have always been
formed in the wet way in the tuff since its deposition. We
should be scarcely justified in declaring this impossible; but
I have never seen any crystals in ordinary volcanic conglomerates of so delicate a form, or in such positions as to make
it incredible that they should have either fallen from the air,
or subsided from water in which they may have more or less
tranquilly floated, enveloped in pumice or light scoriæ, after
ejection from a volcanic vent. When tuffs have been long
exposed to the action of heat by contact with a lava-dyke or
current, or from any other cause, no doubt such new crystal-
lizations may take place; and this was probably the origin
of the rare crystalline minerals of some of the 'peperinos.'
Others were, perhaps, formed of materials ejected after a long
baking in some volcanic focus. The claystone porphyries or
trachytic lavas of early dates are accompanied by conglomerates of very similar character, and were doubtless of similar
eruptive origin *. Some conglomerates buried beneath a volcano may have been themselves more or less altered by heat
—partially fused, perhaps. Hence it becomes difficult to
distinguish such altered tuffs from ordinary lavas. For example, the vitreous trachyte of Ponza has all the appearance of
a partially re-melted pumice-conglomerate. Enigmatical rocks
of this hybrid character are, as might be expected, not unfrequent in all volcanic districts.

The contraction which conglomerates experienced in the

* See Professor Ramsay on the Igneous Rocks of Wales and Scotland,
pp. 175-190; Catalogue of Rocks in the Museum of Practical Geology,
1858.

process of desiccation, combined with a concretionary action, has in many instances given them a divisionary structure, globiform, prismatic, even columnar, or tabular, similar to that assumed by lavas which have consolidated by cooling, and owing to the same cause, *i. e.* the escape of *their* vehicle of fluidity—in this case water. This structure is, however, rarely so perfectly developed in the former as in the latter class of rocks. When fissures were formed by disturbance in conglomerates of this solid character, they have been usually filled (like the veins in marble) by exudation of the finest matter remaining fluid in the vicinity of the crevice, and draining off into it. Many such veins or dykes are observable in the hardened felspathic tuffs of the Phlegræan fields near Naples. Mr. Darwin observed similar veins in the tuff of the Galapagos Isles.

Sometimes the fissures of volcanic tuffs are filled with calcareous spar, or arragonite, occasionally with selenite; not unfrequently with opal, calcedony, or quartz. Bitumen sometimes oozes from them, probably derived from animal or vegetable matter buried within them at the time of their formation. Veins in the hard tuff of Pont du Château (Auvergne) are lined with bitumen, calcedony, and beautiful rosette-shaped groups of rock-crystal.

The solidity as well as the massiveness of these tuffs is, doubtless, owing to the violent mixture of their ingredients with the sea-water, the eruptions which produced them having evidently burst out on a shallow shore. They contain marine shells of recent character. The Monte Nuovo (fig. 35), the nucleus of which is of this hard tuff, was actually seen by eye-witnesses to be so formed by an eruption of mud, *i. e.* volcanic ash and water, tossed up in repeated jets from a low spot on the margin of the sea *. It will be readily seen how such jets

---

* The letter of Pietro Jacobeo di Toledo, printed at Naples in the very year of the eruption (1538), now in the British Museum, says expressly, " The *mud* (fango) ejected was *at first very liquid, then less so,* and in such

would throw off on all sides waves of muddy sediment which must accumulate around the vent into a bank of this matter, in concentric layers, just as we see it in the Monte Nuovo, Nisida, and other crater-cones of the Phlegræan fields, the Galapagos, &c., and be subsequently covered with mantling beds of looser materials of the same kind—the last ejecta of the eruption which fell in a dry state from the air.

Fig. 35.

Monte Nuovo (seen from the environs of Puzzuoli).

§ 15. The accompanying woodcut (fig. 36) presents an ideal

Fig. 36.

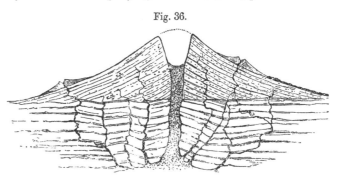

Ideal section of a Volcanic Mountain produced by repeated eruptions.

section of a volcanic mountain after one of its paroxysmal eruptions, which is supposed to have gutted the cone, leaving a great central crater.   The dotted outline is intended for

quantities, that, with the help of the large stones likewise thrown up, a mountain was raised, a thousand paces in height, by the third day."

that of the summit previously to the eruption. It must not, however, be supposed that the alternating beds of lava and conglomerate here represented can ever mantle continuously with any regularity round the whole mountain. On the contrary, each separate current occupied probably but a small segment of the horizontal circular section. In the cliff-sections observable in the interior of the craters of Somma and Vesuvius, and generally of other volcanos, the beds of lava are found to thin out and terminate before proceeding far on either side; many of them having flowed down the outer slopes only in narrow strips, others in wider currents, but none of them probably ever reaching far round the cone. The woodcut, moreover, represents in other respects a greater regularity of structure than is likely ever to be met with in nature; for the multitude of dykes penetrating the central parts of such a mountain, and the inequalities and derangement of the beds by the mutual interference of lateral cones of scoriæ and lava-streams thrown up at different times, must usually occasion

Fig. 37.

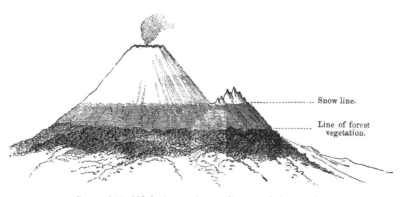

Cotopaxi (17,662 feet), seen from a distance of ninety miles.
(From Humboldt's 'Vues.')

far more confusion in the arrangement of its component parts than is here represented. The section must be accepted therefore as only a suggestive sketch.

Some volcanic mountains, nevertheless, possess a conoidal figure of extreme regularity, owing to the even and uniform emission and accumulation of the ejections from and around the central vent. The cone of Cotopaxi, as represented by Humboldt, is an example of this extreme and beautiful regularity on the largest and most striking scale. Probably few lavas are emitted from this highest peak, but chiefly pumice and ash, which, in the absence of prevailing winds at that great height, would be showered down evenly on all sides (see fig. 37). The truncated summit is " a wall-like annular ridge,"—the crater-margin, no doubt, though for the most part filled up by snow and recent eruptions. The rugged peaks seen to the right (called the Cabeza del Inca) are probably trachytic rock, the remnant of a lateral outbreak of imper-

Fig. 38.

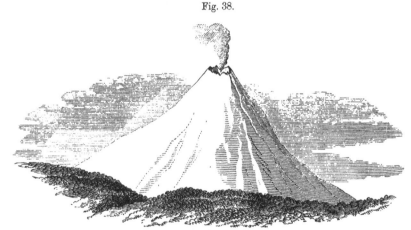

Pic d'Orizaba (17,900 feet high), as seen from the Forest of Xalapa.
(From Humboldt.)

fectly liquid lava; or, it may be, of the circuit of a lateral crater. Humboldt gives no clue to this *.

The Peak of Orizaba (Mexico) is another striking example of the conical figure in a volcano of first-class magnitude (see fig. 38).

* Kosmos, vol. iv., Sabine's Translation, note 454.

The volcanos of Java, which erupt great quantities of mud, *i. e.* pumice-ash violently mixed with water, are equally remarkable for regularity of form. Their conical surfaces are, however, singularly scored, according to Dr. Junghuhn, by radiating ravines, caused by heavy rains acting on the light and friable conglomerate of which they are principally composed.

It must be observed, however, that volcanos which are formed chiefly by the continuous welling-out of very liquid lava-flows from a central orifice are sometimes equally regular in form; for example, the flattened dome of Mauna Loa and other mountains in Hawaii, whose fragmentary ejections are comparatively insignificant.

As a rule, the continued adherence of the eruptive action to a single vent seems to be the main cause of regularity of figure in a volcanic mountain, whether the eruptions are more or less explosive—so long, at least, as no paroxysmal outburst occur to blow a great part of the whole fabric into the air and disperse its materials over the surrounding country. This, however, is an event of not unfrequent occurrence; and the result is, as I have already indicated, and shall proceed to show in greater detail, the formation of a 'crater' of the largest class.

View of Vesuvius from near Sorrento, showing the modern cone rising out of the old crater of Somma.

CHAPTER IX.

ON CRATERS OF VOLCANIC MOUNTAINS.

§ 1. IN the case of simple cinder-cones thrown up by a single eruption, notice has already been taken of the cause and characteristic features of the *crater* or funnel-shaped hollow so often visible either at their summit, or as a breach in their flank, and marking the eruptive orifice. Those conical mountains which are the product of many repeated eruptions from the same vent also exhibit for the most part a central cavity or principal crater at or near their summit. It is, however, liable to frequent changes in form, and is occasionally altogether absent.

These changes, it will be perceived on consideration, are the necessary result of eruptions of unequal violence taking place successively from the same point. For when, after any paroxysmal eruption which has left a crater of considerable dimensions, the activity of the volcano is renewed from the bottom of this hollow with only (as usually happens) a moderate degree of violence, the fragmentary matters ejected will all probably fall within its circuit, and the lavas emitted will accumulate in pools at its bottom. The repetition of such minor eruptions must in this way sooner or later fill up the whole of the vacuity created by the last paroxysm, and cause the matters thrown up, both liquid and fragmentary, to find their way thenceforward over the ridge of the basin—which is, in fact, obliterated.

By this series of operations the crater of a large volcanic mountain becomes replaced by an irregular plain, possessed of a degree of convexity, and often still further heightened by the parasitical cones and pools of lava which the continuance of minor eruptions throws up on its surface—until perhaps their accumulation one over another has built up by degrees a new conical summit on the spot where previously yawned a deep and wide gulf.   In this state of things, the products of all minor eruptions for the most part mantle round the outer slopes of the cone.

If, after this, another violent eruptive paroxysm take place, say by the creation of a deep fissure penetrating into an intensely-heated point of the volcanic focus, its powerful explosions soon break up and shatter the mass of solid and fragmentary rocks which had accumulated within the interior of the ancient crater, and project their fragments to a vast height into the air, whence a part is strewn over the outer slopes of the mountain, while the remainder, which fall back towards the orifice of projection, are again and again driven forth till they are triturated to the finest possible degree of comminution.   The duration of these explosive discharges is usually for days, weeks, months, perhaps even years, as we observed in the sketch given in an earlier chapter of the phenomena of a *paroxysmal* eruption.   The final result of the process is, once more to leave a crater bearing marks of the violence with which it has been torn through the bowels of the mountain, and of a diameter and depth proportioned to the energy and duration of the eruption.

The encircling walls of such a cavity will be more or less perpendicular, and present sections of the beds, both of consolidated lava and conglomerate, through which the aëriform explosions forced their way.   In process of time, however, these walls will often crumble in, and their fragments, forming a talus at the bottom, soften off the abrupt features of this cavity and diminish the steepness of its enclosure.

The horizontal section of this crater of a volcanic mountain, like that of a simple cone, is usually circular, but occasionally elliptical, being lengthened in the direction of the fissure by the enlargement of which it is produced.

§ 2. The phenomena of Vesuvius, during the last century, will serve to exemplify the series of changes described above as incidental to the principal or central vent of every habitual volcano.

In the year 1756, Vesuvius possessed no less than three cones and craters, one within the other, like a nest of boxes, besides the great encircling crater and cone of Somma. Sir W. Hamilton gives a drawing of its appearance in this state (fig. 39).

Fig. 39.

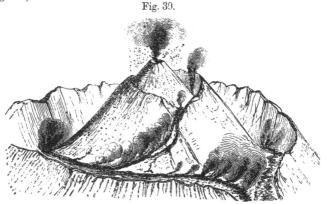

Summit of Vesuvius in 1756, showing cone within cone.
(From Sir W. Hamilton's 'Campi Phlegræi.')

By the beginning of the year 1767, the continuance of moderate eruptions had obliterated the inmost cone and increased the intermediate one, until it very nearly filled the principal crater (fig. 40). An eruption in October of 1767 completed the process, and re-formed the single cone into one continuous slope all round, from the highest point of its truncated, but solid summit, downwards.

An interval of comparative tranquillity followed, until, in 1794, the paroxysmal eruption occurred, described by Breislak, which completely gutted this solid cone, lowered its height, and left a crater of great size bored through its axis. Later eruptions, especially that of 1813, not merely filled up this vast cavity with their products, but once more raised the

height of the cone by some hundred feet. When I first saw
it in 1818, the top formed a rudely convex platform, rising
towards the south, where was its highest point. Several
small cones and craters of eruption were in quiet activity upon

Fig. 40.

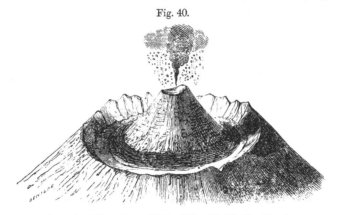

Summit of Vesuvius in 1767. (After Sir W. Hamilton.)

this plain, and streams of lava trickled from them down the
outer slopes of the cone. So things went on until October
1822, when the entire heart of the cone was again thrown out
by the formidable explosions I have so often referred to, and
a vast crater was opened through it, while the cone itself was
found to have lost several hundred feet from its top. In fact,
nothing but an outer shell of it was left (fig. 41). Eruptions,
however, soon recommenced. In 1826–7 a small cone was
formed at the bottom of the crater, and, continuing in activity,
had reached a height which rendered it visible from Naples in
1829, when of course it must have nearly filled up the crater.
In 1830 it was 200 feet higher than the crater's rim; and in
1831 this cavity was completely filled, and the lava-streams
began to flow over it down the outer cone. In the winter of
that year a violent eruption once more emptied the bowels of
the mountain, and left a new crater, which soon began to fill
again from ejections upon its floor; and by the month of
August 1834 this crater had been in its turn obliterated, and

lava overflowed its edge towards Ottaiano.   In 1839 the cone
was again cleared out, and a new crater appeared in the shape

Fig. 41.

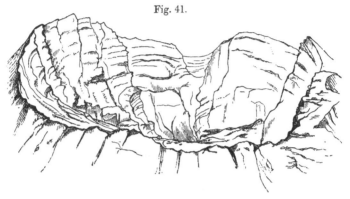

Crater of Vesuvius, a mile wide, after the eruption of October 1822.

of a vast funnel, accessible to its bottom, which for a few
years then remained in a tranquil state.   In 1841, however,

Fig. 42.

Interior of the crater of Vesuvius in 1843.

a small cone began to form within it; a little later, two, and
then three such cones were in activity together there, surrounded by a pool of lava (fig. 42).   These erupted matters
accumulated so rapidly, that in 1845 the top of the interior

cone was visible from Naples above the brim of the great crater, which soon after was completely filled. And the principal cone from that time went on increasing in bulk and height from the effect of minor eruptions, until in 1850 one of a violently explosive character opened two deep craters on its summit, of which I have already spoken. The more recent eruption of May 1855, being confined chiefly to a prodigious efflux of lava from the outer side of the cone, unaccompanied by any extraordinary explosive bursts from the summit, has not altered materially the form impressed upon it in 1850. The two craters are, however, now, or were lately (in 1860), represented by two cones, each with but a slight depression on its summit. And a third has since been formed at a short distance towards the east.

It is thus seen that within the last 100 years the cone of Vesuvius has been five several times gutted by explosive eruptions of a paroxysmal character, viz. in 1794, 1822, 1831, 1839, and 1850 ; and its central craters formed in this manner as often gradually refilled with matter, to be again in due time blown into the air. Meanwhile what remains of the old external crater of Somma is itself becoming choked up by the accumulation of all the lava-streams and fragmentary matter that are expelled towards the northern and outer side of the cone, which is also growing in bulk all round. It would be, therefore, in exact accordance with the habit of this volcano (as of volcanic mountains in general), if at length the ' Atrio ' should be completely filled up, the two cones combined into one, of increased height and bulk, and ultimately, perhaps, the entire conjoint mountain should be blown up by a more than ordinarily violent paroxysm, and the crater of Somma re-formed.

The history of Etna, could we obtain as clear a record of it through a period of some centuries back, would, no doubt, present a similar series of changes. Indeed, it is impossible to look at its present outline from any side without recognizing the fact that the circular margin of the flattish platform upon which the upper cone rests marks the edge of an ancient crater of vast size, formed by the truncation of the mountain, when it rose in a cone of far greater height, by a paroxysmal eruption ; which crater, since its formation, has been filled up to the brink, and moreover loaded with the materials that compose the existing higher cone and enclose its present active mouth. (See figs. 43 & 44.)

Fig. 43.

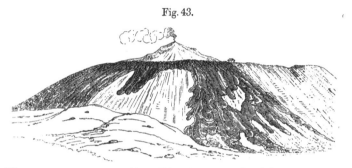

View of the truncated cone of Etna (Mongibello), coated with lava-streams, upon which rest the actual upper cone and crater.   (After S. de Waltershausen.)

Fig. 44.

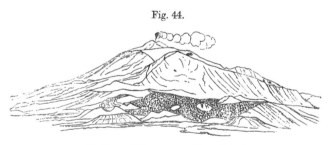

Outline of Etna, as seen from Catania.
(From Mém. Soc. Géol. de France, vol. iv. pl. 2.)

This modern crater itself offers a similar example on a smaller scale.   The process of refilling had already made considerable progress within it at the time I saw it in 1820, a small parasitical cone being then in full activity at its bottom, whence frequent jets of scoriæ tended materially to lessen the depth of the cavity. Since that time it has been more than once reopened and refilled by the explosions of successive eruptions, some of which were of great violence. And the cone itself is meantime growing in bulk, from both inward accretion and outward accumulation.

§ 3. It is evident then, that, besides the operation of time and external forces in damaging or totally destroying the figure of craters, there exists in the volcanic phenomena themselves a tendency to obliterate their forms—a series of causes alternately scooping out the interior of a volcanic mountain, and gradually filling up again the cavity thus produced. Hundreds of eruptions may take place from its bottom before such a great central crater has been filled up to the brim. When, however, this has been effected, the summit of the mountain exhibits a plain, perhaps rising in the middle to a dome, or with one or more minor cones upon it, as in the instances just described of Etna and Vesuvius.

It is the truncation of a volcanic cone, by the blowing off of its apex and the excavation of a crater, which gives to mountains of this character the saddle-shaped or horned profile they so generally exhibit when seen from a distance (see fig. 45) ; and,

Fig. 45.

Characteristic profiles of volcanic cones.

for the reasons given above, it is always uncertain whether, on reaching the summit of such an eminence, a crater may be found, more or less deep, with a minor cone perhaps at the bottom, or a plain, studded with smaller eruptive cones. These latter are similarly truncated, but in their case the apex has never been completed.

According to Professor Junghuhn, a singular external figure

is exhibited by many of the numerous volcanic mountains of
Java, where the erosive action of the tropical rains, to which
their conical surfaces (composed partly of mud-ejections,
partly of loose pumice-ash) are exposed, has scored them all
round with ravines and alternate ridges, radiating with extreme regularity from the central heights like the sticks of an
umbrella. When the mountain has no crater, or only a small
one, these ravines, in the manner of all rain-formed gutters,
widen and deepen downwards, and run out to nothing towards
the summit of the mountain. But when this summit has been
blown off by a paroxysmal eruption, and a great central crater
has taken its place, the upper extremities of the ravines, where
they are cut off by the explosions, form so many notches of
wonderful regularity in the rim of the crater. Though not
equally regular, gaps having this origin will be found in the
encircling rim of many or most of such great craters. They
are to be seen in the crater-ring of Teneriffe, of Palma, of Etna
(the Val del Bove *), &c. The deepest of them will often have
become the drainage-channel of the interior cavity, and been
consequently much enlarged in its dimensions.

§ 4. The eruptions of a volcanic mountain, or habitual
vent of volcanic energy, taking place at intervals from the
same point, the craters will often be formed concentrically one
within the other. The successive craters of Vesuvius have
been always exactly concentric with one another and with the
external semi-crater of Somma. (See the vignette in p. 183.)
Some deviation from this exactness may be expected to take
place at times, through the effectual sealing up of the earlier
orifice, and the opening of a later one upon another point of
the eruptive fissure. Thus in Volcano, one of the Lipari Isles,
a cone with a crater rises within the circuit of a larger crater,
but not quite concentrically. A still later eruption has de-
viated a little from that vent, and thrown up another cone
with its crater (Volcanello) in close proximity to the base of

* See Lyell's Memoir on Mount Etna, Phil. Trans. 1858, p. 63.

the preceding one, but a little further away on the same line. (See figs. 46 & 47.)

Fig. 46.

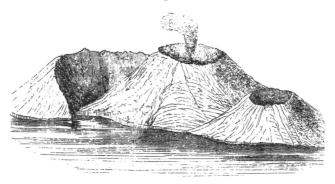

View of Volcano and Volcanello, Lipari Isles.

Fig. 47.

Plan of Volcano and Volcanello.

I shall presently adduce other examples of this very common and characteristic feature of habitually eruptive volcanos.

Geologists have sometimes seemed to overlook the fact, that the changes I have described above are of necessity continually occurring in the figure of such mountains, and have expected to find in every one a central crater, whereas the obliteration of these hollows by subsequent eruptions is as regular a result of the normal phenomena as their formation. Craters also have been spoken of as if they were always connected with lava-streams; these, however, are often emitted from points at some distance from the orifice whence the explosions proceed which produce the crater.

It should also be observed, that the explosions by which a crater is hollowed out from the centre of a volcanic mountain are often immediately *preceded* by the flow of lava in a liquid form from its summit. This consideration should have guarded Dolomieu from supposing that the large crater of Volcano (one of the Lipari Isles) must have been filled to the brim with liquid glass before the small current of obsidian which flowed from its summit in 1786 could have been produced. The emission of this lava was here, as in every similar case in which the abrupt edge of a lava-stream overhangs a crater, antecedent to the excavation of the cavity. As one example I may instance the Solfatara of Pozzuoli, which crater was formed, of course, *after* the emission of the lava-current of Olibano, and of a more recent and smaller current, which, with the crater, probably dates from the year 1198, when an eruption is recorded from this spot.

§ 5. The dimensions of the hollow formed in the manner above described will be determined by the violence and duration of the eruption, and the more or less yielding nature of the walls of the eruptive fissure. Probably the first two conditions will depend much on the depth and temperature of the focus whence the effort proceeds. When these circumstances are favourable to an extreme result, flashing explosions of vapour rapidly enlarge the crevice that communicates with the intensely heated lava below, and not only break up and drive into the air the matters which have during a long previous period accumulated in the old crater and obstructed this habitual vent of the volcano, but even tear their way through, and destroy, the greater part of the volcanic mountain itself, leaving in its place a wide cavity encircled by the ruins of the shattered cone. Examples of such an occurrence have been observed in recent times. Such a catastrophe destroyed, in the year 1638, a colossal cone called the Peak, in the Isle of Timor, one of the Moluccas. The whole mountain, which was before this continually active, and so high that its light was visible, it is said, three hundred miles off, was blown up and replaced by a concavity now containing a lake.

The Isle of Sorca, another of the Moluccas, disappeared entirely in a similar manner during a violent paroxysmal eruption in 1693.

Again, according to M. Moreau de Jonnes, in 1718, on the 6–7 March, at St. Vincent's, one of the Leeward Isles, the shock of a terrific earthquake was felt, and clouds of ashes were driven into the air with violent detonations from a mountain situated at the eastern end of the island. When the eruption had ceased, it was found that the whole mountain had *disappeared*. A hurricane accompanied, and possibly may have been one of the concurrent determining causes of this catastrophe.

In the figure and structure of many volcanos, active or extinct, we may observe proofs of such a paroxysm having not unfrequently taken place. Such I believe to be the origin of all those circular or elliptical cliff-ranges which are so often observed around an active or recently active volcano. They are generally breached at one or more points—sometimes indeed reduced, by time and the action of aqueous or other degrading causes, to mere fragments of the original circuit— the roots or 'basal wreck' of the exploded mountain.

Explosive eruptions of this extreme violence are naturally to be expected to characterize the return of a volcano to a condition of external activity after a long period of apparent repose occasioned by the sustained predominance of the repressive forces. Such an outburst seems to have taken place

in the great eruption of Vesuvius, A.D. 79, by which half of
the old cone of the pre-existing mountain was blown into the
air, and buried Herculaneum, Pompeii, and Stabiæ under its
triturated ruins. The remaining half of the cone exists in the
Monte Somma; and there can be little doubt that the modern
cone of Vesuvius is of a date posterior to that eruption. (See
fig. 48.)

Fig. 48.

Vesuvius half-encircled by the crater-cliffs of Somma, as seen from Sorrento.

The still active volcano of the Isle of Bourbon presents a
remarkable analogy of situation to Vesuvius, rising in a similar
manner to the height of 7500 feet, from the centre of a vast
semicircular enclosure formed by precipitous cliffs, which
consist, like those of Somma, of alternate beds of greystone
lava and conglomerate, whose inclination and direction prove

Fig. 49.

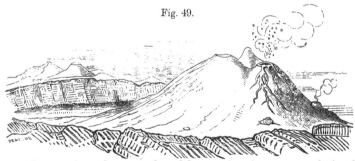

them to have been produced by a volcanic vent, nearly in the
same position as the present one. (See figs. 49 & 50.) In this

case, as well as that of Vesuvius, the appearances are only to be explained by supposing half of the cone of a vast ancient volcano existing on the spot to have been blown up by a paroxysmal eruption of the nature of those described above. This outbreak seems to have been followed by the lengthened phase of moderate and prolonged activity which has characterized the volcano from the period at which the island was first colonized up to the present time.

Fig. 50.

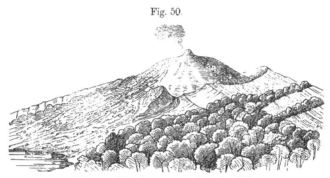

Volcano of Bourbon, as seen from the N.E.

Fig. 51.

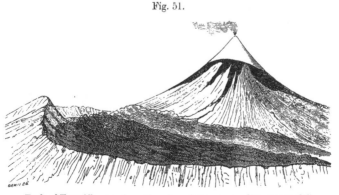

Peak of Teneriffe, seen from the edge of the surrounding crater-cliffs.

Many other volcanos present similar features. In the island of Teneriffe, a semicircular wall of precipitous rocks,

2000 feet in height on some points, and eight miles in its longest diameter, encloses within its limits both the cone of the Peak and that of Chahorra, almost the rival of the former in magnitude, and from which alone eruptions have been known to occur since the island has been inhabited. This

Fig. 52.

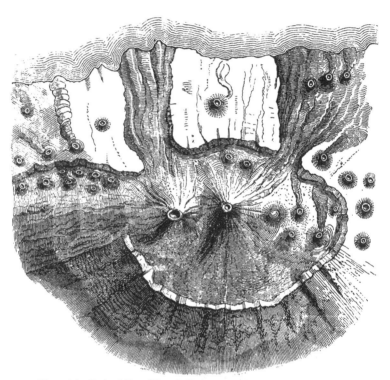

Plan of the Peak of Teneriffe and Chahorra, with the encircling cliff-range.
(From Professor Piazzi Smyth.)

enclosure is evidently a vast crater, formed at some distant epoch, when a volcanic mountain, probably exceeding the Peak in height, and the two existing cones together in bulk, was blown into the air by an intense eruptive paroxysm. (See fig. 52.)

The volcano of Pichincha in Mexico has a double elliptical crater of this character, spoken of by Humboldt as 1400 yards in its shortest diameter, and between three and four thousand feet in depth. Its inner sides consist of tremendous precipices. Many parasitical cones are observed at its bottom, one of which at least is usually in activity, the red-hot scoriæ projected from it being visible at night. This crater was probably produced by the great eruption of 1660, which is said to have launched into the air blocks of twelve feet in diameter, that fell at a distance of eighteen miles (!), and which involved Quito in thick darkness by the falling ashes for days together.

The crater of Bromo in Java is described by Professor Jukes as encircled by a precipitous wall a thousand feet in height. It is four or five miles *in diameter*. From the centre rises a conical mound six or eight hundred feet high, deeply furrowed on all sides, and having a number of subordinate cones and craters, as it were, growing out of it. One of these was belching forth much smoke and steam with a rumbling noise at the time of the Professor's visit *. Many other of the Javanese volcanos, according to Professor Junghuhn, rise "within circular old crater-walls" of equally vast dimensions.

Barren Island, in the Bay of Bengal, east of the Andaman Isles, is another example (fig. 53).

Fig. 53.

Barren Island, in the Bay of Bengal.

This permanently active volcano is a cone about 4000 feet high, rising in the centre of a circular cliff-range, which entirely surrounds it except at one point where the sea has

* Jukes's Manual, p. 291.

broken in. The great crater, of which we have not the exact measurement, must be certainly several miles in diameter, and was, no doubt, produced by a paroxysmal eruption blowing off the summit of a colossal cone.

The Pic de Fogo, one of the Cape Verde Isles, a permanently active volcano, rises, like those of Barren Island and Bourbon described above, 7000 feet above the sea, from within a semicircular crater-range of basaltic rocks 3000 to 5000 feet in height, and has numerous parasitic cinder-cones upon its flanks, two of which were thrown up by eruptions in 1785 and 1799. Many other examples might be quoted, such as Antuco in Chili, and Irasu in Costa Rica, of volcanic cones still in activity, that rise from the centre of external crater-rings of vast dimensions.

Among the remains of extinct, or perhaps only dormant volcanos, such instances are still more numerous, and in many cases equally striking for the magnitude of the area of the encircling crater-walls. Santorini, with its adjoining islets, in the Grecian Archipelago, is often adduced as one of these. I give a section of the group of Santorini (fig. 54), passing in a N.E. and S.W. direction from Thera through the Kaimenis to Aspronisi. The dotted line is intended to represent the

Fig. 54.

Aspronisi.   Thera.

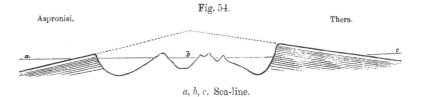

a, b, c. Sea-line.

probable outline of the volcano before the paroxysmal eruption to which I ascribe the production of the crater-basin. The Kaimeni Islands, which now rise from its centre, are the product of recent eruptions (two of them of known dates, 1707 and 1753), and have craters. The three others are partially submerged cones of scoriæ and lava, also of recent origin.

All rise upon the same line of fissure, and correspond very closely, except in size, to the three cones formed within the crater of Vesuvius in 1843, of which a sketch has been given in fig. 42, p. 188.

The island of Nisyros, in the same Archipelago, has a central nearly circular crater, three miles in its longest diameter, its rim being from 2000 to 2300 feet above the sea, which must have also resulted from a paroxysmal series of explosions. The bottom is still in the state of a solfatara. The outer surface of the island is covered with deep beds of pumice-ash, no doubt the triturated fragments of the mountain top which once filled this gulf. The Caldera of Palma, upon which much has been written of late years, is a crater of this character, formed, I have no doubt, by a paroxysmal explosive eruption *; though Von Buch presented it as a type of his imaginary ‘elevation-craters,’ and the occurrence of a central core of trachytic lava-rocks beneath the principal mass of basaltic currents and their conglomerates is supposed by Dr. Daubeny to give some colour to this theory. Sir C. Lyell, in his elaborate and careful examination of this question, decided it against the upheavalists; but seems inclined to ascribe too large a share of the excavation of the Caldera, in my opinion, to the erosive agency of the sea-waves, which there is no reason to suppose ever entered its enclosure. The Barranco will be considered presently. The island of St. Miguel, another of the Canaries, possesses a ‘Caldera’ of equally colossal proportions, which was formed by a tremendous eruption in 1444, recorded by Cabral. The admirable maps of Captain Vidal, published by the Admiralty, give an excellent notion of the configuration of the several great craters of this group of volcanic islands. I have already noticed that of Teneriffe.

The structure of the island of Madeira resembles that of Palma, having also a central circular range of peaks, from

* See Sir C. Lyell's Manual, edit. 1855. p. 499.

which alternating beds of volcanic conglomerate and lava-rock slope on every side with a gentle inclination towards the sea. The interior basin, called the Curral, is unquestionably an ancient crater, of vast horizontal area, but of no great depth, having seemingly been almost filled up since its forma-tion by the products of later eruptions, as many recent-look-ing cones and modern lavas are found within it *.

The island of St. Helena is described by Mr. Darwin as a trachytic volcano encircled by a broken ring of basalt, the area of which measures eight miles by four. The internal cliff-faces are nearly perpendicular, except that they have in some places a flat projecting shelf or ledge cut down in par-allel curves. The outer sides (as in all similar examples) have a moderate slope, to which their component beds of conglo-merate and lava-rock accurately conform.

The island of the Mauritius has a similar great encircling crater-ring of an oval figure, measuring no less than thirteen miles in its shortest diameter! Mr. Darwin describes that of St. Jago, one of the Cape de Verde Isles, as almost iden-tical with this in form, structure, and composition. In both, he says, "the mountains composing the external ring appear to have originally formed parts of one continuous mass. . . . . . At both, vast streams of more recent basaltic lava have flowed from the internal basin through openings in the surrounding hills : at both, recent cones of eruption are scattered round the circumference of the island,—none of them, however, known to have been eruptive within the historical period. . . . . . Both," he says, "resemble the basal and disturbed remnants of two gigantic volcanos, and owe their present form, struc-ture, and position to the action of similar causes †."

§ 6. The horizontal dimensions of some of these ancient external crater-rings are so vast—attaining, as has been seen, five, ten, or even more miles in diameter—as to suggest doubts in many minds of the possibility of their having originated in

* See Lyell's Manual, 1855, p. 518.   † Volcanic Islands, p. 31.

the manner described above, namely by outbursts of vapour exploding continuously for more or less time from a body of subterranean lava in ebullition; since in each such case, however large the area of the crateriform cavity, we must suppose the dimensions of the spherical volumes of vapour, to whose violent lateral expansion as they rose upward, or, rather, as they exploded from the lava-surface, the excavation of these areas is attributed, to have been equally enormous.

If, however, we reflect on the circumstances under which these volumes of vapour were developed,—perhaps at a considerable depth below the mountain, in the interior of a mass of molten rock whose temperature probably far exceeded that of white heat, and under an almost incalculable amount of pressure,—it seems difficult to imagine any limit to their tension, and consequently to the explosive force with which, on reaching the open vent of the volcano, they will have flashed into freedom. The successive explosions of quantities of gunpowder, or the flashes of steam from the mouth of a Perkins's steam-cannon (multiplied many-fold), convey probably but a very inadequate idea of their tremendous force.

Moreover it seems impossible to draw any line of separation between moderate-sized craters (say, for instance, of a mile in diameter, like that of Vesuvius formed in 1822, the origin of which is incontestably that here described, since it was so formed before my eyes, and those of MM. Monticelli, Covelli, and many other competent witnesses) and craters of greater magnitude—five, ten, or even twenty times as large—whose formation in this manner may at first appear problematical, since there are numerous examples of every intermediate size, undistinguishable from the first in form, structure, composition, and every other character except horizontal area.

Again, in the case of concentric craters, the correspondence of the smaller internal with the larger external annular one, as well in figure as in the disposition and structure of the

environing rocks, is usually so complete as to prove conclusively their identity of origin.   When, immediately after
the so often mentioned Vesuvian eruption of 1822, I stood
on the acute ridge of the prodigious crater that had been
drilled through the solid axis of the cone by the gaseous explosions of the previous twenty days, and marked the exact
resemblance of its internal cliff-sections in structure, in composition, and in sweeping curve, to those of the half-encircling
crater of Somma (the Atrio), which were within my view at
the same moment, I found it impossible to doubt that both
the inner and outer concentric craters (no less than the respective cones) owed their origin to similar developments of
eruptive violence, notwithstanding their different dimensions,
the one being perhaps four times the diameter of the other.

But, in fact, the prodigious amount of fragmentary matter
which the paroxysmal eruptions of some volcanos have ejected
in known instances within a few years past—such as those
of Tomboro and Coseguina, already mentioned (p. 170)—an
amount of matter which has been calculated in either case as
equalling three or four times the bulk of Mont Blanc—would
alone lead us to expect the production of cavities of corresponding magnitude in the mountains eviscerated by those
terrific and long-continued explosive discharges.   I have elsewhere * remarked upon the comparatively small quantity of
fragmentary matter ejected by the eruption of Vesuvius in
1822, which only spread a coating of from a few feet to a few
inches in thickness over a circle of four or five miles radius,
and nevertheless left a crater above a mile in diameter, and
more than a thousand—Professor Forbes estimated it at two
thousand—feet in depth.   The infinitely more abundant ejections of such paroxysmal outbursts as those referred to above,
and many other well-authenticated instances †, must neces-

---

* "On Cones and Craters," Quart. Journ. Geol. Soc. 1856, p. 337.

† Take, *inter alia*, the following.   In February 1600, the volcano of
Guayta-Putina, near Arequipa in Peru, threw out during twenty con-

sarily have left, by their abstraction from the heart of the volcano, cavities exceeding that of the Vesuvian eruption in direct proportion to their superior aggregate bulk.   Reckoning then by this scale, it is not unreasonable to suppose that craters of ten or fifteen miles in diameter may have resulted from those tremendous paroxysmal eruptions.   The question, indeed, resolves itself into one of arithmetic.   A very simple calculation demonstrates that a mass of fragmentary matter spread to a depth *everywhere exceeding* ten feet through a circle of twenty-five miles radius, and visibly scattered beyond this to distances of six or seven hundred miles (the facts reported of the eruption of Coseguina in 1835), will amply account for a void left in the volcanic mountain whence all this matter was erupted, exceeding in its dimensions any known crater-basin, and equal to the bulk required to build up within the largest, a conical mountain covering its whole area.

In all the narratives reported of such extraordinary eruptions, we hear indeed of the disappearance of the entire mountain, and its replacement by a hollow basin or lake many miles in diameter.   But we hear at the same time of the dispersion over enormous areas of a proportionately vast amount of fragmentary matter.   I have therefore no hesitation in expressing my conviction that, as in these cases, so also the " external rings " of Santorini, St. Jago, St. Helena, the Cirque of Teneriffe, the Curral of Madeira, the cliff-range that surrounds the volcano of Bourbon, and others of similar form and structure, however wide the area they enclose, are truly the 'basal wrecks' of volcanic mountains that have been blown into the air, each by some eruption of peculiar paroxysmal violence and persistence, and that the circular or elliptical

tinuous days such a quantity of stones, sand, and ashes, as covered the surrounding country to a distance of 90 miles on one side and 120 on the other!   Crops were buried, trees broken down, cattle destroyed, and the roofs of houses broken in by the mass of these ejections throughout this vast area. (A. Perrey, Documens sur les Tremblemens de Terre au Pérou, &c. 1860.)

basin which they wholly or in part surround is a true crater
of eruption.

It must, however, be borne in mind that the process does
not consist in *one*, but in a multitude of repeated explosions.
Some geologists consider these vast cavities to be caused by
the engulfment (*effondrement*) or subsidence of the mountain
at the time of its eruption. Even Sir C. Lyell and Mr. Darwin
appear to countenance this notion, which, however, would
seem to be based upon the "upheaval" fallacy, by which
every volcanic mountain is considered to be merely a hollow
and thin arched crust, which a single explosion is sufficient to
burst like a bladder, its fragments collapsing into the cavity
beneath. That idea is wholly inconsistent with the main
facts of a paroxysmal eruption as above described—the
emission of lava at first from the summit of the volcano—the
usually long continuance of the explosive discharges, and
the prodigious abundance of ejected fragmentary matters
which on all such occasions are observed to be spread over
vast circumjacent areas. It is quite a mistaken view of the
true character of a normal volcanic eruption to compare it (as
M. de Humboldt does) to the single shock caused by the
explosion of a mine, or the bursting of a steam-boiler. I am
not aware of any recorded example of such a *single* shock,
followed by the engulfment of the shattered rocks and immediate quiescence. In all the cases usually quoted in support
of the 'engulfment' theory, as those of Timor in 1638, of
Papandayang in 1772, and Galongoon in 1822, the explosive
discharges are reported to have lasted for *months*, and the
materials of the mountain-tops were *by degrees* blown *outwards*, and spread over the adjoining regions in prodigious
quantities to vast distances. This is, I believe, always the
case. The explosions, having once begun, are *continuous*, for
days, weeks, months, or even occasionally for years,—evidently
proceeding, as has been shown, from a mass of subterranean lava in a state of ebullition, which, having once forced

a communication with the open air, at the weakest point of some fissure broken by its expansive efforts through the over-lying rocks, *blows itself out* through this opening by degrees, although with terrific violence; just as would the boiler of a high-pressure steam-engine, of enormous dimensions and infinite lateral strength, when the valve of the steam-pipe, or an accidental aperture was opened—and *not* after the manner of a *boiler bursting, and discharging all its steam* at once, or of *an exploding mine of gunpowder.*

It is not by one outburst, but by the continued repetition of multiplied *eructations,* caused by the successive development, upward rush, and outward explosive discharge of innumerable vapour-bubbles of prodigious elastic tension, that an eruption is characterized, and it is by their continued action that the more or less solid mass of rock obstructing their escape is—not at one shock, but by degrees, and as the surface of the lava sinks within the vent—broken up and ejected ; many of the fragments falling back repeatedly into the cavity, and being re-ejected, until they are for the most part ground by friction into lapillo (*i.e.* small globular or bouldered scoriæ), or even to the finest powder, which the winds carry to vast distances. This *gradual* process it is that, as it were, eviscerates the mountain, leaving at the close of the eruption, when the ebullition has spent its force, that circular or oval chasm sur-rounded by a ring of steep sloping sides, or precipitous cliffs, which is the well-known form of the larger volcanic craters, and which will be generally of a size proportioned to the violence and duration of the eruption, and to the amount of fragmentary matter thrown out by it and spread over the surrounding slopes, or the adjoining surfaces of sea or land.

The idea, indeed, of the foundering of the summit of a volcanic mountain, by the subsidence into some vast empty gulf beneath, of what was but a hollow crust or roof, is opposed to the characteristic phenomena of normal volcanic eruptions, which are inconsistent with the existence of any vast internal

void immediately beneath such a mountain. Did any such
exist, how is it possible that currents of liquid lava could be
propelled in abundance from the summit, or a high point on
its flank? The gulf which is supposed to swallow up the
mountain would, of course, retain within its bosom all this
fluid matter. The phenomena of eruption seem to attest, on
the contrary, an overflowing abundance and excess of matter
there, solid, no less than fluid and gaseous, that struggles to
find a vent, and having obtained one, the mountain continues,
through a more or less lengthened period of eruption, to discharge the redundant contents of its interior until the plethora
is reduced, and the forces of repression, consisting in the
weight and tenacity of the overlying mass (together with the
weight of the atmosphere, which should not be overlooked),
recover their ascendency and stop any further evacuation. This
is a state of things entirely inconsistent with the existence of
a great internal void capable of swallowing up at one shock
the upper moiety or two-thirds of the mountain.

Certain exceptional cases, in which the process was more
rapid, and subsidence therefore may have played a part, will
be considered presently.

§ 7. It is not intended, of course, to deny that the internal
area of many crater-rings has been enlarged since their original formation by other forces than explosive eruption—especially by denudation. This, indeed, has frequently happened
to such as have been exposed for a long time—perhaps for
countless ages—to the abrading action of the sea-waves and
currents. Under such circumstances, some have probably
been entirely obliterated; some worn down into fractional
segments of the ring they once formed. Santorini is one example. Others may be seen in the Lipari Isles, in the coast
of Naples between the Isle of Procida and Misenum, in the
Ponza Isles, and, indeed, in every cluster of volcanic islands
and insular rocks.

A very common form assumed by such fragmentary segments of a volcanic cone as have been long exposed to the denuding action of the sea, is that of the Isle of Ventotiene, in the Mediterranean, near Naples (see fig. 55), where a thick

Fig. 55.

Island of Ventotiene, N. of Ischia. A mass of greystone, with stratified tuff above it.

bed of trachytic lava, underlying strata of fragmentary ejections, has, through its superior solidity, been preserved above the level of the waves, while the remainder of the subaërial cone has been washed away. A disciple of the elevation-crater theory, and indeed any casual observer, would be apt to conclude, from the mere outline of such a mass, that the lava-bed had been upheaved in bulk, and had carried up the conglomerate strata to their present angle of inclination: whereas I believe that in this, and innumerable other analogous instances to be seen wherever waves beat against a volcanic coast or island, both the lava and the overlying stratified beds are in the position in which the first originally flowed, and the latter were spread, upon the outer slope of a volcanic cone, whether submarine or subaërial. Mr. Darwin, in his description of St. Helena, justly remarks on the vast amount of degradation which many insular volcanic mountains have evidently suffered from the action of the sea-waves. Portions, he says, " of the basaltic ring whose segments still nearly encircle that island have been wholly removed through areas of several square miles, leaving precipitous cliffs on either side from one to two thousand feet in height. There are also ledges and banks of rocks, rising out of profoundly deep water, and distant from the present coast between three and four miles, which can be traced to the shore, and are found to be the continuations of well-known great dykes. The mind recoils from an attempt to grasp the number of centuries of exposure to the Atlantic waves necessary to have ground into mud and dispersed the enormous cubic mass of hard rock which has been pared off the circumference of this island." He then contrasts with these evidences of external waste in St. Helena—confirmed by the equally worn aspect and relief of its internal surfaces, denuded dykes, pyramidal cones of clinkstone, and no traceable streams of lava or scoriæ-cones—the comparative freshness of the next island, that of Ascension, where "the surfaces of the lava-streams are glossy, as if just poured out, their boundaries well defined, and traceable to perfect craters, no dykes perceivable, and the circumference of the island low, and scarcely at all eaten into by the waves." Yet Ascension has not been in eruption since its first discovery, three or four centuries back! Similar contrasts are often to be found within the limits of the same island, where an old eruptive vent has been deserted for one removed to some distance; for example, in the Isle of Bourbon, two-thirds of which are occupied by the remains of an early volcanic mountain long since extinct and deeply eaten into by aqueous erosion, the remaining third consisting of the products of the still active volcano.

The lesson taught by these observations is worth the notice of geologists engaged in the study of the relations between the earlier trap-rocks and the sedimentary strata which they penetrate or overlie. Many a massive dyke, or bed of such rock, may have been the remnant of a once extensive and bulky volcanic island, worn down to a mere stump, or to a few segments of its hardest rocks, by the secondary or tertiary seas which afterwards enveloped it in their sedimentary deposits.

§ 8. Much change likewise must have been often occasioned

in the encircling walls of a crater by earthquake-shocks toppling down its precipitous cliffs, partially filling up the cavity
with taluses, and, in conjunction with atmospheric and aqueous
erosion, smoothing down the superficial outlines of the external ranges, or cleaving a passage through them for the exit
of rain-torrents or lake-waters that may have collected in
their basins.

The last-mentioned influence is worth dwelling upon, as it
affords the explanation of a not unfrequent feature in the
larger and more perfect crater-basins, namely, their exhibition
of a principal breach or opening on one side, through which
their drainage is usually effected. Of such lateral breaches
the great Baranco opening into the crater of the Isle of Palma
(Canaries) is an example often referred to. The Val del
Bove, on the flank of Etna, the Cirque of Teneriffe (see p. 198),
the Curral of Madeira, the cliff-range of Bourbon, and many
others that might be mentioned, are instances of vast craters
breached in the same way on one or more sides. In nearly all,
I believe, the direction of the principal breach corresponds with
that of the longer axis of the crater's horizontal figure—in
other words, with the fundamental fissure by whose local enlargement it was presumably formed. The widening of this
fissure through a considerable length upon the occurrence
of the paroxysmal eruption to which the crater is owing, is in
itself a highly probable circumstance—in complete conformity
with the statements already brought forward, of the visible
production of some great cleft through the flank of a volcanic
mountain on such occasions. Should a chasm of this kind
remain open for a lengthened term—that is to say, not blocked
up by lavas or tuff-deposits from subsequent eruptions—giving
outward passage to the rain-torrents that fall or collect in
the interior of the basin, especially in tropical countries, or
to the deluges caused by the melting of snow or ice in
colder regions or on lofty volcanic mountains, it is inevitable
that the opening must be greatly enlarged by the under-

mining and washing away of its banks. In some of the instances mentioned above, there is patent proof of the vast scale on which denudation has thus cooperated to excavate the gorge which drains the great crater. At the seaward opening of the Val del Bove, enormous accumulations of alluvial volcanic matter have overspread the plain *. In the Baranco of Palma similar eluvial conglomerates reach a thickness of 800 feet, and " bear testimony," in Sir C. Lyell's opinion, " to the removal of a prodigious amount of materials from the Caldera by the action of water †." They are interbedded with lava-streams, probably the result of eruptions from the bottom of the Caldera subsequent to its formation. Certain remnants of basaltic lava-beds and conglomerate strata within the Caldera itself are also, perhaps, due to these eruptions.

In Teneriffe, one of the Barancos opening out of the great crater, through the Valley of Taoro, is said by Professor P. Smyth to have been enlarged to three times its previous width in the space of a few hours by a single debacle proceeding from a violent storm of rain, or water-spout, that fell upon the mountain on the 6th November, 1829. It undermined the cliffs of hard lava-rock and conglomerate on either side to a great depth, and spread their ruins over a wide devastated area of the plain beyond, carrying much also into the sea ‡.

In the Mont Dore, the Valley of Chambon is a 'Baranco' of similar character; and the immense masses of conglomerate still remaining at its opening into the main valley of the Allier, at Nechers, Pardines, and Issoire, attest the vast force of aqueous denudation by which it was excavated. We must bear in mind how peculiarly liable to the erosive action of torrents are the loose and fragmentary strata of which vol-

---

* See Sir C. Lyell's Map of Etna (Phil. Trans. 1859, pl. xlix.), copied from the Atlas of S. v. Waltershausen.

† Manual, 1855, p. 508.   ‡ Piazzi Smyth's Teneriffe, 1859.

canic mountains are chiefly composed, while even the harder
lava-beds that alternate with these are full of vertical joints,
causing them to be easily displaced by frost and undermining
currents of water *.

The description given by Dr. Lauder Lindsay of the chief
features of an eruption which took place in May of last year
(1860) from the volcano of Kotlugja in Iceland may assist us
to form some faint conception of the vast changes occasionally brought about in the physical configuration of a large
tract of country by such occurrences within the course of
a very brief time. The eruption was preceded, as usual, by
local earthquakes. Then a dark columnar cloud of vapour
was seen to rise by day from the mountain, and by night balls
of fire (volcanic bombs) and red-hot scoriæ, to the height (it
would seem) of 24,000 feet—being seen from a distance of
180 miles. Deluges of water rushed from the heights, bearing
along whole fields of ice, and rocky fragments of every size,
some vomited by the volcano, but in great part torn from the
flanks of the mountain itself, and carried to the sea, there to
add considerably to the coast-line, after devastating the intervening country. In truth, no more powerful causes of superficial change on the face of the earth can be conceived than
such paroxysmal eruptions from snow-covered volcanos. For-
tunately they are but " few and far between," as well in time
as in space, or the surface of the globe would be uninhabitable.

* Sir C. Lyell at one time was inclined to ascribe, I think, an undue
amount of influence to the action of marine waves and currents in the
production of many craters, even going the length of giving them the
name of ' Craters of Denudation' (see his paper on the Structure of Volcanos, Quart. Journ. Geol. Soc. 1849). He has since, however, I believe, modified this view considerably, although perhaps still inclined to
refer the larger class of craters to subsidence, assisted by denudation,
rather than to explosive eruptions, which I consider in every case to be
their true *origin*;—denudation, of course, as admitted above, having in
certain cases enlarged their internal area, and indeed often (as in the
instance of Ventotiene given above) worn away the greater part of the
enclosing walls.

On the whole, I consider it unquestionable that, whatever modifications they may have subsequently undergone from the causes last referred to, the origin of all the larger craters, no less than of those of inferior magnitude, is to be sought in the outburst of aëriform explosions.

§ 9. *Crater-lakes.*—I make no exception even of such wide saucer-like hollows, surrounded by comparatively low elevations (always, however, composed of volcanic strata with a quaquaversal outward slope), as the lake-basins of Bracciano, Bolsena, Laach, and others that might be named.

With regard to those of smaller size, as, for example, the Lakes Ronciglione, Nemi, Albano, and others in the Roman States, there can be no reason for doubting their explosive origin any more than in the case of the neighbouring Monte Albano, where a central cone and crater (the Monte Cavo) rises within a larger external crater-ring, both enclosures being breached towards Rome, in which direction they have sent forth streams of lava, and both, of course, formed in the same manner. The crater-ring of Rocca Monfina, north of Naples, is, I believe, likewise of eruptive origin, although there, as in Astroni, a central boss of trachyte has led some to believe it an ' elevation crater.'

Whether the bottom of a crater should be occupied by a lake will very much depend, as has been already noted, upon the ash and other fragmentary matters of which its conglomerates are composed being of an aluminous nature, so as to form a paste capable of *holding water*. Hence such lakes chiefly occur among the volcanos that have erupted trachytic lavas in which felspar predominates. For example, the crater-lakes of the Phlegræan fields, Agnano, and Averno are formed in pumice-tuff. Most of the small lake-craters, or Maars, of the Eifel, which emitted generally basaltic lavas, were produced by eruptions that broke through strata of shale belonging to the carboniferous period, which, when shattered into dust, forms a clayey lining to the cavity equally impermeable

to water.   In the volcanic country surrounding Auckland, in
New Zealand, there occur, together with cinder-cones of the
ordinary character, numerous small circular crater-lakes,
which, like those of the Phlegræan fields, owe their origin to
eruptions on the shore of the sea, but through beds of tertiary
marl; consequently the ejected ashes, being tumultuously
mixed up with calcareous mud and water, were likely to
form a retentive clayey paste.   These New Zealand craters
have generally one or more minor cones within them, the
product of explosions subsequent to their formation.   Each
has also usually given issue to a current of greystone lava*.
They vary in diameter from a mile or more to only a few
yards.   Their marginal heights are often very low, rising but
little above the plain around, though the surface of the water
within the basins is much lower, as may be seen in the section
of their usual form given by Mr. Heaphy (fig. 56).

Fig. 56.

The low subconical banks that surround these and similar
small crater-lakes consist, of course, of the fragmentary
matters ejected by the explosive eruption which produced the
cavity.

In the case of some of the larger saucer-like basins, such as
Laach, Bracciano, and Bolsena, which measure from three to
eight miles in diameter, it is probable that a volcanic hill,
composed of the matters ejected by preceding eruptions,
existed there and was broken up when the basin was formed,
and its skirts covered by their fragments.   These, being of a
light and loose texture (pumice or felspathic ash), would be

* Heaphy, Quart. Journ. Geol. Soc. 1860.   Mr. Heaphy's admirable
drawings of the New Zealand volcanos, in the possession of the Geological
Society, are well worthy of examination.

likely to form a series of curved ridges, of no great height or steepness, which time and meteoric denudation will have still further smoothed down; and such is the general character of those basins.

§ 10. Where, however, the crater-lake has been hollowed out of granite, basalt, or other hard and massive rock, the encircling banks are mostly abrupt and precipitous; and the sections they exhibit of the rocks through which the cavity has been drilled show few or no signs of disturbance. Of this character are the Lacs Paven, Guéry, Servières, Bouchet, Tavana, St. Front and others in Auvergne. The lake of Gustavila in Mexico, described by Humboldt, and of which he gives an engraving (fig. 57), is apparently of the same kind, as well

Fig. 57.

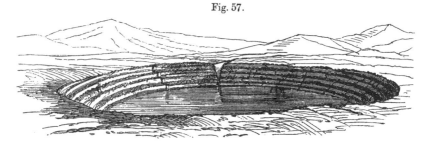

Lake of Gustavila (Mexico).   (From Humboldt's 'Vues des Cordillères.')
(N.B. The terraces round the interior slopes are artificial.)

as those "numerous small lakes of volcanic origin" in Central America, described by Mr. Squier.

One of the chief characteristics of this remarkable class of crateriform hollows (*pit-craters* they have been called) is that, however deep their interior may be, their borders often rise but little—in some cases scarcely at all—above the level of the surrounding country. Nevertheless that repeated explosive eructations attended their formation is proved beyond question by the fact that beds of scoriæ, volcanic bombs and lapillo, or of pumice and ash (matter which could only have been thrown up by explosive bursts from a surface of liquid lava), are found in every instance (I believe without exception) scattered around or on one side of them; although the bulk of such ejecta appears frequently insufficient to account for the mass of matter which must once have filled the cavity. There would seem, therefore, in these cases reason to believe in the subsidence of the remainder into some void beneath.

It is not improbable that the powerful explosions, of evidently short duration, to which many of these pit-craters are clearly owing (as shown in the small amount of scoriæ and volcanic ash ejected), proceeded from some proportionately extensive cavity, some disk-shaped vapour-bubble, formed on the surface of a subterranean pool of highly liquid lava by the collection of volumes of steam supplied from yet lower depths, in the manner suggested in an earlier page (p. 43). The increased tension of the vapour, as it received further additions from beneath, or as its temperature was augmented, may be supposed to have brought on at length a sudden rupture of the overlying rocks, and the violent escape of the confined steam in one or a few tremendous belches from the opening so broken, followed by the subsidence into this gulf of the ruins of the shattered rocks.

We have seen (pp. 79–81 *supra*) that, on or near the surface of some extremely liquid lavas, bubbles of large size, many yards in diameter, are found. Dana saw such in vast numbers in Hawaii, and Darwin in the Galapagos Isles. The volumes of steam that inflated these bubbles rose, no doubt, from the bottom of the lava-bed, if not from the chimney itself of the volcano, and expanded as they rose, both from the pressure upon them diminishing, and by the union of many bubbles in their ascent. If then we suppose such volumes of vapour to have risen from still greater depths through the widest part of a fissure filled with lava of extreme liquidity and specific gravity, and communicating with a molten mass beneath—previous to the opening of any outward avenue of escape—it is reasonable to believe that they would collect

into a colossal bubble or blister on the surface of the elevated
lava.   Further, it is very conceivable that a body of highly
liquid lava so forced up through a deep fissure may sometimes
spread horizontally to a greater or less distance among the
rocks it penetrates—partly by mechanical force, partly, perhaps, by melting the more fusible beds with which it comes in
contact.   Evidence of such intrusions of lava between strata
is far from unfrequent wherever denudation has disclosed the
internal structure of volcanic tracts of all ages.   And where
this has happened, the volumes of vapour rising from beneath
may have sometimes collected on the surface of such a subterranean pool of lava in one or more immense flattened bubbles, like the air-bubbles that are often formed beneath ice.
Such bubbles or blisters would necessarily have an approxi-
mately circular outline.   Nor does it appear incredible that
they may, in extreme cases, have spread horizontally to the
dimensions of even some miles in diameter.   Should the
tension of the steam contained in such a disk-shaped cavity
increase by additions of vapour, or of heat, or both (as must
be the case unless some other vent is opened, on the assumption of continued increments of heat from beneath to the
lower lava-mass with which it communicates), the sum of the
forces of resistance above must be overcome at length ; and
should the lava-surface be at no great depth, the overlying rocks
may be broken up, throughout a wide, more or less circular
superficial area, corresponding to that of the exploded bubble.
Their shattered fragments would probably subside into the
void thus opened, without any very violent outward explosions (much in the manner of the chalk cliffs blown up recently at Seaford), and only a few partial spirts of vapour and
ejected scoriæ may take place from the lava beneath through
their interstices, its ebullition being choked by these accumulated fragments after the contents of the cavity had escaped *.

* While these sheets are going through the press, I have received the
account of M. Daubrée's experiments " on the capillary infiltration of

Or, perhaps, the lava itself may be rapidly discharged through some other orifice affording an easier issue, opened by the shock of the explosion or simultaneously with it.

The latter supposition receives some support from an examination of a still rarer class of craters in which such a lateral escape of lava unquestionably plays a part; I mean those which are seen in numerous points of the Sandwich Isles, and of which Kilauea in Hawaii is the best-known and in all respects most striking example.

§ 11. In this instance, a mountain rising nearly 14,000 feet above the sea, of the shape of a flattened cone or dome (Mauna Loa), has been formed chiefly by the repeated outflows of a highly liquid lava boiling up and over the lips of a central vent at its summit. It is still frequently in eruption, and possesses at present a crater of considerable size. Another much larger crater (Kilauea) has at some time or other been broken through the flank of the mountain at the distance of sixteen miles, and at a level 10,000 feet lower. It is to the phenomena of this last crater that attention has been chiefly attracted, partly through its comparative facility of access, but principally, no doubt, from their very striking character. It is an immense chasm, of an irregularly elliptical figure, of varying depth, and some three miles in its longest diameter. Its internal walls are precipitous cliffs, for the most

water through porous rock-matter, although opposed by a strong counter-pressure of elastic vapour," in which he suggests that the gradual penetration of the surface-waters of the globe in this manner to subterranean cavities at no great depth, within bodies of heated lava, may explain many of the phenomena of volcanic action; and he especially refers to the crater-lakes of the Eifel, &c.

I will not here attempt to discuss the question of the source from which the aqueous vapour in lavas proceeds, which may be noticed in a future page; but I may here say, that M. Daubrée's ingenious suggestion only professes to account for the access of aqueous vapour to the *surface* of a subterranean body of lava,—not for that intimate permeation of its entire mass to extreme depths by steam, which the phenomena of volcanos indicate, I think, with certainty. (See p. 39, *supra.*)

part composed of horizontal beds of black rock; and at the bottom—at times 1000 feet or more below the rim—is usually seen a lake of more or less liquid and incandescent lava, but covered in great part with a hardened crust, through which in several places the molten lava may be seen to boil up like water in a spring, and moderate discharges of vapour take place, throwing up scoriæ, and forming as many small cinder-cones. This surface does not, however, long retain the same level, but sometimes rises to the brim of the crater; at others it sinks out of sight, the entire caldron being emptied of its contents by discharge through some fissure opened at a still lower level in the flank or near the base of the mountain. After a time, this duct being closed, the lava rises again within the crater. The subsidence often leaves one or more irregular shelves or 'black ledges' of solid rock within the outer crater-ring at different levels, marking the height to which the molten matter had previously reached. According to Professor Dana, this series of phenomena has been seen in operation several times within the last quarter of a century; the lava sometimes rising so high as to overflow the brink of the cavity, sometimes subsiding again to the depth of 1000 feet or more, when the internal reservoir has been 'tapped' by the opening of lateral issues at lower points on the slope of the mountain. The awful grandeur of the scene presented by the caldron of seething lava, when looked on by night, can scarcely be exceeded by the volcanic phenomena witnessed at any other spot on the globe; and from the moderate character of the explosive eruptions of scoriæ from this vent, the crater may be approached, its shelves of lava walked upon, and its phenomena inspected with very little personal risk.

The upper and central crater of the mountain (Loa), 10,000 feet higher, resembles that of Kilauea in every respect, except that its flows of lava burst from cracks near the outer edge of the hollow, or a short way down the outer slopes, and it does not appear to discharge its contents by a subterranean duct at any much lower level.   Indeed the fact of the column of lava maintaining itself at this extraordinary height proves the solidity and compactness of the framework of the mountain, which is not surprising, since its slopes rise at the low angle of from 4° to 8°.   The crater is 8000 feet in diameter, circular, and from 500 to 800 feet in depth.   Like the pool of Kilauea, it is encircled by a shelf or terrace of hardened lava within a wider crater of irregularly elliptical figure.   The cliff-walls that bound both the outer and inner crater are vertical, and show horizontal beds like those of Kilauea.   An eruption took place from this crater in 1843, which did not in any manner affect the phenomena of Kilauea.   The lava continued to flow from near the summit for ten weeks.   Scoriæ do not seem to be thrown out from either of these craters on a large scale at present. In the year 1789, however, powerful ejections of this nature are described by the natives as having produced darkness at mid-day, and destroyed by their fall several men in an army which was then on its march across the country.   It is evident that such an eruption would account for the formation of either of these craters in the usual manner.   Once opened by aëriform explosions, we may suppose them to have been re-peatedly filled by the welling-up of lava from the vent at their bottom, and emptied again by a tapping through ducts at a lower level, in the manner above described.   The fissure which gave lateral issue to the lava of Kilauea in one of its latest eruptions (1840) was at first narrow and small.   It was gradually lengthened downwards and enlarged, the lava flowing success-ively from openings at a lower and yet lower level, as in the cases of Etnean eruptions already mentioned.   The extreme length of the fissure (which could be traced through the disturbance of the surface-rocks above) was twenty-five miles, and the lava-stream reached a distance of forty miles, covering fifteen square miles of surface, and averaging ten or twelve feet in depth.   Its discharge emptied the whole lower pit of the crater, calculated to hold 15,400,000,000 cubic feet of molten matter! *

These are stupendous phenomena, but yet quite in accordance with the normal laws of ordinary volcanic action which we have deduced from the more common circumstances of eruption. Professor Dana observed several thousand smaller pit-craters in the same island—now extinct—but probably produced in the same manner as the greater active craters. They are usually in the close vicinity of scoriæ-cones from 200 to 1000 feet in height; which proves that in each instance very considerable aëriform explosions occurred, and these probably produced the crater. Mount Rea has nine such cones on its summit, each 500 or 600 feet high. The peculiar character of the phenomena of the Sandwich Islands' volcanos seems to be due to the extreme liquidity and viscidity of their lava. These qualities are attested by its very vitreous texture and filamentous or ropy forms, indicating likewise an intense temperature, to which probably is owing the rapidity of its escape by the fissures opened through the lower flanks of the mountain, and also the fact of such large surfaces of molten matter remaining so long in a state of incandescence as are occasionally seen in the remarkable lava-pool of Kilauea.

§ 12. I have said that the phenomena of these Sandwich Islands' craters might throw some light on the origin of the class of 'pit-craters' previously described, which are drilled through pre-existing rocks, and in which the bulk of fragmentary matter ejected by aëriform explosions does not bear its due proportion to the size of the cavity. I suspect that in many of these cases there has been a 'tapping'—i. e. an escape by some channel opened at a lower level—of the mass of molten lava, the rise of which to (or nearly to) the outer surface of the earth had produced the crater by sudden and violent explosions of no long continuance. Certainly, in the cases of Lakes Paven, Guéry, Tavana, Bouchet, that in which

* Dana, American Journal, 1850, vol. ix. p. 383.

the Fontaulier rises above Montpezat, and many other pit-craters which I have personally examined, there are evident signs of the escape, contemporaneous with their production, of abundant flows of lava from lateral openings at some little distance from, and below the level of, the crater.

The accompanying cut (fig. 58) gives a sketch of one of

Fig. 58.

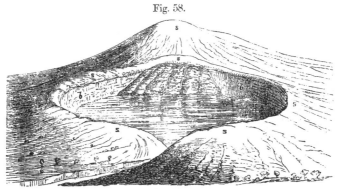

Lac Paven, at the foot of Mont Chalme (Mont Dore).   s, Scoriæ ; b, Basalt.

these—the Lac Paven (Mont Dore).   It is about a mile in cir-cumference, and lies at the foot of the cinder-cone called Mont Chalme, visible in the centre of the view, which rises to a height of 600 feet from its base.   A lower bank of scoriæ also encircles the hollow on the edge of the steep cliffs which almost everywhere encase it.   These last are composed of basalt—a portion of one of the earlier currents from the Mont Dore—through horizontal beds of which nearly 100 feet thick the crater has been excavated.   At the opposite side of Mont Chalme is a breached crater which has given issue to a copious current of basaltic lava that rapidly descends in a north-westerly direction down a narrow valley to the town of Besse, which is built upon it.   The surface of this lava-stream sinks very much in the middle below its lateral banks, having a concave cross section, indicating that the central and lower mass continued to flow down, as in a covered channel, long after the surface and sides had been consolidated. Another very extensive lava-current stretches from the base of Mont Chalme to the east, and appears likewise to have flowed from thence. At its presumed source there is a small depression having at the bottom an orifice opening into a subterranean vaulted cavity of unknown dimensions, but certainly of considerable depth, called the 'Creux de Soucy.' It seemed to me probable, from a close inspection of the localities, that there had been in this case, at the time of the eruption of Mont Chalme, a 'tapping' of a copious pool of lava which had risen within the crater of Paven, and occasioned the explosions by which it was hollowed out; and that the rapid subsidence of this body of lava by lateral efflux had emptied both that cavity and the vaulted cavern of the lava they contained. The depth of the waters of Paven is now, according to M. Ramond, about 300 feet.

I incline then to the opinion that in all these cases the lava having risen at an intense temperature up some fissure into proximity with the open air, the vapour imprisoned in its upper portion—perhaps having the form of a disk-shaped bubble or blister of large dimensions—flashed out in explosions of such sudden vehemence as to shatter the adjoining and overlying rocks throughout a circle of corresponding superficial area, at the same time projecting upwards a certain amount of scoriæ and lapillo; but that the formation of a lateral opening at the same time—perhaps by the same shock—at a lower level quickly emptied the flue of its liquid contents, bringing its explosions to an early termination (or rather transferring them to the new vent), and allowed the greater portion of the shattered rocks to subside without further disturbance into the void left by the escape of the lava. This explanation is in accordance with what we know of the occasional rapid shifting of the eruptive phenomena from one vent to another on the same fissure, and appears to satisfy all the conditions of the problem. Mr. Squier describes a great number of circular or elliptical crater-lakes in Central America as " surrounded by abrupt cliffs of blistered lava-rock and scoriæ, from 500 to 1800 feet high." They have rarely any visible outlet for their water, which is usually salt and bitter. The level of the lakes, and still more their bottom (since they are often of great depth), is far below the surface of the surrounding country. Two of these lakes, Slopango and Amatitlan, are of prodigious area; the first measuring twelve miles by five, the latter thirty by ten or fifteen *. In the absence of more precise information, it is difficult to give an opinion as to the origin of these last two basins, the dimensions of which seem almost to place them in a class apart. It would be very desirable to ascertain whether any ravines or valleys in their neighbourhood have been deluged by lava which may have found a vent there from within them. We have no evidence on this point. But Mr. Squier describes all these lakes as generally lying at the foot of some great eruptive volcano. The largest, Lake Amatitlan (in the province of Guatemala), closely adjoins two immense volcanic mountains that rise some 10,000 feet above it, one of which, Atitlan, was in eruption in 1828, and again in 1833, and ejected vast quantities of stones and ashes that covered the coast for many leagues. Another adjoining volcano broke out paroxysmally in 1766, and overwhelmed with its ejections villages at the distance of nine miles. The volcanic forces are indeed developed on so vast a scale in this district, that there would seem no reason for attributing its crater-lakes (however large their area in some few instances) to any special or ano-malous origin, other than the violent explosions of steam-bubbles to which I have referred the great crateral basins of other localities.

§ 13. It would appear, however, that in some cases the eruption of volcanic matter is accompanied by the subsidence not only of the column of lava which had risen within the

* Squier's 'Mexico,' 1850, p. 270.

Q

vent, but also of the neighbouring surface-rocks themselves.
Several of the cinder-cones of New Zealand, as described by
Mr. Heaphy, have been thrown up on the margin of the sea
exactly upon a line of fault in the tertiary strata whose upcast
forms the sea-cliff, and show a clear synclinal depression of
the elsewhere horizontal beds, on either side towards the
eruptive vent (see fig. 59).   Mr. Darwin describes several

Fig. 59.

Breached cinder-cone near Auckland (New Zealand).
(From Mr. Heaphy's paper, Quart. Journ. Geol. Soc. 1859.)

similar examples in St. Jago (Cape de Verde Isles), and gives*
a drawing of one (Signal-Post Hill) resembling precisely in
character that represented in Mr. Heaphy's sketch.

Such subsidence is not in itself an improbable consequence
of the outward discharge of a quantity of lava and of elastic
vapour from beneath the fractured surface-rocks.   It is, however, certain that in many, perhaps the greater number of
volcanic tracts, the reverse has taken place; that is to say,
the superficial levels of the country adjoining active or recently extinct volcanic vents have been elevated rather than
depressed.   I may mention, as examples, the whole western
coast of Italy, the southern and eastern coasts of Sicily, the
Atlantic groups of Iceland, Madeira, the Azores, the Cape de
Verde Isles, Ascension, and St. Helena.   Among the numerous

* Volcanic Islands, p. 9.

volcanic islands of the Pacific, marine deposits or coral-banks, covered by or alternating with volcanic matters, have been raised hundreds of feet above the sea, evidently at periods more or less coincident with their eruptive eras. Indeed, even in the instances just given of local subsidence immediately adjoining some volcanic vents, the general line of coast simultaneously suffered considerable elevation, and it is only by comparison that the strata underlying the cones appear to have sunk. There were probably in such cases minor local oscillations of level accompanying the eruptions of these districts, similar to those which have been so minutely traced by Sir Charles Lyell, Mr. Babbage, and others, in the shore of the Bay of Puzzuoli, adjoining the Temple of Serapis—the result in that instance being a general rise of the coast, interrupted by local depressions on a small scale.

It should, however, be remarked, that while the elevation of sea-beaches, or marine deposits, is obvious at once to the eye, proofs of depression are by no means so readily discoverable, unless in those very rare instances where sunken forests or works of human art are observable beneath the waves. Whatever portions of the earth's surface may have so subsided (and not risen again above the sea-level) are thenceforth usually lost to sight. It is only within a very few years that observation has been directed to the possibility of this occurrence; so that the rarity of known examples of the kind cannot be accepted as any indication of their general non-existence. I shall have to recur to this subject in a future page.

§ 14. *Tendency of volcanos to shift their principal point of discharge.*—So long as a volcanic mountain preserves on the whole a conoidal figure, it is obvious that its eruptions must have remained constant for the most part to the central vent. Occasionally, however, as has been shown, the ancient habitual issue at the summit of the volcano is deserted, in favour of some new aperture, more or less distant, on the flank or at the foot of the mountain, which consequently loses by degrees its regularity of form,—a new cone forming itself around the new habitually eruptive orifice. It is reasonable in this case to suppose that the original aperture has been effectively— perhaps permanently—sealed up by the immense weight and cohesion of the matters accumulated above it, and the subterranean volcanic energy forced to open a fresh channel at some weaker point, probably on the prolongation of the primary fissure in the overlying rocks.

Etna offers an example of such a shifting of the chief eruptive vent of the volcano. This, at some earlier period, appears to have existed at a point within the upper basin of the Val del Bove, from which a general quaquaversal dip of the beds observable in its cliff-sections may be traced. This point Sir C. Lyell calls the axis of Trifoglietto. The formation of the great irregular crater of the Val was probably the result of the last paroxysmal eruption from that centre. The more recent and still active vent of Mongibello (the summit of Etna proper), whence numerous eruptions of more or less violence have occurred within the historical era, is distant some four or five miles to the north-west of the old axis.

It is noteworthy, however, that the latest considerable eruption (that of 1852–3) sent forth a prodigious and almost unexampled torrent of lava from two or three fissures opened in the immediate vicinity of Trifoglietto, and near the centre of the old caldron of the Val. The summit-crater of Mongibello was, it is true, in explosive eruption at the same time; and, indeed, the vents opened along a fissure radiating from thence downwards, showing that the lava within the volcanic focus was still in communication with the atmosphere through that central duct. I must refer my reader for further details on this interesting eruption to Sir C. Lyell's able paper*.

In Madeira likewise a twofold axis of eruption is very

* Phil. Trans. part ii. 1858.

clearly to be traced in the structure and external forms of the great mountain-masses. The Isle of Bourbon presents another instance of the kind. A vast central crater is environed by the summit-heights of the north-western half of the island. It has been, to all appearance, extinct from a very early period, the rocky masses around having suffered a great amount of degradation, probably as much the result of earthquakes—to which that part of the island is much subject—as of aqueous erosion. The habitual point of eruption has since evidently shifted its position more than once in the same (south-easterly) direction: first, to a distance of about fifteen miles—the Plain des Calaos—where a flattened dome of considerable magnitude was formed, still studded with cones and craters ; next, to the site of the actually eruptive volcano, which rises in a dome or cone to the height of 6000 feet from the centre of a vast semicircular cliff-range—the walls of a prodigious crater, formed by some explosive paroxysm of extraordinary violence, and which subsequent eruptions have nearly filled up to the brim with the existing volcano. (See figs. 49 & 50.)

So, again, the recent volcano of Teneriffe has an evident treble axis : that of the Peak proper, which is not known to have been active since the island was occupied, though its vitreous lavas have a very recent aspect, and its small summit-crater emits sulphureous vapours ; and those of the two mountains closely attached to it on either side—Chahorra on the west (whence all the recorded eruptions have taken place—the last in 1798), and Monte Blanco (now wholly extinct) on the east. All three are on the same line—the longest diameter of the old environing crater-ring.

In Java, according to Junghuhn, several of the larger volcanos have had two or more centres of eruption—always ranging on the same line. One in particular, called Gede, shows a regular cone 9326 feet high, sloping at an angle of 30°, and truncated, like Etna ; while a twin cone adjoins it, somewhat less lofty, and probably earlier in point of age, since it has suffered much waste, and has on one side a deep valley comparable to the Val del Bove—most likely therefore an ancient crater.

In truth, such a shifting of the eruptive vent along the line of the original or principal disruption of the solid surface-rocks, as the first-formed vents become overloaded with erupted matters, or as some accident of fracture may determine, is so probable an occurrence, from what we know of the general nature and conduct of volcanic energy, that we should expect to find it (as is the case) a very common feature in all districts that have been the seat of such phenomena.

When the principal crater of an habitual volcano has been in this manner deserted, it often remains for some time in the condition of a solfatara, through the continued emanation of acid vapours from the residuum of heated lava left in the chimney of the mountain. These vapours find their way by percolation through the pores of the solidified rock, or crevices too narrow to permit any fresh intumescence below. Meantime the mass of lava within and beneath this sealed-up vent is most likely cooling down by degrees. But the process may last for centuries; as has been the case with the solfatara of Puzzuoli, which, from the white colour of its altered rocks, was called λευκογαιος by Homer. The small crater of the Peak of Teneriffe is in this condition, as was just stated. Many of the craters of the Javanese volcanos are solfataras.

These languid emanations of vapour appear to be chiefly confined to the trachytic volcanos; perhaps because their lavas, when solidified, are so much more porous than basalt. Their action upon these rocks, and the various saline and earthy substances derived from it, have been already noticed.

§ 15. Before quitting the subject of Volcanic Craters, a word should be said on the remarkable resemblance presented by the surface of the Moon to some of the volcanic tracts of our Earth which are most plentifully studded with crateriform hollows surrounded by subconical banks. The analogy is so close, that it is impossible for a moment to doubt the volcanic character of the lunar enveloping crust.

" The generality of these craters," says Sir John Herschel, "present a striking uniformity and singularity of aspect. They are wonderfully numerous, occupying by far the larger portion of the Moon's visible surface, and almost universally of an exactly circular or cup-shaped form, foreshortened, of course, into ellipses towards the limb. But the larger have, for the most part, flat bottoms within, from which rises centrally a small steep conical hill. *They offer*, in short, *in its highest perfection, the true volcanic character*, as it may be seen in the crater of Vesuvius, or in a map of the volcanic district of the Campi Phlegræi or of the Puy de Dôme. And in some of the principal ones, marks of volcanic stratification, *arising from successive deposits of ejected matter*, may be clearly traced with powerful telescopes. In Lord Rosse's reflector, the flat bottom of Albatignius is seen to be strewed with blocks not visible by inferior telescopes ; while the exterior of another (Aristillus) is all hatched over with deep gullies " (like the Javanese volcanos described by Junghuhn) " radiating towards its centre."

I subjoin, for the purpose of comparison, maps of two districts, one of Terrestrial, the other of Lunar surface (figs. 60 & 61). The latter is taken from the environs of the crateral mountain Maurolychus ; the former from the Phlegræan fields adjoining Naples, and includes Vesuvius. From some of the loftiest and largest mountains of the Moon (not represented in the woodcut) there stretch outwardly on all sides numerous radiating lines, reflecting a brilliant light, and therefore elevated like causeways above the intervening shadowy hollows. These are probably either lava-streams that have flowed to great distances from the central eruptive heights, or dykes protruded upwards in vertical ridges from radiating fissures— a mode of eruption already noticed as characterizing some

trachytic and phonolitic lavas. For example, the volcano of
Pichincha is described by Humboldt as a lofty long-drawn
ridge leading up to a still higher crater—a description precisely

Fig. 60.

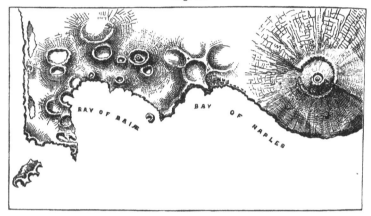

Map of Terrestrial Volcanic surface—The Campi Phlegræi.

Fig. 61.

Map of Lunar surface—Mount Maurolychus.

corresponding to some of the lunar mountains. The clinkstone ridge stretching north from the high centre of eruption

of the Mezen (see p. 135) is another example. Some of these great dykes or currents, indeed, proceeding from the higher mountains, traverse wide areas of the Moon's surface, crossing the largest of those dusky hollows, resembling dried-up sea-basins, which form so conspicuous a feature in the face of the planet. One thing seems certain, namely, that the lunar surface is no longer eruptive—at least, that its volcanos have been quiescent for centuries, since no change has been observed by astronomers in its mountains. It bears, indeed, the aspect of a burnt-out globe, once imbued with volcanic life and intense outward activity, probably with seas and an atmosphere, now dried up and extinct—at least as respects the hemisphere which alone we are able to contemplate, and which, by an excentricity in its centre of gravity, seems fastened irrevocably towards us by the powerful attraction of our larger and denser planet.

Compared with those of the Earth, the lunar craters are far more numerous (in proportion to the respective areas of the planets) and more generally distributed, their interior is deeper, their diameter greater, and their external banks shallower. They most nearly resemble the pit-craters and larger crater-lakes I have described as rather exceptional in their characters. Judging from this analogy, I suspect them to have derived these peculiar features from the explosions of vapour that produced them breaking through a surface of soft and semi-liquid matter in successive bubbles, whose bursting would throw off all round a concentric ridge formed of repeated layers of this substance—as was remarked of the very similar crater-cones of the Phlegræan fields and of New Zealand, which were so formed on a shallow sea-shore.

If a shallow pan (a common frying-pan) be filled to the depth of an inch or two with plaster-of-paris mixed with water in which glue has been dissolved (to prevent too rapid setting) to the consistency of paste, and placed on a stove, so as to cause the water in its lower stratum to boil with some violence, the bubbles that break repeatedly from its surface, following one another in succession upon the same points, leave at last, when all the fluid has evaporated, numerous deep circular pits, with a low bank of matter around each, so closely resembling those on the face of the Moon, that it is difficult to resist the conviction of the latter having originated in some very analogous process, however different the scale of operations.

Graham Isle (or Ile Julie), thrown up by a submarine volcano, off the west coast of Sicily, September 1831.

CHAPTER X.

SUBAQUEOUS VOLCANOS.

§ 1. As yet our attention has been confined to the phenomena of those volcanic vents which open at once in the atmosphere; but it must not be forgotten that eruptions are liable to take place on any point of the globe's surface, and therefore as well on that part which is covered by permanent bodies of water as on that which is dry.

Indeed, from the far greater extent of the former, which exceeds the latter in the proportion nearly of three to one, we might expect the number of subaqueous eruptions in the same proportion to exceed those that take place in open air. It must, however, be remembered that the repetition of eruptions from the same vent will sooner or later have the effect of raising the apex of a submarine volcanic mountain above the level of the sea, and that, with probably the greater number of habitual submarine vents, this limit of elevation has already been reached, and they have become visible as insular volcanos.

It is very seldom that an opportunity is afforded of observing a volcanic eruption from the bottom of the ocean, or any inland sea. The resistance created by the greater density of the medium and its refrigerating influence are likely to prevent the explosions of steam, in most instances, from taking place at all—at least from reaching its surface, and at all times from attaining any great height above that level, and, consequently, being visible at any considerable distance. It is only to the crews of vessels casually passing near that such an opportunity can occur. We must not expect, therefore, to be possessed of many accounts of such phenomena.

§ 2. Instances, however, are not wholly wanting, and the details of what has been observed on these occasions lead us to conclude that the volcanic phenomena take place very much in the same manner from the bottom of the sea as from the open surface of a continent, subject only to modifications produced by the lower temperature of the surrounding medium, and the greater external pressure, caused by the weight of the overlying column of water, which in this case becomes one of the elements of the repressive force. In fact it can scarcely be otherwise, since we know that the instant the cone of the submarine volcano raises its peak above the waves, it enters into the class of subaërial volcanos, and the nature of its activity is in all respects precisely the same as that of a continental vent.

The principal instances which have been witnessed of submarine eruptions are :—

1. Several successive eruptions off St. Michael, one of the Azores; the first on record in 1638* : two other eruptions took place from the spot in 1691 and 1720; the last produced an island six miles in circumference, which, however, before long disappeared. In 1812 it was re-formed at a spot previously 300 feet deep. The new island was taken possession of by Captain Tillard for the British Government, and named Sabrina. It was 150 feet in height above the sea-level, and about a mile in circumference. The explosive jets of scoriæ lasted for six days, during which its growth was watched. In the course of a few years it had been, like that previously formed there, wholly washed away by the waves, and replaced by deep water.

2. An eruption, continued at intervals for five years, gave

* Sanderson's Hist. of Charles I.   Mémoires de l'Académie, 1721.

birth to the Isola Nuova off Santorini, in the Grecian
Archipelago, in 1707–12*.   It measures four miles in
circumference at the sea-level.   Santorini itself is reported by Pliny to have been produced in the same
manner B.C. 236, as well as two other neighbouring islets,
Jera and Thia, now the Great and Little Kaimenis.   All
three rise in the centre of the vast broken crater-ring of
Santorini, already described (p. 200).

3. An island was thrown up at some distance from the coast
of Iceland during the violent eruption from Skaptar Jokul
in 1783.   This likewise has since disappeared.

4. A new volcanic island was formed in the Aleutian group
near Unalashka in the spring of 1796, called Bojuslaw
by the Russian hunters†.   It was at first 250 feet high;
but in 1816, from the continuance of its activity, it had
increased to a height of 3000 feet, and a circumference
of twenty miles.

5. .The Isle Julia (otherwise called Graham's Isle), off the
south-west coast of Sicily, thrown up in 1831, on a spot
where 100 fathoms of water existed a few years before.
(See vignette, p. 234.)   It disappeared shortly after.

6. On the coast of New Granada, near Carthagena, in
1848, a column of fire and smoke rose to a great height
from the sea, continuing for several days.   A small island
of lapillo and black sand was found there after the close
of the eruption ‡.

7. M. Daussy and, after him, Mr. Darwin have collected
various reports tending to show the existence of a considerable tract of volcanic sea-bottom beneath the Atlantic,
nearly midway between Cape Palmas on the west coast of
Africa and Cape St. Roque on the east coast of South
America; that is, *in the narrowest part of the ocean* between the two continents.   Earthquake-shocks have been

* Hist. de l'Académie, 1708. Humboldt, Pers. Narr. vol. i. p. 448.
† Kotzebue.   ‡ Comptes Rendus, xxix. p. 531.

repeatedly felt by vessels passing over this disturbed area, which measures about nine degrees from east to west, and three to four degrees from north to south. The sea has been observed to be powerfully agitated in the absence of wind, muffled sounds heard from beneath, vessels struck as if by a bank or rock, columns of smoke seen to rise, and scoriæ or pumice found floating in abundance there. In some spots, islets of sand or cinders have been seen to rise above the water-level, but have subsequently disappeared. In some, shallow soundings are met with; in others none can be obtained.

On all these occasions columns of smoke (steam, mixed with ash) by day, and flames (jets of red-hot scoriæ) by night, were seen to rise from the sea, which was considerably agitated, discoloured, and heated to such a degree as to kill numbers of fish.

At length the dark-coloured rocks showed themselves above the sea-level. These, in the cases of Santorini and the North Pacific Island, were lithoidal (lava) as well as fragmentary, and the island produced by those eruptions has consequently remained firm; in the other instances, the fragmentary cone alone appears to have risen above the water-level, and the action of the waves and marine currents upon such loose matters soon mined and degraded them, reducing the island to a submarine shoal.

The reports of actual observation in these instances, indeed, lead to the opinion, that a volcanic eruption taking place from a subaqueous vent *at a moderate depth* below the surface of the water, conducts itself in a manner very little, if at all, different from a subaërial one. The same explosions of aëriform fluids are observed : rocky fragments, ignited scoriæ, and their comminuted ashes are thrown upwards; the heavier, in falling, accumulate round the vent into a cone with a central crater, while the lighter are borne to a distance by the tides and currents of the sea. Lava probably issues and spreads over the subaqueous bottom, seeking the lowest levels, or accumulating upon itself, according to its liquidity, volume, and rapidity of congelation—following, in short, the same laws as when flowing in open air.

Humboldt and Von Buch have both declared their belief, that in submarine eruptions the strata previously forming the bottom of the sea are uniformly elevated in mass, and that positive eruptions do not take place from the vent until these strata have been raised above the surface-level of the water. Their opinion on this point has been followed by several other continental geologists, as also by Dr. Daubeny. This supposition is not, however, warranted by any observations, since, in all the cases already mentioned, the only rocks which showed themselves above the sea-level were uniformly lavas, either lithoidal or fragmentary, evidently the products of the eruption by which they were raised *.

§ 3. It is true that, at considerable depths, the vast augmentation of the repressive force occasioned by the pressure of the column of water above the vent must proportionately im-

---

* It is remarkable how closely the relation by Seneca of the production of the island of Hiera corresponds with the phenomena which in the present day we are led, by a fuller study of similar facts, to believe to characterize such submarine eruptions :—" Majorum nostrorum memoriâ, ut Posidonius tradit, cum insula in Ægeo mari surgeret, spumabat interdiu mare, et fumus ex alto ferebatur. Nam primum producebat ignem, non continuum, sed ex intervallis emicantem, fulminum more, quoties ardor inferius jacentis superum pondus evicerat. Deinde saxa revoluta, rupesque, partim illæsæ, partim exæsæ et in levitatem pumicis versæ. Novissime cacumen montis emicuit. Postea altitudini adjectum, et saxum illud in magnitudinem insulæ crevit."—" Within the memory of our ancestors, as Posidonius relates, when an island rose in the Ægean Sea, the sea for a time bubbled up, and smoke [vapour] was emitted from its depths. At first it threw out flames " (red-hot stones probably), " not continuously, but spirting up at intervals, like lightnings, as often as the vigour of the fire underneath overcame the weight above. Then stones were driven out, and rocky masses, partly uninjured " (by the fire ; i. e. fragments of pre-existing rocks), "partly eaten into and converted to light pumice. Lastly, the summit of a hill appeared ; and afterwards increased in height, till that rock grew to the size of the [existing] island."

pede the ebullition of the lava to which any fissure of escape penetrates. It is therefore probable that, at depths exceeding several hundred feet, there will occur no development of steam on the large scale, and that either the rocks forming the sea-bottom are elevated by the intumescence of the lava beneath them, until they reach the point at which the tension of the confined vapour can overcome the pressure arising from the superincumbent column of water, or lava alone may issue, and be piled in a dome-like mass, up to the same point, whence alone eruptive explosions can occur. But it should be recollected, that since the weight of the column of water at any depth must be added to the sum of the repressive forces by which any eruption whatever at that point is resisted, so, when the expansive force of the confined lava does at length overcome this, the temperature and tension of the confined interstitial vapour must be proportionately intense.

It is also probable that the vapour which escapes from the lava, even at moderate depths, will be instantly refrigerated, by contact with the colder strata of water through which it passes in its ascent, and *condensed* (as in the refrigerator of a steam-engine) ; so that, until the submarine volcanic hill has risen to within a short distance of the atmosphere, *no volumes of vapour will rise to the surface,* and consequently no visible aëriform discharges will announce the occurrence of eruptions in the depths of the ocean.

Torrents of lava will probably be poured out, rock-masses fractured, elevated, and, perhaps, their fragments pushed up to a certain height, by every such eruption, and much disturbance occasioned in any sedimentary beds in course of deposition on the floor of the sea. But until the accumulation of these matters shall have raised the orifice of the mountain to such a height that the aqueous vapour it discharges is no longer wholly condensed by the pressure and refrigerating action of the surrounding water as it rises, no other appearance of its eruptions will be visible from without than what consists in the partial discoloration and disturbance of the water above the volcano.

When therefore the eruption becomes visible, and the steam produced below is discharged from the surface of the sea, and particularly when it is seen to throw up scoriæ, it may be safely argued that the summit of the submarine volcano is at no great depth, and that only a short continuance of eruption is required to raise it above the sea-level, and create a new subaërial volcanic island, more or less permanent, according to the more or less solid nature of its materials, and the power of the marine waves or currents to which it is exposed *.

§ 4. The mineral or saline compounds which, in greater or less quantities, always accompany the aqueous vapour evolved from a volcanic vent, will, on its condensation, mingle with the waters of the ocean, and add to the ingredients of the same nature with which they are impregnated, and which, it may be conjectured, were originally derived in great part from the continuance through countless ages of this source of supply.

§ 5. *Disposition of products of submarine volcanos.*—Though the volcanic phenomena seem to be essentially the same, whether acting from a subaqueous vent or one in open air, yet the difference of the medium into which the eruption breaks forth must considerably modify the *disposition* of the substances produced.

Fragmentary rocks.—As has been already said, a large portion of the matters thrown up will accumulate immediately around, into a cone ; but those which, when the orifice approaches the surface of the water, are scattered to any height by gaseous explosions, and particularly the lightest and finest of the fragmentary matters, will remain suspended for a considerable time in the agitated fluid, imparting to it a turbid

---

* In a subsequent chapter I shall refer to Mr. Mallet's view of the action of submarine volcanos as the cause of earthquakes—a view from which I am obliged to dissent.

colour, and upon the cessation of the disturbing causes be gradually and evenly deposited, over a large surface of the bottom of the ocean, in the form of sedimentary layers. Those composing the upper part of the cone itself will also probably before long be carried off by marine currents and likewise spread over extended areas.

Thus, no doubt, were formed the stratified and conchiferous tuffs which cover the maritime plains of Western Italy, the Campagna di Roma, and the Terra di Lavoro; and which have penetrated the proximate valleys of the Apennines, then maritime creeks or estuaries, but since raised above the sea by a general elevation of the western shore of the Peninsula. Pumice, it is well known, floats upon water; when, therefore, the fragmentary ejections are of this nature, they may be driven by winds or currents often to very great distances, and deposited on the shores towards which the currents set.

If the waters beneath which the eruption occurs are impregnated with calcareous matter, the tuffs formed in this manner will have a calcareous cement, and will contain seashells, &c.; or will occasionally alternate with the limestones or calcareous sandstones deposited by the surrounding ocean.

In this manner were produced the calcareous and conchiferous Peperinos of the Veronese, Vicentino, and the Euganean Hills; of Southern Sicily; of the freshwater basin of the Limagne d'Auvergne; of the Deccan; of the Pampas plains of S. America, and numerous other localities; where basalt, as well as calcareo-basaltic conglomerates, are found to alternate repeatedly with regular strata of limestone.

When the comminuted and pulverized ejections of a volcanic eruption taking place at or about the water-level are of such a nature as to 'set' when mixed with water, which we have seen to be the case with felspathic ash, it appears that the matter constituting the cone, being, from the vicinity of the vent, mixed by frequent agitation with the surrounding water into a thick mud or kind of mortar—perhaps influenced like-

wise by the intense heat emanating from the volcanic orifice
—subsequently assumes a greater consistency and solidity
than the ash which was deposited as sediment at a distance
from the volcanic mouth, or showered upon the surface of the
cone itself, after its rise above the water-level.

Thus the tuff which constitutes the lower part, or nucleus,
of the eminences surrounding the submarine craters of the
Phlegræan fields is sufficiently solid to be worked as a building-stone; while the intervening flat spaces consist of loose
tuff, identical in composition with the other, and differing
only from it in the incoherence of its materials.   These beds
of loose tuff are found also on the surfaces of the harder
cones.   The difference of solidity between these varieties ap-

Fig. 62.

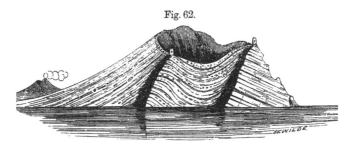

Natural section of a tuff-cone, the Cape of Misenum.

pears to consist in the former having *set*, or effected a degree
of attractive cohesion, during its desiccation.   Indeed, the
escape of volumes of steam from a volcanic vent but a little
below the surface of a body of water must probably take place
by the belching up of many large bubbles of thick mud, which,
as they burst, will throw off the clotted matter on all sides in
great circular waves, likely to form, when subsequently solidified, those thick beds of pumiceous tuff, usually showing a quaquaversal dip from the eruptive axes, often arranged in curved
anticlinal ridges, which we meet with near Naples, and in other
sites of felspathic littoral eruption.   The Cape of Misenum
(fig. 62) is an example of the summit of a cone formed by

such submarine explosions.   The tuff of which it is composed
has considerable solidity, as indeed is proved by the resistance
it still offers to the waves of the Mediterranean.   The little
insular cone of Nisida, at the extremity of the promontory of
Pausilipo, shows an almost identical composition and structure
where the banks of its breached crater are denuded.

§ 6. *Lithoidal products of subaqueous eruptions.*—There can
be little doubt that the same laws will influence the disposition of lavas produced below any body of water from a subaqueous volcanic orifice, as when emitted in open air.   They
will in the same manner spread laterally beneath the cover of
a scoriform envelope, with a rapidity and to an extent proportioned to the expulsive force, their fluidity, and the permanence of that fluidity, as well as the accidents of level in the
surrounding surfaces.

Of these circumstances, the first, viz. the expulsive force,
is only so far influenced by the pressure of the superior
column of water, as to need a greater effort, and therefore,
perhaps, a higher temperature in the lava, to balance the
corresponding increase thereby caused in the total amount of
the repressive force.   Increased temperature may increase the
fluidity of the lava, at the moment of its emission from the
vent; and the *permanence* of this character is probably
rather augmented than lessened by the nature of the medium
with which the liquid lava is brought in contact,—the surface
being instantaneously consolidated, and the upward rise and
escape of the vapour prevented by the refrigerating influence
of the cold water, and by the intense external pressure acting
on the surface of the current.

Thus we should expect lava-beds produced at the bottom of
the sea to exhibit a greater lateral extension compared to their
thickness than those which have flowed under the pressure of
the atmosphere alone, and that this extension should be proportionate to the depth of the column of water they supported.
This reasoning would seem to be confirmed by the generally

great horizontal dimensions of those lava-beds which are considered as of submarine origin, such as the bulk of the earlier trap formations; *e. g.* those of Iceland, Ferroe, Ireland, the Hebrides, Germany, &c., which usually occur in parallel, widely-extended sheets.

Again, it must be expected that currents of lava which have flowed at great depths under water will present comparatively few scoriform parts. This character has, in fact, been generally observed amongst the older submarine traps. It may, however, in part be owing to the denuding influence of the waves or currents.

Vesicles, on the contrary, may be expected to abound in the *interior* of the lava, whenever its liquidity was sufficient to permit the agglomeration of the vapour into parcels—the bubbles expanding as the lava flows on—while, for the reason mentioned above, very few will make their escape by rising outwardly. The Flœtz-trap formations, as the earlier volcanic rocks were formerly called, are frequently vesicular, or, rather, amygdaloidal.

We have already noticed, in a former chapter, the occasional occupation of such cellular cavities and pores in a lava-rock, subsequently to its consolidation, by crystallized minerals, deposited from water penetrating these interstices ; and it is obvious that a bed of lava, existing for a long period under the surface of the sea, and exposed to an intense pressure, will be far more liable to the percolation of water in this manner through its substance, than when consolidated in air. As those elastic fluids which remain in it subsequently to its consolidation are slowly condensed by refrigeration, they tend to produce a vacuum in the cavities they occupy, and oppose no resistance to the penetration of water from above, which is urged to descend, not by its own weight alone and that of the atmosphere, as in the case of a subaërial rock, but by that of the whole supported column above in addition.

It is also obvious that the various mineral substances con-

tained in the water of the sea, by entering into new combinations with the silex and other substances volatilized or dissolved
in the condensed vapours of the lava, may occasion the crystallization of new minerals within the vesicles and minute interstices of the rock; and this may serve to account for the
numerous varieties of zeolites and other minerals that characterize the amygdaloidal basalts and traps, of which the
greater number certainly have been produced by subaqueous
volcanos. The slowness with which these substances are
separated from their aqueous menstruum is probably the
cause of the great regularity and perfection of their crystalline
form*.

§ 7. In the case of subaqueous eruptions, it will rarely be
possible to distinguish whether they take place from a new or
an habitual vent. Both circumstances, no doubt, will occur,
as in subaërial volcanos.

The eruptions that threw up the Kaimenis, already mentioned more than once, burst from the centre of the crater of
a volcanic mountain, itself for the most part of submarine
origin. Those off the coast of St. Michael may be supposed to
have proceeded from a subsidiary or lateral vent of that great
insular volcanic mountain.

On the other hand, the Phlegræan fields, as the volcanic
district of Pozzuoli and Cumæ, near Naples, has been called,
present an instance of numerous submarine eruptions, each
from a fresh point on a shallow shore. Some of the cones left
by these phenomena are very regular and entire; others present lengthy circuitous ridges, embracing crateriform basins,
of which some contain lakes. They are all formed of an indurated felspathic tuff, containing occasionally sea-shells and
fragments of wood, both unmineralized, and are covered by
strata of loose tufaceous conglomerate, similar to that which
is generally dispersed over nearly the whole neighbouring
plain of Campania, and which appears to have been deposited

* See Professor Phillips's Address, Quart. Journ. Geol. Soc. p. li, 1859.

there at a time when the Mediterranean washed the foot of the nearest Apennine. Mr. Heaphy describes a very similar district in the neighbourhood of Auckland in New Zealand*.

Some knowledge as to the conduct of submarine volcanos may be derived from observation of such of their products as have, by subsequent elevatory action on a wide scale, been raised above the sea-level. Such examples are numerous among the coral islands of the Pacific. The flat basaltic platforms of the north and south coast of Iceland, of the Ferroe Isles, of the north-east of Ireland, of the north-eastern part of Teneriffe, and of numerous other localities, tend to show that the conduct and disposition of lava, when issuing from a submarine vent, is, as has been supposed above, very much the same as when poured out on dry land. The chief difference would seem to be, first, that it flows more uniformly and spreads more widely over a flat surface, probably from retaining its internal heat longer under the pressure of the water than of the atmosphere only. 2ndly, That either a less proportion of conglomerate, or fragmentary matter, is usually ejected from a subaqueous than from a subaërial volcano; or that it is more widely spread and less thickly interstratified with the contemporaneous lava-beds. Both of these explanations may well be true; for in deep water it is not likely, as has been already shown, that aëriform explosions should be able to throw up much scoriform lava. And when the vent has risen near enough to the surface to permit of such explosive ejections, they will be just about the level at which the abrading and degrading force of the ocean-waves and currents, and their influence in widely distributing all loose materials within reach, are at their *maximum.* On these grounds we should expect in such cases to find very much what we do observe in formations of this presumed origin, namely extensive platforms of basaltic or trappean rock, in repeated beds, overlying one

* Quart. Journ. Geol. Soc. August 1860.

another, with perhaps little or no scoriform matter intervening, and stratified conglomerates or ash-deposits in other
neighbouring districts.

Mr. Darwin, in his work on South America, describes a large
geographical area of that continent to the east of the chain of
the Andes as composed of metamorphic schists, clay-slates,
and plutonic rocks, which at one time must have formed the
floor of the ocean, and were then covered by vast streams of
lava (claystone and greenstone porphyries), together with
alternating piles of angular fragments of similar rocks—all
ejected from submarine volcanos, and apparently, from the
compactness of the rocks so formed, in deep water. This
volcanic formation was subsequently covered by gypseous
deposits of the age of our chalk, mingled with the products of
contemporaneous volcanic eruptions. And on some points,
especially in Chili, these beds were again loaded, in the later
tertiary period, with a vast pile of volcanic submarine tuffs
and lavas, previous to the final elevation of the continent
above the water-level, or the opening of the great subaërial
volcanic range of the existing Cordilleras*.

Professor Ramsay also gives a very similar description of
the igneous rocks interbedded with, and breaking through, the
Silurian strata of Wales and Shropshire†. He distinguishes
with great sagacity between those lavas which were contemporaneous with the strata in which they lie—as shown by
their having altered the underlying stratum, while that deposited above them is unaffected—and those intrusive lavas
which were apparently injected between earlier-formed and
solidified strata, as shown by the overlying bed being as much
altered as the one beneath. The latter disposition appears to
be confined to the hornblendic lavas (greenstone); the former
prevails among the felspathic traps. Is this owing to the

* Daubeny, p. 503. Darwin, South America, p. 237.
† Catalogue of Rock-specimens in the Museum of Practical Geology,
1858, p. 174.

greater liquidity of the former class better enabling them to penetrate the seams of the stratified rocks among which they were forced? Professor Ramsay speaks of these intrusive beds of lava " running often *for miles* directly along the strike of the strata, and then breaking suddenly across it." These extensive intrusions took place, it appears, in the slate-rocks, which might be expected to split more readily along the planes of stratification than across them.

§ 8. When at length the summit of an habitual volcanic cone is permanently elevated above the surface of the sea, it enters into the class of subaërial volcanos, and produces one of those volcanic islands which are of such frequent occurrence in the Atlantic, Pacific, and Indian Oceans.

But it is not solely by the accumulation of erupted matters that this increase of height can take place. It may be produced, without any addition of bulk to the mountain itself, by its elevation in mass from below, occasioned by a general expansion of the subterranean bed of lava, raising, together with the volcanic cone, a greater or less superficial extent of the neighbouring strata forming the pavement of the sea.

This has probably happened to the coast of Italy, just described. The volcanic cones of the Phlegræan fields have certainly sustained a change of level, relatively to the surface of the sea, since their formation, without having received any addition to their bulk by subsequent eruptions. Marine shells of existing species are found in Ischia at a height exceeding 1000 feet above the sea. Again, the flanks of Etna are in many parts formed, to a very considerable height, of alternating beds of lava and tertiary limestone full of shells.

We are thus led to distinguish between islands or tracts of volcanic origin which were produced solely by subaqueous eruptions, and subsequently elevated above the sea-level (with or without any superincumbent calcareous or arenaceous strata) by successive expansions of the subterranean lava-bed (which I have called plutonic action), and those islands or tracts which owe their progressive rise above this level chiefly to the gradual accumulation of matter expelled during repeated eruptions from an habitually active vent, and in which this process has continued long after their emerging into open air.

§ 9. The latter will have all the characters of a volcanic mountain, viz. a pyramidal or conoidal form, and a structure of alternate beds of lava-rocks and conglomerate, with occasionally interbedded strata of marine sediments about their base, all dipping more or less rapidly away on all sides from the central summits.

The former class of volcanic formations will seldom exhibit this regularity of form and structure. The lava-beds which are produced at any depth under the sea-level will probably spread, as we have observed, to a greater lateral extent than those emitted in open air; and, again, the abrasive action of the waves or marine currents will generally wear down and disperse the accompanying fragmentary ejections to a greater distance from the vent. Hence a submarine volcanic mountain will not present by any means so decidedly conoidal a form as one thrown up in open air, but will be far more depressed and flattened, and composed of comparatively horizontal beds.

Again, the forcible elevation of such a mass by subterranean expansion will probably still further derange its figure; and, in lieu of a gently sloping conical mountain, its outline will present vast plateaux with little or no inclination, enclosed by precipitous cliffs, or separated from each other by deep fissure-like chasms, and composed of alternate beds of lithoidal and fragmentary lava-rocks, exhibiting the marks of their submarine origin as well in their great compactness of texture, in the absence of scoriform parts, and in the numerous amygdaloidal minerals they enclose, as in their peculiar disposition and general forms. Of such an upheaved submarine volcano the Ferroe Isles present a typical example.

The island of Teneriffe exhibits, on the other hand, both these volcanic formations geographically united. It consists chiefly of an immense volcanic mountain, of which the central peak reaches an elevation of 12,200 feet above the sea. In form this mountain approximates to an oblong cone, although its regularity has been greatly disturbed by the products of numerous lateral eruptions. The principal crater evidently once occupied the centre of the ellipsoidal base, and corresponded with it in shape and dimensions. Its enclosing walls, or what remains of them, form a vast elliptical range of cliffs, within the area of which rise the more recent cones of the Peak and Chahorra. But this immense mountain constitutes little more than two-thirds of the superficial extent of the island. On its north-east side, and in the prolongation of the longer axis of the ellipsoid, projects a remarkable headland, differing from the other part totally in form and structure, having a flat and terraced outline in lieu of a conical one, and made up entirely of vast horizontal beds of basalt and basaltic conglomerate, while the rocks of the central volcanic mountains are wholly trachytic. This basaltic district is clearly a submarine volcanic mass, elevated above the sea-level subsequently to its production, since which time the proximate trachytic volcano has alone remained habitually active, and continues so to the present day.

The island of Palma, another of the Canaries, offers an excellent example of a perfectly regular and complete insular volcanic mountain. Its form is perhaps as near an approximation to the cone as the disturbing causes which must necessarily influence the figure of such a mass will ever permit; the outline of its ground-plan is nearly circular; and it rises from the coast on all sides, at first by gently inclined slopes, which become gradually more and more rapid, till they terminate in the ridge which forms the summit of a steep range of cliffs encircling its central crater. This deep cavity is drained by one outlet, or Baranco, doubtless originally a radial fissure produced during some violent eruption. The ravines created by numerous other rivulets diverge like rays from the central heights. A few parasitical cones are sprinkled on the surface of the outer slopes, each of great regularity, and productive of its peculiar lava-current*. The Great Canary Island shows a very similar figure and composition.

The island of Tristan d'Acunha, in the middle of the Southern Atlantic, appears to be an insular volcanic mountain of equal regularity. Captain Carmichael, who visited it in 1816, describes it as of a conical figure, nine leagues in circumference, and 8000 feet high, with one large central crater a mile round, having a lake at its bottom 150 yards in diameter. The lava and scoriæ are basaltic, and of a fresh aspect†.

St. Paul, or Barren Island (see fig. 53, p. 199, *supra*), Fayal in the Azores, and the Pic de Fogo, one of the Cape Verde Isles, are regularly conical eminences, the summits of submarine volcanic mountains. In fact, such appears to be the general form of those insular volcanos which have been produced by habitual eruptions from a central vent, since its elevation above the sea-level, of which numerous examples are to be found among the Archipelagos of the Pacific.

Where two or more habitual vents were situated so near to one another as to bring their erupted matters into contact, the island will, of course, be formed of a string of such conoidal mountains; and many instances of this likewise are to be met with among the volcanic islands of both the Atlantic and Pacific Oceans.

* Palma was selected by Von Buch as a typical example of an elevation-crater, or upheaved volcanic mountain. There is, I think, every reason to believe this view to be erroneous, and that it owes its origin, on the contrary, to quaquaversal flows of lava and ejections of fragmentary matters from a central vent. Sir C. Lyell gives a map of this island in his 'Manual.' The fact of the bottom of the great crater having a boss of trachyte in no degree argues the upheaval of the encircling basaltic beds. It is a circumstance common to many volcanic craters, *e. g.* Astroni, Rocca Monfina, and others. † Trans. Linn. Soc. vol. xii.

Even the entire island of Java, a great part of Sumatra, Celebes, Formosa, the Philippines and Kurile Isles, Japan, the promontory of Kamtschatka, and many of the largest islands of this part of the globe, appear to consist of one or more rows of volcanic mountains thrown up from vents produced upon the same or parallel fissures in the crust of the globe.

No doubt, contemporaneously with, or in the intervals between, the eruptions of such habitual volcanos, local elevations of the neighbouring surfaces, by the force of subterranean expansion, will have often occurred; and hence strata of limestone, sandstone, and even of the schistose rocks and granite itself, may form a part of these islands, without controverting their volcanic origin for the most part. " If," says Mr. Darwin, " archipelagos are in course of time, by long-continued action of the elevatory and volcanic forces, converted into mountain-ranges, it will naturally result that the inferior primary rocks will often be uplifted and brought into view."

As examples of islands, composed chiefly of volcanic products, which appear to have risen from below the sea, solely by subterranean expansion, without having been since augmented in height or bulk by the products of external eruptions, I have already named the Ferroe Islands, and the trap islands amongst the Hebrides, and may add those of Ponza, Zannone, and Palmarola, on the coast of Italy.   It is, however, not always easy to distinguish upheaved islands of this character from such as are only the remnants of subaërial volcanos, the greater part of whose mass has been worn away by the action of the sea.   When, for example, a cone of the latter class, composed chiefly of scoriæ, with perhaps but one great lava-current, has been long exposed to the violent degrading action of the waves, the whole of the former, consisting of loose conglomerates, will often have been swept away into the depths of the sea, and only the latter (the lava-bed), or a portion of it, left above water, together with such overlying

strata, perhaps, of ash or scoriæ as it may support.   I have
already given (fig. 55, p. 209) a sketch of one such island, whose
outline must remind all geologists who have visited volcanic
archipelagos, of many similar instances—instances that are
to be found among the islands of the tertiary as well as
of more modern seas.   By many geologists these insulated
remnants would be considered examples of basaltic or trachytic rocks upheaved from depths below the sea ; whereas
it is often possible that their level has never varied at all
since their original deposition.   Some of these subaërial
volcanic islands have numerous superficial strata of calcareous sandstone, which help to induce the persuasion of
their being of submarine origin.   The lime, however, is, I
believe, derived from countless generations of land-shells (*He-
lices*, &c.) which have lived and died on the surface, and has
been carried down with the rain-water among the arenaceous
beds of tuff, or, in many cases, drift beach-sand, which it has
cemented into a rock often hard enough to be used as a
building-stone.   Such a sandstone of recent terrestrial origin
coats nearly all the small volcanic islands of the Mediterranean.   Mr. Darwin found it in those of the Atlantic and
Pacific likewise.   The mollusks probably obtained the lime
(through their vegetable food) from the volcanic rocks, which,
especially the trachytes, usually contain a portion of this
mineral—even now, as we know, largely produced from the
interior of the globe by calcariferous springs.

These considerations seem to show that it is only when
undoubted marine sedimentary strata are found overlying, or
interbedded with volcanic formations, we can confidently assert
the upheaval of the mass above the level at which it was
originally produced.

Iceland, like Teneriffe, presents a combination of both characters.   The high mountains called ' Jokuls,' with which this
island abounds, mark the site of so many subaërial volcanos,
which continue still, or have lately been, in occasional activity ;

while two large portions of the island on the north and south consist of flat plateaux, composed of repeated beds of basalt and basaltic conglomerate (trap-tuff), bearing marks of a submarine origin.

In the Isles of France and Bourbon a similar contrast is observable,—the former having all the distinguishing characteristics of a submarine volcanic formation, elevated bodily since the cessation of the eruptive phenomena; the greater portion of the latter presenting the appearance of an ordinary volcanic mountain, produced by repeated extravasations of lava from two or three habitual sources above the sea-level,—one of these vents remaining in continual activity to the present day.

Where a volcanic mountain has been elevated in mass by inferior expansion, it will, of course, be accompanied in its rise by the recent beds of marine formation which had been deposited upon it; and even strata of earlier date may occasionally be raised with it by the same expansive force acting upon a more considerable extent of surface than that which supports the volcanic mountain.

Thus the northern extremity of the Ile de France presents a flat plain, composed of a calcareous rock full of recent madreporites and other coralline bodies, overlapping the volcanic rocks which make up the remainder of the island. The neighbouring islets of La Platte and Les Colombiers offer a similar conformation.

So, again, in the cluster of islands lying eastward of Java, the greater part of their surface is composed of coralline beds, unmineralized, and exactly similar to those which constitute the neighbouring reefs still in progress of formation below the level of the sea. These beds visibly rest upon volcanic rocks, which have in this instance, no doubt, been elevated in mass, together with the coral reefs which had been built upon their highest eminences.

The island of Pulo Nias, on the western coast of Sumatra, which is seventy miles long and twenty-five wide, exhibits the same fact throughout its whole extent, with this exception, that the coral-beds rest on stratified sandstones and limestones of older character. The nearest coast of Sumatra, however, is volcanic; and the elevation of these extensive recent coral-beds to a height of some hundred feet is most probably owing to subterranean expansions coëtaneous with the volcanic phenomena of the neighbouring island *.

There seems, indeed, great reason to believe the almost innumerable coral-islands of the Pacific and Indian Oceans to be generally based upon the summits of submarine volcanic mountains. Their usually circular or elliptical figure, however, is not always to be attributed to their being built upon the circular ridge of a submarine crater, since Mr. Darwin has ingeniously demonstrated that this figure would be equally the result of the operations of the madrepore worms around a gradually subsiding central island or rock of any kind. A large number of such islands, however, forming an entire (and, according to Mr. Darwin, a locally distinct) class, are clearly not sinking, but rather rising. For by the ordinary process of growth, so well described by him, the coralline mass cannot attain, at the utmost, to any greater elevation than a few feet above high-water mark; whereas, in the numerous archipelagos of these vast oceans, a very great number of islands, which are shown by their composition and structure to have been originally formed as coral reefs, rise much above this level, attaining frequently from two to three hundred feet in height, and occasionally much more.

Those, however, which have reached this height have been observed generally to consist of a substratum of some lava-rock, supporting the coralline beds; and it is a common remark of navigators that they are subject to frequent and violent earthquakes. All these facts combined make it difficult to doubt that these islands are the summits of subaqueous volcanic mountains, which, when raised to within a certain

* Dr. Jack on the Geology of Sumatra (Geol. Trans. 2nd ser. vol. i.).

distance of the surface of the ocean, are immediately occupied by the remarkable zoophytes which elaborate coral for the site of their erections; and this distance is not considerable, perhaps never exceeding a hundred feet, these animalcula requiring light, and the movement of the sea-waves, as it would appear, for their support.

Subsequently, by subterranean intumescence, acting probably by repeated shocks, the mountain is more or less elevated, and, by the continual growth of fresh coral on its shelving shores, is progressively augmented to a considerable size.

Mr. Darwin, distinguishing, as has been said, those coralline islands of the Pacific which are, or have recently been, undergoing elevation from those which are sinking, places in the latter category all the coral 'Atolls' and islands which stud the central basin of the Pacific—where, if his view be correct, a vast area of the earth's surface is, and has long been, slowly subsiding. In either case, however, he admits, as I understand him, the probable volcanic origin of nearly all these rock-foundations of a coralline superstructure*.

The Carib Isles offer another striking example of this association of elevated and eruptive rocks. The line of islands which lies farthest to the west, or "under the wind," is described as consisting solely of recent volcanic cones, strung together at greater or less distances; while the eastern islands are formed of calcareous strata, and frequently of recent coral, supported on a foundation of trachyte and other volcanic rocks. These last therefore must have been elevated, together with the overlying strata, by subterranean expansion.

I shall presently advert to the general character of this elevatory action.

* "Coralline Islands." From the report of Mr. Sawkins, it appears that the island of Tongataboo, one of the Friendly group, during a recent earthquake, was depressed on the N.E. side, so as to cause the sea to encroach on it for two miles inland, while the western coast rose several feet; thus showing the irregular action of the elevating force.

CHAPTER XI.

SYSTEMS OF VOLCANOS.

§ 1. THE occasional shifting of eruptions to fresh vents —probably along the line of the original fissure—is, as we have seen, one of the most ordinary characteristics of volcanic action. It occurs on the smallest scale, as when a train of lateral vents open in succession one below another upon the flank of a volcanic mountain. And that it has continually occurred on the largest must be evident to any one who inspects a map (or still better, a globe) upon which the geographical position of the several volcanos (active, and dormant or extinct) is accurately marked. The general linear arrangement of these eruptive points on bands which traverse large segments of the earth's surface is very obvious; and although there are several which appear to be insulated, or to form independent groups, yet, since these rise chiefly as islands from the ocean, it is not impossible that there may exist between many of them some connecting links of submarine volcanic action, unobserved as yet, owing to causes already adverted to, which preclude us from the knowledge of what phenomena of this kind are going on below the deep waters. On this account it is, I think, unnecessary to infer (as Von Buch has done), from an apparent want of connexion among such isolated volcanos, or groups of volcanos, any difference between the circumstances of subterranean energy to which they are owing, and those which gave birth to the linear volcanos. If we are justified in presuming that the great trains of vents—such, for example, as that which all but encircles the Pacific Ocean—indicate the existence of extended fissures, or lines of dislocation, across the earth's crust, through which subterranean igneous matter has for a vast series of ages been habitually protruded, we may reasonably believe that other eruptive fissures exist beneath the ocean, connecting some of the more dispersed localities of volcanic outbreak by intermediate vents, which must remain concealed from observation unless the volcanos should raise their peaks above the water-level, and so enter into the subaërial class.

§ 2. That a volcanic eruption occurs as a consequence of the production of a linear rent or fissure through the solid overlying rocks seems, indeed, to be the normal rule, on whatever scale the phenomena develope themselves. Examples have been given of such rents, both of early and recent age, which traverse the earth's surface for distances of from 50 to 200 miles. The smaller fissures into which lava has forced its way from beneath, whether it obtain an issue outwardly or not, usually seem, before long, to be sealed up by its solidification, and rarely reopened. The main or primary fracture through which any considerable eruption has taken place probably remains for a long time filled with lava still fluid—at least at its central axis, or at some widest point or points—and in a condition, therefore, to give ready passage to fresh eruptive matter on the occurrence of any new impulse from beneath, or a further widening of the fissure by any commotion propagated from a distant quarter. Moreover there will, no doubt, always have been several adjoining and parallel—in some cases transverse —rents, formed simultaneously by the same dislocating effort —some of them mere cracks, or solutions of continuity— which, although not sufficiently opened at first to admit the intumescent lava, will yet have weakened the cohesive force of the rocky masses through which they were broken, so far as to cause the next successful effort of subterranean expansion to take effect upon one or more of these fractured lines of least resistance rather than in any new direction.

The whole tenor of the volcanic phenomena goes to support

this view,—eruptions habitually breaking out at or about the same spot, or on a prolongation of the same superficial line which unites other vents, or on a parallel or transverse line at no great distance.

Mr. Darwin remarks of the volcanic vents of the Pacific and Atlantic Archipelagos, that they have been generally formed in two, three, or more parallel rows, or in others transverse to these; and in each of these systems there is seldom more than one orifice at a time in activity *.

Whether the eruptive action should be continued from the same point, or be shifted to some fresh opening at a greater or less distance, will, no doubt, be mainly determined by the comparative amount of resistance offered at that part of the earth's crust which overlies the spot where the tension of the heated subterranean matter is at its maximum. Within what limits of horizontal or, indeed, vertical distance an habitually eruptive vent may relieve this tension by drawing off the increment of heat which occasions it, so as to obviate any eruption at a neighbouring point, must remain uncertain. There is, however, every reason to believe that a connexion (or partnership) of this kind does exist between neighbouring localities.

Some geologists seem to suppose that in such cases the connected vents ought to exhibit their sympathy by simultaneous activity. But it is rather the contrary that should be expected—namely, that a period of eruptive activity in one vent of the same connected system should concur with a phase of repose in the others. And this anticipation seems to be justified by well-known facts. During the last eighteen centuries, for example, while Vesuvius has been frequently active, the neighbouring seats of ancient volcanic energy—Ischia, Ponza, the Phlegræan fields, and Rocca Monfina—have been, with rare exceptions, completely tranquil. And the exceptions themselves tend to confirm the rule, since they occurred (viz. the eruption in Ischia of 1302, and that of the Monte

* Volcanic Islands, p. 128.

Nuovo in 1385) during periods of secular inactivity in the neighbouring vent of Vesuvius. Such contiguous vents may be therefore considered as belonging to the same subterranean focus, for the relief of which one chimney at a time suffices; just as an eruption at a distant point on the lower flank of Etna, or any other great volcano, relieves the central or axial focus in a great degree from its plethora—almost as completely, perhaps, as if the discharge took place from the summit. Humboldt was of opinion that " the whole mountainous part of Quito might be considered as one volcano, occupying 700 square leagues of surface, of which the volcanos of Cotopaxi, Chimborazo, Antisana, Tunguragua, and Pichincha are subsidiary vents. So the group of the Canary Isles may be justly reckoned as one volcano, and the whole island of Iceland another.

§ 3. Nevertheless a certain amount or, rather, kind of independence seems to be maintained between the separate foci or subterranean sources of eruptive energy belonging to one volcanic system—nay, even to one volcano. This was observed on the smallest conceivable scale by M. Deville, who saw, on the summit of Vesuvius, in 1856, two minor cones, each with its crater, within one of which a pool of incandescent lava continually bubbled up at a white heat, while the bottom of the other, at a lower level by at least 300 feet, was empty,—showing, as he justly observes, the extreme localization of the ascending lava in separate, though closely adjoining, fissures or 'chimneys.' I witnessed something similar in the crater of Stromboli (already described) in 1820, namely, two orifices within a few yards of each other, one of which was overflowing with liquid lava, the other apparently empty, and productive only of continuous windy belches. Bory de St. Vincent's description and engraving of the Cratère Dolomieu, on the Volcano of Bourbon, presents an exactly parallel example. Again, in the interior of the singular pit-crater of Kilauea (Hawaii), Professor Dana tells us that lava has been seen to flow from orifices in its perpendicular walls at a height of 200 feet above the open bottom of the pit. A still stronger fact of the same kind is, that the adjoining crater of Mauna Loa is not unfrequently in eruption, and emitting streams of lava from its summit, 13,760 feet above the sea, while the open crater of Kilauea upon its flank, at a lower level by nearly 10,000 feet, and at a distance of only sixteen miles, continues as tranquil as usual, its pool of liquid lava showing no sympathy with that which is at the time rising, under a prodigious impulse from beneath, in the neighbouring axial vent of the same mountain. "How," asks Dana, "if there were any subterranean channel connecting the two vents, could this want of sympathy exist? How, according to the laws of hydrostatic pressure, can a column of fluid stand 10,000 feet higher in one leg of the siphon than in the other?" He concludes that "volcanos are no safety-valves, as they have been called; for here are two independent and apparently isolated centres of volcanic activity, only sixteen miles distant from each other, sustained in one and the same cone *."

Dana, however, in this reasoning overlooks the distinction between the flow of heat, and that of a heated liquid. Undoubtedly it is clear that in all these cases there could be no immediate *fluid* connexion between the two adjoining fissures of discharge, or the enormous hydrostatic pressure of the liquid lava must have forced it to rise in both to the same level. The irregularity and narrowness of one of the two fissures may, indeed, to a certain extent have impeded and strangled the upward ascent of the lava to the same level as in the other leg of the supposed siphon; but it is more probable that, by refrigeration or pressure, causing local consolidation, *liquid* communication is completely cut off in such cases between the two neighbouring fissures. Still more, of

---

* Proceed. American Assoc. 1849. See also Sir C. Lyell's 'Principles,' 1853, p. 553.

course, is this likely to be the case between two comparatively distant orifices or vents belonging to the same system, each of which may have its independent focus of more or less liquid lava, in a condition of greater or less tension, according to the accidents which have determined the flow of heat towards it, and consequently its expansive force and tendency to eruption.

It is obvious that, though the connexion between neighbouring volcanic vents, whether on a small or large scale, may not be such as to admit of the direct flow of liquid lava from one to the other, as between the two legs of a siphon, they may nevertheless be connected in such a manner as to admit of the flow of heat by convection, through solid or semi-liquid intervening matter, so that its escape from one by external eruption may to some extent relieve the tension and consequently diminish the expansive force and liability to eruptive action of the other.

§ 4. In an earlier chapter it was shown that the phenomena of active volcanos demonstrate the continued accession of increments of caloric from some unknown source to the mass of lava (or the material whence lava is derived) beneath every habitual vent. And it was suggested as probable that it is the tension (or expansive force) thus occasioned, as portions of this mass pass from a solid to a fluid or fused state, accompanied by a great amount of dilatation, which—at length overpowering the resistance arising from the weight and cohesion of the rocks above—rends them asunder, with more or less violent shocks, and injects some of the fissures so formed with intumescent mineral matter ; and that should the upward pressure of this matter force a way through some weaker part of one of the fissures into approximatively free communication with the atmosphere, it enters there into eruptive ebullition.

If this view be correct, it would seem to follow that, on the cessation of any such expansive process, from the outward escape of the redundant heat through a volcanic spiracle, the *residuum* of molten or liquefied lava in the vent, or in what may be called its focal reservoir beneath, may be reconsolidated—its temperature being no longer so high as to keep it liquid under the pressure to which it is subject. In this condition it may be expected to remain until either the transmission of fresh increments of heat from beneath or from either side may occasion its renewed effervescence, and reproduce the same series of phenomena. Or the sudden reduction of the pressure to which it is exposed, through the penetration of a fissure into it, owing to movements communicated from a distant quarter, may bring about the same result. In this way subterranean masses of lava may (as was before suggested, see p. 123) be repeatedly fused and liquefied, or at least softened and again reconsolidated, with more or less of change, perhaps, in their mineral character.

We know, indeed, with certainty so little of what is going on beneath the immediate surface of the earth, that conjecture has the field to itself. It is a mere conjecture which, according to the theory accepted by many geologists, assumes that surface to be but a hardened crust enveloping a fluid and molten nucleus. Mr. Hopkins has shown* conclusively, that though we may be warranted in supposing the globe to have been once so far in a fluid state (at least to some depth from the surface) as to have derived its oblate-spheroidal figure from its rotatory movement, yet it is quite consistent with this hypothesis, as well as with the well-known fact of its actual increase of temperature from the surface downwards, that it may now be solid, not merely at the centre, but *throughout*. Too little is known of the comparative force of the antagonistic influences of high temperature in resisting the solidification of mineral matter, and of great pressure in promoting it, to enable us to solve such a problem. The experiments which have been recently conducted by Messrs. Hopkins, Fairbairn, and Harcourt, for the purpose of throwing

* On Theories of Elevation, &c. (Brit. Assoc. Report, 1847, p. 47).

light on this question, have as yet, it seems, owing to the difficulty of conducting them, and other impediments, presented no decisive results.

Mr. Hopkins, however, very justly remarks that—upon the assumption (a very probable one) that solidification began first at the centre of the globe, through condensation by pressure, and at a later period on the surface, through the escape of heat into the surrounding void, by the formation of a solid frozen crust, when the tumultuous circulation of fluid matter within the still liquid portion had ceased, and cooling by conduction alone took place—"the globe must have necessarily passed through a condition in which a solid exterior shell rests on an imperfectly fluid and incandescent mass beneath." He goes on to say, "Whether the exterior shell and solid nucleus are *now* united, or are separated by matter still in a state of fusion, it is impossible to infer" from any *à-priori* reasoning derived from the above-mentioned facts.

The phenomena of earthquakes have, indeed, sometimes been adduced in support of the notion that the surface of the earth is but a thin solid crust resting on a fluid sea of molten matter beneath, the undulations of which, when some cause disturbs it, occasion those of the crust above. Mr. Mallet has sufficiently shown the untenable character of this view of seismic shocks, which are rather of the nature of a vibratory jar propagated through solid rocks, and afford no support to the idea of an internal fluid.

The phenomena of volcanos, it is true, as we have seen, indicate the existence of certain masses of subterranean mineral matter immediately within or beneath them, at times in a state of greater or less liquidity—perhaps of fusion. But the non-correspondence in point of level of the columns of lava that have risen up closely adjoining channels (even within the area of one volcano, or of one crater), to which I have lately adverted, proves conclusively that the reservoirs of liquefied matter with which these several proximate ducts communicate are not always in *fluid* communication with one another.   It seems therefore certain that the condition of even imperfect fluidity in the matter underlying an habitually active volcano is local merely, and temporary, varying from time to time according to circumstances of temperature and pressure; and these circumstances must of necessity be liable to great local changes and irregularity through differences or accidental changes (which we know to be continually taking place) in the position, bulk, weight, cohesion, and consequent resistances of the rocks above, independently of variations in the rate at which heat may be supplied from beneath.   We have seen that the cooling of lava within a fissure effectively seals it up, and repairs the rent through which an eruption perhaps once took place;—how, too, the accumulation of erupted matters, solid and fragmentary, tends to augment the local sum of re- sistances to the expansive energy of the focus beneath; which, moreover, is itself lessened, for a time, by every escape of heat in its eruptions. We can therefore easily understand that any particular focus or reservoir of lava from which an eruption has proceeded may be wholly consolidated, while at the same time a neighbouring portion of the same general subterranean mass of mineral matter is undergoing a gradual increase of temperature and consequent tension, and approaching or even reaching to fusion, and so preparing for an eruption, which may break out on a small or a large scale, and at a greater or less distance, without affecting, to outward appearance, the neigh- bouring recently active (or perhaps still only dormant) vent. The activity of the one focus may be readily supposed to draw off, through intervening solid matter, some portion of the caloric whose increase tends to cause the ebullition of the lava within the other (and so far act as a safety-valve), without doing this fast enough, owing to peculiarities of structure, texture, and composition in the intermediate rocks, wholly to prevent its occasional effervescence and eruption.

§ 5. This view may not only be true as respects portions of

subterranean matter and volcanic vents laterally adjacent, but probably is so likewise as to foci at different levels one below the other.   For it has been shown by facts that the moderately tranquil activity of a volcano continued for a considerable period does not prevent the occurrence of a paroxysmal eruption from the same vent, but proceeding from a focus apparently at a greater depth beneath the mountain, whose increasing temperature and expansive force have not been sufficiently lowered by the slow conduction of heat through the intervening rocks to effect its full relief by means of the eruptions that have occurred from higher-seated and less heavily weighted foci.

§ 6. If, then, we endeavour to arrive at some definite idea of what is going on beneath the crust of the globe in volcanic localities (especially along the course of the primary fissures of eruption), it would seem highly probable that there may exist there separate parts, or pockets as it were, of more or less intensely heated mineral matter, at greater or less depths and horizontal distances; some, perhaps, comparatively cooled down by past expansion; others gradually acquiring, by increase of temperature, that extreme tension which will, sooner or later, enable them to overcome the resistance caused by the weight and cohesion of the solid overlying rocks, and obtain relief either by dislocating and uplifting a considerable area of the latter, or by a volcanic eruption.   Between the different parts, heat (of an intense character, such as to maintain the whole mass in a state of extreme tension) may be everywhere circulating by conduction, seeking an equilibrium— which is as much a law of its nature as it is of water to seek a level.   And as the visible result of these subterrestrial changes, earthquakes and eruptions may take place from time to time along and in the vicinity of those lines of dislocation which have become the habitual channels for the outward discharge of internal heat; that is to say, within which volcanic chimneys have already established themselves.

No doubt, it is *possible* that the deepest of all these presumed reservoirs of volcanic force may be in communication with a widely extended belt, or even continuous shell, of molten or liquid mineral matter intervening (as suggested by Mr. Hopkins in the above-quoted passage) between the nucleus of the globe, solidified by compression, and its outer crust, hardened by cooling through radiation of its heat into space. At all events, the supposition (in itself so probable) of the existence of such a continuous envelope at a former period will best account for the very general dispersion of volcanic vents or eruptive fissures over the whole surface of the globe, and for the occasional appearances of connexion or interdependence between the several existing fissures or habitual issues, although at wide distances from one another.

But it is not unreasonable to suppose that, even in that inferior belt, the intervals between the different principal fissures of volcanic eruption may long since have been completely solidified, and admit at present of the tranmission of heat only by conduction (whether outwardly or laterally)—not by circulation, or actual transit of the heated matter. And these intervals would in such case either preserve at present a stationary condition, or obtain relief by the comparatively slow upheaval of large areas, accompanied by little outward display of energy in the shape of either earthquake or volcano, or, perhaps, by a series of paroxysmal elevations at distant intervals. Or, on the other hand, some of them may be undergoing subsidence from the loss of heat, for the outward escape of which favourable avenues are afforded in some neighbouring quarter.

Such areas may be plausibly assumed to correspond with those extensive tracts of our globe which separate the more active bands of seismic convulsion and volcanic activity, but which yet, there is reason to believe, experience more or less of oscillatory vertical movement; sometimes by a gradual and tranquil process—a sort of creep—such as is now believed to be raising the extreme north of Norway and Siberia, as well as of America, at the rate of a few feet in a century, and similarly depressing the bed of the Baltic, the coast of Greenland, the central area of the Pacific, and a part of the Indian Ocean * ; sometimes by a paroxysmal effort, or series of efforts —such as would seem to have upheaved the mass of the Alps and Pyrenees by several thousand feet, and that of the Himalayas and the plateau of Thibet to a still greater height, since the deposition of the early tertiary strata.

* Mr. Hopkins, in his Anniversary Address to the Geological Society of 1853, p. lxx, speaks of "movements of depression, acting slowly and continuously during long periods of deposition of sedimentary matter, as proved by the immense observed thickness of such stratified matter, reaching at times to 20,000 to 40,000 feet. That such movements have taken place follows as a necessary consequence of the law of distribution of organic beings, which asserts that each class of marine animals can only flourish within comparatively small limits of depth. There may also have been slow and continuous elevations ; but of such movements we have not the same demonstrative proof as of the continuous movements of depression."

The force of this evidence, however, has been somewhat modified by the recent discovery of many species of mollusks attached to electric cables which have been taken up after lying some time on sea-bottoms at depths of even 5000 or 6000 feet. Nevertheless, that large superficial areas have been subject to depression continued through long periods of time and in every geological age, is a fact which no geologist now thinks of disputing.

CHAPTER XII.

RELATION OF PLUTONIC TO VOLCANIC ACTION.

§ 1. In support of the hypothesis advanced at the close of the last chapter, we have, in the first place, the well-known evidence of mines and artesian wells to the fact that the temperature of the crust of the globe increases everywhere in a very rapid ratio from the surface downwards, varying from one degree in 50 to one in 100 feet of vertical depth, and consequently that a large amount of heat is continually in course of out-ward transmission from within this envelope through the superficial rocks, and the waters that permeate or cover them, into surrounding space. Secondly, we have the phenomena of volcanos, proving, as has been shown, that, besides this, another considerable amount of heat is continually effecting its outward escape—with less regularity, but with equal con-stancy—by the exhalation from within of heated vapours and thermal waters, and the eruption of incandescent lavas. The continuance of these phenomena, through every past age of the globe, proves the accession of continual increments of caloric from great depths within its interior to the mass of lava, or the materials from which lava is elaborated, that underlies the outer hardened and comparatively cooled crust.

It would seem, indeed, that the outward transmission of internal heat by these two combined modes is insufficient for its discharge as rapidly as it is supplied from within, inasmuch as a third collateral order of phenomena, the plutonic—*i.e.* the occasional upheaval of large areas of the solid surface of the globe, accompanied in some cases, perhaps not sensibly in all, by earthquakes—attests the frequent expansion (only to be

accounted for by increased temperature) of extensive underlying masses of matter.   But it is possible that the elevation
of some tracts may be compensated by the proportionate depression of others at a greater or less distance, and therefore
that these oscillatory movements may be the result rather of
the lateral shifting of the flow of heat from one mass of subterranean matter to another neighbouring one, than of its
positive increase on the whole.   Such a lateral diversion of
the outward flow of heat we may presume to be caused (as
was suggested in the first edition of this work *, and more
fully argued by Mr. Babbage in his notice on the Temple of
Serapis †) by the deposition, over certain areas, of thick,
newly-formed beds of any matter imperfectly conducting
heat, like sedimentary sands, gravels, clays, shales, or calcareous mud, by which the outward transmission of heat
being checked, it must accumulate beneath, while a portion
of it will pass off laterally to augment the temperature of
mineral matter in neighbouring areas; just as the water of
a spring, if its usual issue is blocked up, will accumulate in
the fissures or pores of the rock containing it, until it finds
another vent on either side and at a higher level.   Owing to
this increase, the resistance opposed by the overlying rocks
in that quarter may be, sooner or later, overcome, and their
elevation brought about, through the dilatation of the mineral
matter beneath.   Their upheaval may take effect by violent
and sudden jerks or paroxysms ; or gradually and slowly, like
a creep : but in either case it must be accompanied by their
dislocation and the production of cracks, faults, and fissures
through them ; the rending asunder being probably accompanied in each case by a jarring vibration propagated through
their horizontal extension to considerable distances, corresponding to those sudden undulatory shocks of the surface
which we call earthquakes.

* 'Volcanos,' ed. 1825, p. 30.
† Geol. Proc. vol. ii. p. 72 ; and Ninth Bridgewater Treatise, p. 200.

Such rents will be, for the most part, vertical—that is, at right angles to the strain of the surface-rocks. Many of them (the greater number probably), especially when the expansion which occasions the elevation takes place at great depths, will be mere faults, *i. e.* separations of continuity, allowing massive portions of the solid beds affected to rise above the level of the rest, the two sides of each cleft remaining in close contact and tightly pressed together, though perhaps more or less shifted in relative level. Others will be wedge-shaped; that is, open at one end and closed at the other. And of these, some will *gape* downwards, towards the heated and expanding mass beneath; others upwards, towards the outer atmosphere.

We have seen, in an earlier chapter (p. 48), that there will be a tendency to the production of outwardly opening fissures of this last kind towards the central parts of the elevated area, the lower portions of which will be kept forcibly closed by horizontal compression; and that, on the other hand, the fissures formed on the margins of the upheaved area will tend to open downwards, being similarly closed above. These last alone, then, will be injected by the rush of heated and liquefied matter from beneath, which may or may not, according to the nature of the fractures, force its way into communication with the atmosphere and enter into eruptive ebullition as a volcano. Or it may remain for some time in a molten con-dition at a greater or less depth beneath the outer surface, as one of those localized pockets or focal reservoirs which have been already spoken of, in a state of extreme tension, ready to break out at any time into external eruption, should any new commotion of the confining rocks so far moderate the com-pression to which it is subjected as to enable it to rise through some newly opened or widened fissure into the vicinity of the atmosphere. On the other hand, the expanding matter below the central axis of dislocation in the elevated area, though upheaving the solid rocks above in anticlinal or tilted masses, will not only have to sustain their entire weight, but will be

resisted in its upward progress by the *jam,* or horizontal compression, which must tend forcibly to squeeze together the
opposite sides of any fissure formed in that situation*.

§ 2. The results, then, of such a local change of temperature
would seem to be, first, the dilatation—whether or not amounting to fusion—and consequent upward pressure and bodily
rise of the expanding matter beneath the centre or medial
line of the area affected, but without producing its outward
extravasation there; and, secondly, and at the same time,
the upward rush and (sooner or later, probably) the external eruption of portions of this heated and fluidified matter
through fissures formed towards the margin of the elevated
area, and ranging in parallel lines on one or both sides
of its central axis of maximum upthrust. The latter process
will not necessarily be accompanied by any elevation or
even disturbance of the rocks through which these fissures
are broken—indeed is more likely to be attended by their
subsidence than their elevation, because the escape of much
molten matter and hot vapour at these points may admit of
the partial sinking of some adjoining area of the overlying
rocks, their enormous weight being no longer supported to
the same degree as before. That such subsidence does, in
fact, sometimes take place is shown by many known instances
where the upper stratified rocks in the vicinity of an eruptive
fissure or volcanic vent, instead of dipping outwardly from it,
are found to dip inwardly towards it,—this dip evidently
originating in the depression of the parts adjoining the fissure
on one or both sides as the lava from beneath escaped
outwardly.

Mr. Heaphy gives an example of this kind from the New
Zealand volcanos in the neighbourhood of Auckland. (See
fig. 63, and Quart. Journ. Geol. Soc. xvi. p. 245.)

Mr. Darwin describes and delineates a perfectly similar fact,
in the island of St. Jago†.

* See note, pp. 50 & 51, *supra.*   † Volcanic Islands, p. 9.

T

Fig. 63.

Tertiary strata dipping synclinally towards a volcanic vent, in a cliff-section
near Auckland.   (C. Heaphy.)

To what distance the relief so caused by the escape of
erupted matters may be propagated horizontally along the
underlying heated zone, in such degree as to admit of the
actual lateral transfer of fluid matter under the squeezing
pressure of the overlying rocks, must be doubtful.   But it is
quite conceivable that this effect may in some cases extend far,
and cause the depression of overlying areas more or less remote.   And even beyond the limit at which fluid movement
ceases, heat may be expected to pass away by conduction, and
slow subsidence to occur.

The first class of results will be signalized outwardly by
earthquakes, and *upheaval* of the superficial rocks affected in
a conical or dome-shaped mass, or more probably in a great
anticlinal ridge ; the latter by the *eruption* of lavas and gaseous
explosions, and occasionally the slow *subsidence* of neighbouring areas.

Sir J. Herschel * exclusively attributes these alterations of
level in the earth's crust to "changes in the incidence of
pressure on *the general substratum of liquefied matter* which
supports the whole." I think I have adduced sufficient grounds
for doubting that the substratum is generally liquefied, and
for the belief that the phenomena of elevation and depression

* Phys. Geogr. p. 116.

may be equally explained by the supposition of the transference, not of *liquid* matter, but of *heat* from one part of a *solid* substratum to another, and that the cause of this transference is not so much variation of pressure, as of obstruction to the outward escape of heat, owing to the accumulation or diminution of sedimentary strata above—more or less imperfect conductors of heat. It is, however, conceivable that the lateral flow of heat thus occasioned may be often accompanied by the fusion of those layers of matter through which it finds a way; in which case pressure will operate to some extent in the mode suggested by Sir J. Herschel, to transfer bodily quantities of the fluidified matter from beneath an area where sedimentary matter is accumulating towards another which offers greater facilities for its escape, either through habitually eruptive fissures, or by intrusion into lines of previous dislocation. This view of the subject seems to be that of Mr. Babbage, which Professor Phillips, in his able Presidential address to the Geological Society in 1859, spoke of as "securely founded on the inevitable effect of the vertical displacement of the interior surfaces of equal temperature by every removal of matter from the land and its deposition in the sea—masses of the earth's crust being in consequence changed in bulk and changed in place." Indeed, without the continual transmission of fresh heat from below, varying its upward course according to the varying conductivity of the overlying matters, it would seem that, in place of continual oscillations, both the internal isothermal planes, whether liquid or solid, and the outer terraqueous surface of the globe, under the influence of meteoric abrasion, must long since have arrived at a uniform level, and thenceforward remained undisturbed.

§ 3. The relation here presumed between the plutonic and volcanic forces will account for such facts as the precise coincidence of the eruption of 1835 in the island of Juan Fernandez with the great earthquake which affected the coast of Chili at a distance of 300 miles, on the east, and raised it by several feet, at the same instant of time*; and other instances already mentioned (see p. 7) of the coincidence of volcanic eruptions with violent earthquakes affecting neighbouring regions.

It is yet more strongly supported by the very general fact that eruptive vents are found to occur in strings or lines at some distance on one or both sides of the principal mountain-ranges of the globe, and preserving a decided parallelism either to their axes of *maximum* disturbance, or to the outlines of the nearest elevated areas; while parallel depressed areas are often observable on the other side of the train of vents, where subsidence may be presumed to have operated contemporaneously with the rise of the upheaved area and eruptions from the volcanic fissures.

It is, I think, impossible to cast a glance at any map of the globe on which the sites of volcanic development are indicated without being struck by the truth of this generalization, which I ventured to proclaim in the first edition of this work in 1825, and which has been since admitted, and attributed to the cause to which I then assigned it, by MM. Humboldt, Von Buch, Darwin, Lyell, and other geologists, as well as by Sir John Herschel, in his recent volume on Physical Geography†.

* So, too, the three great volcanos of Chili—Osorno, Minchinmado, and Orcovado—immediately opposite the Isle of Chiloe, were observed by Mr. Douglas (who at the time of this great earthquake was residing there) to break out into violent eruption at the moment of its occurrence. They remained eruptive for several months. Again, in the account of the earthquake of 1822, at Valparaiso,—"At the instant of the shock two volcanos in the neighbourhood of Valdivia burst out suddenly, with great violence and noise, illuminating the sky for several minutes, and as suddenly subsiding to a quiescent state."

The inhabitants of the whole coast, says Mr. Darwin, are firmly convinced of an intimate connexion between the suppressed activity of their volcanos and the more formidable tremblings of the ground (Geol. Trans. Ser. 2, vol. v. p. 616).

† "Trains of volcanic eruptive vents are, as a general fact, parallel to, or, as in the Andes, crown the crests of, mountain-ranges or raised continental shores" (Darwin, 'Volcanic Islands,' p. 129). See, too, Her-

For example, throughout the entire length of North America, the train of volcanos that, with occasional intervals, closely borders its western shore maintains a very exact parallelism with the not far distant elevated axis of the Rocky Mountains—the back-bone of that continent—and its prolongation southwards through Mexico.   In South America, the immediate coast-range of the Cordilleras is itself eruptive, and has been so through all past geological periods up to the present day—from the commencement of the deposition of those vast sedimentary formations which are seen to form a great system of parallel elevated ridges and terraced plains, flanking on the east in close proximity the lofty chain of volcanic vents along the entire north and south extent of the continent.   Most remarkable, too, is the fact, that precisely at that medial spot of the two Americas where the breadth of elevated land is reduced to a narrow isthmus, and its height to a few hundred feet, between the 10th and 20th degrees of N. latitude, we find both an extraordinary development of volcanic activity on the western, and on the eastern side of the sunken basin of the Caribbean Sea another parallel N. and S. chain of active vents —almost the only instance of eruptive action on the eastern side of the continent from Baffin's Bay to Cape Horn.   Here elevation and eruption have evidently taken place in inverse proportion to each other.

Again, that prodigious volcanic band which borders the Pacific Ocean on the west, from the northern extremity of

schel's 'Physical Geography,' p. 115; and Von Buch, 'Iles Canaries,' p. 664.   The complete accordance of Baron Humboldt in this view of the relation between the plutonic and volcanic forces may be seen from the following passage in the last volume of his 'Kosmos' (p. 415, Sabine's translation) :—"I am inclined to believe that islands and coasts are only richer in volcanos because the upheaval effected by internal elastic forces is accompanied by depression of the bed of the adjacent sea; so that an area of elevation borders on an area of subsidence, and *at the limit between these areas* great and profound clefts and fissures are occasioned."   This is precisely the view embodied in the text of the first edition of this work (pp. 194-207).

Kamtschatka southwards, follows, though at a considerable
distance, the general trend of the principal coast-lines and
elevated mountain-ranges of Eastern Asia, repeating indeed,
on some points, its lines, whether curved or straight, in a most
striking manner; as, for instance, in that portion which,
threading the Andaman Isles, Sumatra, Java, Flores and
Timor, and turning northwards through the Moluccas and
Philippine Isles, forms an advanced breastwork encircling and
parallel to the neighbouring elevated coasts of Siam, the Malay
Peninsula, Borneo, and Cochin China. And the Australasian
volcanic embranchment through New Guinea and New Caledonia to New Zealand pursues the exact curve of the high
lands of the coast of Australia adjoining it on the west.

On the other hand, the interval between these two great
lines of eruptive activity fringing the opposite continents of
Asia and America, is occupied by the vast depressed basin of
the Pacific Ocean, great part of which is believed, with apparent probability, to have been undergoing for ages a continuous process of subsidence.

Take another example. A belt of seismic and volcanic
disturbance borders on the south the broad chain of heights,
or rather platform, that composes Chinese and Russian Tartary. Taking a westward direction from the valley of the
Ganges, where it may be said to continue that of the west
coast of Burmah, just referred to, it is prolonged, through
Central India and Cutch, into Persia. The groups about
Lake Van, Ararat, and Elbourz continue it towards Asia
Minor, Syria, and the Archipelago, whence it ranges along
the whole length of the valley of the Mediterranean, many of
whose islands and coasts are, or have been, the sites of volcanic eruption, maintaining, in this long east and west course,
a general parallelism to the direction of the elevated ridge
which constitutes the spine of the Old World—across its
entire breadth from Cape Finisterre to China, through the
Pyrenees, Alps, Carpathians, Caucasus, and Himalayas. Thus,

also, there is every reason to believe that, during the period which witnessed the rise of this ridge from below the ocean (probably by a series of paroxysmal efforts)—carrying up, on its solid crystalline axes, the tertiary marine deposits now found there at elevations of some thousand feet above the sea, and throwing them often into vertical folds,—volcanic outbursts were contemporaneously taking place from distant but parallel fissures of eruption, ranging north and south across Central France, and along the whole western coast of Italy to the extremity of Sicily (parallel to the Alps of Dauphiné and Cenis, the Maritime Alps, and the Apennines); and east and west (parallel to the main range of the Alps and Carpathians) across the whole plain—then probably a shallow sea-bottom—of Germany, Bohemia, and Hungary, from the Eifel to Moldavia, and thence eastwards along the northern base of the Central Asiatic platform.

So, again, westward of Europe, we may more than suspect the existence of an irregular curving belt of volcanic development, partly still submarine, partly subaërial, threading the several volcanic groups of Iceland, the Azores, Madeira, the Canaries, the Cape Verde Isles, St. Paul, Ascension, and St. Helena,—a belt which rudely corresponds with the outline of the adjoining European and African coast, and is bordered on the other side by the vast depression of the Atlantic valley. Even within the limits of our own little islands, we may trace a parallelism in the early volcanic outbursts of the Hebrides and the north of Ireland, on the N.W., with the crystalline plutonic axis of Scotland, Wales, and Devon; while to the south of the Grampians another parallel volcanic adjunct shows itself in the old trap-dykes and eruptive lavas that range from the Friths of Forth and Tay across the island to Westmoreland.

§ 4. No doubt, innumerable exceptional irregularities must have occurred. Accidental differences in the position of the points of greatest and of least resistance in the overlying

rocks will almost everywhere have modified the general tendency to parallelism between the lines of maximum dislocation
and elevation and those of outward eruption, so as to hinder
their complete accordance.   Moreover, transverse fissures,
taking a direction more or less perpendicular to the primary
lines of dislocation, have evidently (as might be expected from
the theory of their production *) in some places given rise to
transverse axes of elevation and fissures of eruption.

As one striking example of the latter fact, I may point to
the volcanic chain of the Aleutian Isles, which branches off
nearly at right angles from the two great north and south
bands bordering the Pacific Ocean on the east and west.
Another is observable in the transverse chain of the Mexican
volcanos, and more than one instance among those of Central
America.

In the European system, the elevated axes of the great
Uralian North and South range, and of the Dauphiné and
Maritime Alps, the Apennines and Cevennes, cross the general
direction of the Alps and Carpathians nearly at right angles ;
each of these mountain-chains being accompanied by a parallel
belt of volcanic rocks.

This transverse intersection of parallel volcanic and plutonic ranges may be observed on a still smaller scale in the
region of Rome and Naples, where one volcanic band passes
through Vesuvius, Rocca Monfina, and the Alban and Umbrian
groups, parallel to the main ridge of the calcareous Apennines,
and another connects Mount Vultur, the Phlegræan fields,
Procida, and Ischia, crossing the first in a line ranging south-
west, and precisely parallel, in its turn, to the embranchment
of elevated Apennine limestone which forms the promontory of
Castel à Mare and Amalfi, and the island of Capri.

§ 5. Sometimes the relative local conditions of the antagonistic forces of expansion and resistance will have caused the
primary and transverse lines of maximum dislocation to com-

* See note, p. 50, *supra*.

bine into a curve. That of the Alpine range, from the Julian on the extreme east to the Maritime Alp on the west (doubling back, indeed, still further to the east again in the Northern Apennines), may be referred to as a well-known example. Another may be seen in the great curve of the Carpathian primary axis from Moravia eastwards, and then round to the south, where it still almost blocks up the valley of the Danube and joins the Balkan, enclosing the whole of Hungary in its sweep. We may compare to these curved dislocations that remarkable circular sweep, noticed above, of the volcanic train of the Pacific round the east coast of Borneo, itself apparently the extremity of a spur from the Thibetan platform forming the elevated axis of the Cambodian peninsula.

Sometimes the force of upheaval, instead of being concentrated on a line, and so giving rise to an axial upthrust, will have been spread over a wide area, and caused the elevation of an extensive tract, the strata of which will have preserved more or less of their horizontality, though probably traversed by numerous faults, and here and there perhaps squeezed up into undulating folds by accidental irregularities of pressure. Of this I may instance, on the largest scale, the enormous plateau of Central Asia, and the vast plains of Siberia and European Russia north and west of it. In not a few cases it is allowable to suppose that, instead of an upthrust axis tilting the overlying horizontal strata on both flanks, one side of the main primary fissure has been alone, or principally, elevated, leaving the area on the other side comparatively unmoved, or retaining its original level. Of this disposition it would not be difficult to find examples. Indeed, Mr. Symonds, in his admirable paper on the Malverns, suggests this very theory to account for the peculiar phenomena of that upraised tract*. And the entire continent of South America may be instanced as an example of the same kind on the largest scale,—the great, nearly straight, north and south

* Quart. Journ. Geol. Soc. 1860.

primary fracture, along the west coast, having in this case given issue to a vast series of volcanic outbursts, accompanying, through a long term of ages, the progressive upheaval of the wide sloping plateaux that form the basins of the Orinoco, Amazons, and Plate Rivers, and the Pampas of Patagonia, together with the axial ranges of Brazil, on the east, while on the west the bed of the Pacific either remained stationary or subsided.

This general parallelism of the lines of external eruption with those of maximum upheaval, and their proximity to areas of presumed depression, cannot be looked upon as fortuitous. Some general cause must have occasioned so general a fact.

§ 6. The rare occurrence of active volcanos within the interior of the raised continental tracts seems to confirm the view here taken as to the nature of this cause. It moreover accords with and explains the usual structure of the axial ranges of such elevated tracts observable wherever denudation has sufficiently disclosed it—viz. a central core of hypogene crystalline rock—granite, syenite, or porphyry—which has evidently been thrust up from beneath, and has carried up, and shouldered on either side, the schistose and sedimentary strata that overlaid them. The absence, among these axial crystalline masses, of glassy or vesicular lavas, scoriæ, pumice, ash, or any other of the characteristic products of subaërial or submarine volcanos, proves the different circumstances under which they were extruded*.

§ 7. Now what was the character of this plutonic matter at the time of its upthrust? The opinion of M. Scheerer, of Christiania, upon this subject, formed from close and mature study of the great development of granite in Scandinavia, may be thus epitomized. After proving by analysis that water is combined chemically with the crystalline minerals

---

* It would be convenient if the term 'eruption' were confined to volcanic action, and that of 'expulsion' or 'extrusion' employed to designate the upthrust of plutonic matter unaccompanied by aëriform explosions.

of granite in proportions reaching even to ten per cent. in some, he concludes that all granite formed at one time a kind of watery paste—' *une bouillie aqueuse*,' or moistened magma, into the composition of which hydrates of silex, of alumina, and other bases entered ; that it occupied in this state a very much larger space than in its actual solid condition ; that it seems to have been intensely heated, under a degree of com- . pression sufficient to prevent the vaporization of the water, the result being that the solid atoms already separated by the heat would be still more separated (or tend to separate them- selves still further) by the interposition of high-pressure steam, which would greatly add to the liquidity of the mass. This condition of the granite, he says, although it may be called a state of fusion, is not one of simple igneous fusion, and the results on its cooling would be proportionately different. The crystals of felspar and others not containing water would crystallize first—the mica, which contains much water, pro- bably next—and the silex, which the heated water would longest hold in solution, last. This silicate in its liquid state moreover would fill the shrinkage-rents or other crevices formed in the granite as it consolidated, giving rise to quartz- veins, &c.* M. Elie de Beaumont fully adopts these views of M. Scheerer, which are identical with those put forward in the first edition of this work, and are, I believe, now generally accepted by geologists.

§ 8. If now we reflect on the probable condition of the upper layer of the plutonic granitoidal matter at the time of its upthrust against or through the superincumbent solid crust—crystalline or granulated in its texture (the process of crystallization being perhaps commenced but not completed), yet so far liquefied as to penetrate the finest crevices of the rocks against which it is pressed—at an intense temperature, as proved by its metamorphic influence upon them—and

* " Sur la nature plutonique du Granit et des Silicates crystallins " (Bulletin, 2ᵉ sér. iv. p. 479 *et seq.*).

subject itself to an enormous squeeze between the upward pressure of the expanding mass beneath, and the downward pressure occasioned by the weight and cohesion of the overlying rocks, as well as to a violent more or less horizontal compression or frictional shove from either side towards the central fissure of dislocation and upheaval—I think we may perceive, as a necessary result, that the dragging movement impressed under these circumstances upon the more or less solid crystalline particles composing this upper layer must have forced, or led, them (according to the degree to which the process of crystallization had proceeded) to take that laminar arrangement which in an earlier chapter was shown to be produced in the felspathic lavas by similar conditions of lateral movement under great pressures, and which, acting upon mineral matter of the triple composition of granite, will transform it into a rock resembling the typical character of gneiss, or 'laminated granite,' and, if continued, would be likely (as shown also in the case of the felspathic lavas) to crush and contort the rock so laminated into those capricious zigzag folds which characterize the crystalline schists. Whatever crevices were formed during this process (as would especially be likely at the extreme angles of flexure) must be instantly occupied either by the expanding granitic magma from beneath, or the more liquid siliceous juice with which the semi-consolidated gneissic matter itself was permeated, thus giving rise to the veins of granite or of quartz so frequent in these rocks*.

---

* Professor Rogers describes the gneiss of North America as having " a larger crystalline grain in the lower beds, where it passes into granite, lesser crystals in the middle zone, and a finer grain and minuter lamination in the upper. In these last, any insulated felspar-crystals have the form of lenticular knots " (which is precisely the effect that would be produced by a dragging squeeze). " It is penetrated by numerous dykes and veins of granite, and also of serpentine, which terminate in the gneiss, and in the neighbourhood of which it is usually much contorted." (Geology of Pennsylvania, p. 70.)

That the gneissic rocks were crystallized and solidified, or nearly so, before their injection and elevation by these contemporaneous veins, is clear from the fact of their having split asunder to admit of the penetration, as likewise from the acute zigzag foldings into which they were crumpled at the time of their protrusion. The inferior granitic matter evidently never found its way in these axial regions up to the open air while retaining a temperature high enough to cause the ebullition of the water contained in it. Its effervescence was probably stifled by the weight of the overlying masses, and especially by the crush and jam of its own upper laminated and consolidated layers towards the throat of the axial fracture. The woodcut (fig. 64) may serve to give an imper-

Fig. 64.

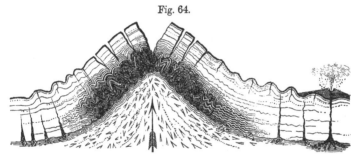

Ideal section of a mountain-chain elevated by the upthrust of a granitic axis, of which the upper layer is laminated and crumpled by the friction and oblique pressure which it undergoes.

fect idea of the supposed conditions of upthrust of such an axial wedge of granite.

The planing action of aqueous denudation, and the subsequent deposition of sedimentary strata or conglomerates upon the worn surface, have in most cases more or less obliterated or obscured the signs of this struggle. They may, notwithstanding, be recognized by observant eyes wherever natural sections of a mountain axis sufficiently disclose its structure. The granitic axial wedges were themselves probably for the most part consolidated long before they reached the position in which we now find them, since they are seen to have been repeatedly fractured and penetrated by dykes of varying mineral matter, whose intrusion was, no doubt, synchronous with the successive steps of their elevation.

Such repeated intrusions must have caused the proportionate lateral as well as vertical dilatation of the axial mass, and shoved off the elevated strata on either side.

§ 9. It is true that an axial injected mass of liquefied granite, cooling by outward conduction of its heat, would, as it consolidated, lose about one-sixth of its bulk, and sustain proportionate shrinkage-rents more or less at right angles to the cooling surface. But these, I presume, would be immediately filled either by injection from beneath, or by exudation of the finer crystalline matter, or silicate, from the sides of each fissure; thus maintaining the full horizontal dimensions of the mass at the expense of its height: so that when fresh expansions beneath fracture it again, and inject new matter into its veins, it will still grow in bulk laterally, and continue to thrust off still further on either side the strata that lean against it, and perhaps compress them into parallel folds.

It does not, however, seem at all unlikely that some of the more extended masses of the kind were expelled at a high temperature and in a condition of imperfect liquidity (such as we have recognized in many of the large-grained granitoidal trachytic lavas), and spread over the bottom of an ocean too deep to admit of any gaseous ebullition. I refer particularly to such masses as the old granitic platforms of Central France, of Brittany, of Devonshire, of Scandinavia, &c. In some spots, as in the Val di Fassa, there is an evident overlapping of fossiliferous strata by extravasated syenite or granite *.

* The syenite of Skye has been partly *protruded* in huge semi-solid bosses (thrusting aside the disrupted strata, which it has altered to a remarkable degree) and partly *erupted* in masses which overlie shales of the Liassic age without disturbing them, but insinuating itself into cracks,

A crust would, no doubt, harden on the upper surface of all such extruded crystalline masses, which, as the process of cooling, and consequent contraction, proceeded downwards, must have formed a sort of arch, pressing with a powerful lateral thrust upon the sides of the fissure it occupied, or whatever rocks bordered its flanks. The latter would be thus again forced outwards and more or less crumpled.

§ 10. The repetition of upthrusts, and intrusions of heated mineral matter into fresh ruptures from beneath, must, sooner or later, force the bulk of the crystalline axis, in a solid boss or wedge, entirely through the higher folded and laminated layers, as well as the overlying rocky strata ; both of which would be partly carried up with the rising axial ridge, partly tilted and shouldered aside into positions where their own gravity must assist the tendency of the lateral pressure of the intrusive wedge, and cause them to slide off on either side in fractured blocks, or waving undulations parallel to the main axis, according to their degree of induration, or softness, or slipperiness at the time. Fresh injections of intrusive granite into the axial mass taking place from time to time would have the effect of driving these lateral strata further and further aside, occasioning, after the manner of landslips, an infinity of fissures, faults, corrugations and other disturbances, varied by such accidental resistances as the lateral movements might encounter. Owing to these irregular influences, the corrugations may be formed but partially parallel to the primary axis of elevation—which, indeed, itself may (as has been already observed), from similar causes, lose its rectilinear direction and become curvilinear or otherwise distorted.

If, as seems not at all improbable, the lateral corrugations should reach down to depths at which the mineral matter is

and conforming exactly to all the inequalities of the stratification. There is a difference in the texture, says Mr. Geikie, of the 'disrupting' and the 'overlying' syenites. The former is coarser in grain, the latter finer and more felspathic (Geikie on Skye, Quart. Journ. Geol. Soc. 1847, p. 14).

in a molten state, or in such a condition of tension as to intumesce upon its relief, to a certain extent, from pressure, this result would be likely to take place at the base of the anticlinal curvatures, while, on the other hand, any cracks that may be formed at the base of the intervening synclinal folds would, owing to their gaping downwards, be instantly injected, giving rise to the trap-dykes, or intrusive wedges of igneous rock, which we frequently observe in such positions (see fig. 65).

Fig. 65.

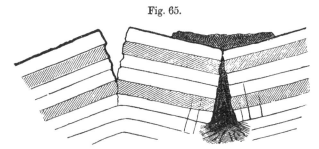

Example (of frequent occurrence) of plutonic rocks at the junction of anticlinal, and volcanic at that of synclinal, strata.

The snap caused by the creation of each of such fissures would, I imagine, occasion those sensible superficial vibrations through the adjoining and overlying rocks, which we call earthquakes.

§ 11. I observe that Professor Rogers attributes the parallel undulations of the superficial strata—which, as remarkably exhibited in the two typical examples of the Appalachian and Jura chains, he admirably describes—to the effect of a wave of translation impressed on liquid matter beneath, and propagated laterally in the manner of a sea-wave, raising the thin and imperfectly consolidated crust in wrinkles as it moved forward *.   I do not think it necessary to suppose either the

* " The wave-like structure " (of the Appalachians, &c.) " is caused by an actual pulsation in the fluid matter beneath the earth's crust, propagated in the manner of great waves of translation."   " The oscillation of the crust produced by an actual *floating* forward of the rocky parts is the cause of an earthquake." (Rogers, Geology of Pennsylvania, p. 886.)

action so sudden and single, or the crust so thin, or the liquid substratum so generally distributed beneath it, as this theory requires, which is, in fact, only the old one that ascribes the undulatory movements of earthquakes to the transmission of waves along the surface of a fluid underlying the thin solid crust of the globe. I have already said that I agree with Mr. Mallet in believing the vibratory undulations of earthquakes to be transmitted through solid matter entirely, and not to require for their production any hypothetical fluid substratum. It seems to me that the lateral and parallel foldings of the superficial strata on one or both sides of an elevated range may be amply accounted for by the conjoint influence of the two causes which I have here indicated : viz., first, the horizontal compression to which (on the principle exemplified in a breaking beam) the upper layers of an elevated area will be subjected towards its marginal limits (see fig. 4, p. 49) ; secondly, the immense lateral thrust that must be consequent on the tilting up of the overlying strata at a high, or, indeed, any angle, and the repeated extrusion through them of axial wedges of hypogene matter, which, as they were consolidated and began to subside, would (as explained above) force outwards the lateral abutments. The strata acted on, it should be remembered, were mostly at the time, as shown by the sharpness of their flexures, in a soft and pasty state, permeated, no doubt, by water, and formed of clayey micaceous and other slippery substances—therefore liable to be stretched and crumpled up even under the influence of their own gravity alone, as they slid away laterally towards a lower level, in the manner of a landslip.

The very similar but more intensely crumpled and zigzag foldings of the inferior strata, or, rather, of the upper laminated granitic matter beneath these (gneiss and the crystalline schists), I attribute, as already stated, to the horizontal compression or jam sustained (as in the beam) by these lower layers in the neighbourhood of the axis of dislocation, aided by the frictional drag of the wedge of granitic matter pressing powerfully in an oblique direction against them in its struggle to rise and force itself through the narrow throat of the fracture (see fig. 64).

This theory I prefer not only to that of Prof. Rogers, but also to the one suggested by Mr. Hopkins, as I understand it —namely, that the horizontal compression evidenced in the crumpling-up of the lateral strata on either side of an upthrust axis was caused by the mutual pressure of great angular masses of elevated rock, *collapsing in consequence of the outward escape of vast volumes of vapour,* the expansion of which beneath had occasioned their previous elevation*. I do not believe that there is any evidence of such an escape of vapour having accompanied or followed the elevation, whether paroxysmal or gradual, of the plutonic axis of a mountain-chain. It is, on the contrary, as has been remarked above, the absence of such outward escape of vapour which mainly distinguishes plutonic from volcanic action. On the other hand, I cannot conceive the upthrust of such an axial wedge unaccompanied by the several conditions of horizontal compression and consequent crushing in both the upper and lower layers of the uplifted mass, which I have indicated above†.

§ 12. With this important exception, I agree in much of the view taken by Professor Rogers as to the effects of an axial upthrust upon the strata thrown off on either side—the ' keying ' of their flexures by the intrusion of molten matter

* Brit. Assoc. Report, 1847.

† I am not aware that this view of the crushing action of the process of upheaval, at these distinct positions, has been pointed out by any other writer on geological dynamics. I suggested it in the first edition of this work, but it does not appear to have been noticed by subsequent authors. Had Mr. Hopkins intended, in the above-cited passage, to refer only to the lateral thrust of an arched crust subsiding as the intruded fluid matter beneath was undergoing refrigeration and consequent contraction, I should fully agree with him in considering this to be *one* of the modes in which the crumpling of the lateral strata may be accounted for. (See p. 287, *supra.*)

from beneath, and its consolidation there—the peculiar steepness of the foldings on the side towards which the shove takes place, the crests of the waves breaking over, as it were, as they advance—their gradual flattening and widening as they recede from the centre of disturbance towards the unaffected areas on either side—the general parallelism of the greater undulations to the general trend of the chief neighbouring upthrust axes— and the occasional disarrangement of this parallelism, or occurrence of transverse undulations occasioned by cross-movements (like the waves of a chopped sea), perhaps at different epochs—as also the 'cleavage' of such stratified masses as are composed of inequiaxed particles, by the repeated transmission of squeezing pressures in planes parallel to the strike of the wave.

In the Rocky Mountains, the Andes, the Himalayas, and the Alps, lateral undulations, characterized by these features, are observable on the largest scale. The Cordilleras of Chili, according to Mr. Darwin, present eight or more parallel, highly tilted anticlinal ridges in succession, fifty to sixty miles apart,—intrusive volcanic rock having frequently risen through the base of the synclinal troughs. Each of these ridges is half as high as Etna. Those of the Himalayas, from the observations of Captain Strachey, appear to be of even more stupendous magnitude. He finds there a general parallelism of all the principal ridges, of the intervening valleys, or lines of drainage (which are also the lines of the chief faults or ruptures), of the strike of the elevated strata, of the succession of stratified deposits, and of the lines of igneous eruption or intrusion. The great axial heights (the snowy peaks of the Himalayas) are of granite, which in its upthrust has shouldered off the schists and strata on either side. The vast undulatory ridges of the Jura range, thrown off to the north by the elevation of the Central Alps, are well known, and have been often described. In our islands many similar series of parallel waves of strata are familiar to geologists.

§ 13. The replications of some strata, especially the schistose, are so numerous and repeated, as to suggest their having undergone a more extended lateral movement than it is easy to account for. But it should be recollected that the same pressure which crumples the strata into these folds also flattens and extends them in the direction of their planes. The distortion of some fossils in slates shows that the squeeze they underwent often doubled the original extension of each layer, of course reducing the thickness of each in the inverse proportion. A stratum therefore of shale, for example, originally measuring a mile in horizontal extent, might, during the process of replication, be so far attenuated that it would cover somewhat more than two miles if unfolded, and yet the two vertical planes between which compression took place may have approached each other in only a very small degree in a horizontal direction,—the increased extension of the squeezed strata having been gained entirely in the opposite (*i. e.* a vertical) direction—which will be usually that of least resistance. This extension of squeezed strata in the direction of their planes will be proportionate to the facility with which the particles slip or slide past one another, and hence is greatest in the micaceous shales, and least in the calcareous or coarse arenaceous rocks.

Wave-like plications, indeed, could only take place at all when the strata were at the time in such a condition of softness or composed of such slippery materials as would allow of their yielding to the force that operated on them in curved or zigzag flexures. When they were so rigid as to break rather than bend, the effect upon them of the transmission of the horizontal pressure, or undulatory impulsion, would be to break them up by generally vertical fissures into separate masses, and often to elevate the side of each fissure whence the shock travelled (or, in the case of oblique fissures, the side towards which they lean) above the opposite one—owing to the well-known tendency of separate masses, when in contact and subjected to pressure oblique to the resisting surface, to be displaced by a sliding motion (see p. 51). Hence it is that, as we pass from the axis of an elevated range over moderately inclined strata of this character, we usually find the abrupt faces of the greater faults fronting us, and, unless planed away by subsequent denudation, forming so many escarpments or steps to be surmounted in our path. There are faults in the Appalachians of 1000 feet perpendicular. The tendency of formations of limestone and sandstone, especially the former, when too much indurated to admit of folding, to break up under elevatory shocks into rectangular masses of which the strata still preserve their horizontality, is exemplified in many districts. As one remarkable instance I may mention that called Les Causses, in the old French province of the Cevennes—an extensive elevated platform of horizontal cretaceous and oolitic strata, cut up, by intricate and narrow clefts nearly a thousand feet deep, into separate blocks bounded by precipitous cliffs. Another is to be seen in the magnificent mountain masses of dolomitic limestone in the Tyrolese and Carinthian Alps.

The numberless successive movements to which the same area has often been subjected are shown by the many instances observable of faults seen to stop at certain levels; traversing, that is to say, the palæozoic and stopping at the secondary rocks; or the latter, and not the tertiary.

Indeed, the various modes of displacement to which stratified masses have been subjected, under the joint influence of the elevating force and their own gravity, multiplied by the ever-changing accidents of resistance occasioned by their peculiar composition, structure, condition, and position at the time of each shock, the number and force of these, and the shiftings of the points from which they proceeded, must necessarily have been so numerous as to defy classification or description. All the great mountain-ranges of the globe present examples of the confusion resulting from these complicated conditions of mechanical disturbance. But though the results, taken in detail, are often difficult of explanation—only a fractional amount of the operating causes being discernible at any one spot—there generally reign throughout sufficient order and uniformity of character in them to enable geologists to ascribe their phenomena primarily to the up-thrust of some great central axial wedges of crystalline hypogene rock—granite or its congeners.

Extreme distortion and replication are, however, generally confined to the vicinity of such an axial ridge throwing off layers of laminated granite and the overlying strata on either side. The conditions of resistance or of subterranean expansion have occasionally caused the elevation of a wide area in such a manner as to preserve its horizontality more or less completely—the primary fractures being confined to its marginal limits. Or the superficial rocks have been raised up on one side only of a primary fissure, leaving their continuation on the other side undisturbed, or nearly so. Such variations will necessarily have affected in a corresponding manner the position of the fissures of volcanic eruption as well as of plutonic disturbance; bringing the two classes, for example, in the latter case perhaps to coincide—as seems to have occurred with the great axial dislocations and eruptive fissures of the American continent—instead of having a considerable interval between them, as is generally the case in the European and Asiatic examples*.

§ 14. *Theory of Earthquakes.*—In the above suggestions I have referred the sensible undulatory movements of the earth's surface, which we call earthquakes, to the snap and jar occasioned by the sudden and violent rupture of solid rock-masses,

* Sir J. Herschel (Phys. Geogr. p. 302) inclines to ascribe the crumpling of strata to the subsidence of slimy sediment in hollows of the sea-bottom during its tranquil deposition. This theory fails to account for (indeed, is inconsistent with) these general facts :—1st, the increase of corrugations as an axis of upheaval is approached ; and 2ndly, their parallelism to it. Moreover there still remain faults and fractures to be accounted for. And it is evident that the same disturbing movements which so broke up consolidated strata, must have crumpled them up when in a soft state.

and perhaps the instantaneous injection into them of intumescent molten matter from beneath. Mr. Mallet, in his able
Report on Earthquakes*, on the contrary, sees, in submarine
eruptions of volcanic matter, the chief agent in the production
of the more violent earthquakes. He thinks " an eruption of
igneous matter taking place beneath the sea must open large
clefts or fissures in its rocky bottom, through which water
gains access to the ignited surfaces of lava beneath." The
water remains, he thinks, " at first in the peculiar state which
Boutigny calls spheroidal, until the lava-surface is cooled down
to the point at which repulsion ceases, and it comes into close
contact with the heating surfaces; then a vast volume of
steam is evolved explosively, and blown off into the deep and
cold water of the sea, [where it] is instantly condensed; and
thus a sort of blow and impulse (or several) of the most tremendous sort is given at the volcanic focus, and, being transferred outwardly in all directions, *is transmitted as the earthquake-shock*," &c.

In this view of the result of a submarine eruption I cannot
concur. It appears to me, like all the theories which ascribe
volcanic action to the penetration of meteoric or oceanic water
to some heated or metallic nucleus below the earth's surface,
to be arguing in a vicious circle (like the old Eastern fable
which makes the globe rest on an elephant, the elephant on
a tortoise, but omits to say what the latter rests upon); for in
these theories, the first action, or the originating cause of the
whole series, is the formation of fissures in the earth's crust.
But what causes these? Not (according to the theory) volcanic
action; for that is itself occasioned by the fissures admitting
sea-water. Not earthquakes; for they are, *ex hypothesi*,
themselves the results of subaqueous volcanic action. Well,
then, the originating cause being still to be sought in that
inferior expansive force which produces the fissures, why
may we not believe that the production of these fissures—*i. e.*

* Brit. Assoc. Report, 1850, p. 79.

the violent and sudden rending asunder of the solid rocks which form the surface of the earth, many miles probably in thickness, and whether below the sea or not—may, by the jarring vibration caused in the overstrained rocks as they snap apart, propagated on either side through their continuous masses in undulatory pulsations, be the true cause of the earthquake?—and should any of these fissures, opening downwards, penetrate so far into a heated lava-mass below as to give rise to its ebullition by relieving it from the excess of pressure that confines its intensely elastic gases or interstitial vapour—of a lava-dyke or a volcanic eruption?

I cannot but think the more reasonable theory to be, that the volcanic eruption is thus brought on by the same primary cause as the earthquake, namely, the expansion of some deeply-seated mass of mineral matter, owing to augmentation of temperature or diminution of pressure.

This, in fact, accords with what Mr. Mallet himself says in another part of his Report*:—" There is more than a mere vaguely-admitted connexion between the earthquake and volcano, as heretofore commonly acknowledged—so vaguely, that the earthquake has sometimes been stated as the *cause* of the volcano, and sometimes the volcano of the earthquake, neither view being the expression of the truth of nature. They are not in the relation to each other of cause and effect, but are both unequal manifestations of a common force under different conditions." In this view I need hardly say that I fully concur; but it appears to me, on the grounds dwelt on above, inconsistent with the notion advocated by Mr. Mallet himself, that the earthquake is the *result* of a submarine volcanic eruption.

§ 15. *Progression of Plutonic action.*—The analogy of *volcanic* phenomena, which present occasional paroxysms, but generally exhibit minor developments at successive intervals, or a continuous moderate activity, leads to the belief that the

† Fourth Report on Earthquake Phenomena, p. 67, in Brit. Assoc. Report, 1858.

action of the *plutonic* forces likewise has probably been sometimes paroxysmal, sometimes gradual (frequent moderate
efforts alternating with intervals of quiescence)—sometimes
continuous, but slow and comparatively tranquil—like a creep.
Such a view of the conduct of plutonic energy is in perfect
harmony with the observations of geologists upon the numberless examples of changes of level and disturbances in the
superficial rocks. It is to be remarked, as Mr. Hopkins says
in his paper on the Lake-district (1848), that dislocation and
elevation are not necessarily produced to an equal extent by
the same plutonic action. " Great dislocations may have
been the result of more violent, and great elevations that of
more continued or more frequently repeated action of the
elevatory forces." It is nevertheless true, as has been said
above, that the greatest amount of dislocation is *usually*
observable in the most elevated tracts of the earth's surface—
the mountain-ranges of the globe—and that as we recede from
them, the fewer and the less violent in general are the derange-
ments visible in the superficial rocks, such as faults, dykes,
vertical or high angular tilting, contortions of strata, &c.

Mr. Darwin expresses the opinion that mountain-chains
are only subsidiary phenomena to wide continental elevations,
effected very slowly and by repeated shocks, with intervals of
rest; every shock being attended with one or more fractures,
and the injection of lava or fluid stone of some kind from
beneath into them, followed by its cooling and consolidation.
" In the Conception earthquake of 1835 there were 300 successive shocks ; so the fluid stone must have been pumped into
the axis by as many separate strokes *." He believes that the
ridges of the Cordilleras were upheaved by steps as slow, probably, as those which create a mountain by the successive accumulation of the matters erupted from an intermittent volcano.

---

* Darwin " On the identity of the force which elevates continents with
that which occasions volcanic outbursts " (Geol. Trans. 2nd ser. vol. v.
p. 610).

§ 16. *Metamorphism.*—During all these alternating periods of convulsion and repose, the internal heat is, no doubt, passing, both by conduction and convection, through whatever rocky masses intervene between the intensely heated matter beneath and the external surface of the globe, accompanied by water, steam, or gases, conveying chemical agents of various kinds. By this means many metamorphic effects must be produced on those rocks. As a general rule, the signs of metamorphism and disturbance are met with together, and in similar proportions. M. d'Archiac, in the Preface to the volume of his ' History' for 1853, justly calls attention to the fact of consolidation and tendency to metamorphism in the sedimentary strata of mountain masses, exhibited by the hardening of the limestones, their assuming certain peculiarities of colour and frequently subcrystalline and even saccharoid textures, the conversion of the laminated marls and sandy clays into schistose beds, and the indurated and compact character of the sandy elements. On the other hand, the continuations of *the same beds,* when forming horizontal table-lands or extended plains composed of conformable and undisturbed strata, exhibit entirely distinct mineral characters, being comparatively unsolidified, and putting on very different features of colour and texture *. These different characters of the same set of rocks—according as they have been disturbed and crumpled up, or remain almost in their original repose—are exhibited in formations of all ages, and lead to the inference that the greater amount of metamorphism is mainly due to the greater energy of dynamical causes †.

It is, then, only what might be anticipated, that the class of rocks which have generally sustained the greatest amount of

---

* See, in proof of this, Murchison's description of the horizontal low plateaux of soft clay, or sand, which in part of Russia represent the Silurian system ('Siluria,' chap. ii.), while in the Ural they have been converted into crystalline schists, quartzite, and granular marble.

† Address of Professor E. Forbes to the Geological Society, 1854, p.76.

derangement, namely the foliated or schistose crystallines, and which are also usually found in closest proximity to the intrusive granitic matter, have suffered also the greatest amount of metamorphism—to such an extent, indeed, as to be no longer recognizable as strata of arenaceous, calcareous, or argillaceous sediments, which they perhaps were before they were exposed to intense heat, permeated by hot water or steam conveying chemical reagents of great activity, possibly fused and reconsolidated under vast pressures and powerful internal movements. There is nothing improbable, or irreconcileable with the results of experimental research, in the idea that, by exposure to such influences, strata of ordinary clay, sand, silt, or calcareous mud, may be eventually changed into crystalline limestones, slates, mica-schist, gneiss, or granite itself. The so-called metamorphic rocks may be readily believed to have had such an origin; and we may conceive this process of extreme metamorphism to have been going on from the beginning of time, and to be still in progress in the depths of the earth's crust, creating fresh supplies of molten or liquefied and intensely heated granitic matter out of the inferior beds of sedimentary deposit from the superficial waters, after they have been covered by many thousands of feet in thickness of similar strata, and perhaps depressed to depths where the subterranean igneous energy is at work in sufficient intensity.

§ 17. It is, however, quite unnecessary to suppose (although the doctrine is maintained by a large class of geologists) that the peculiar laminar structure of the chief crystalline (or metamorphic) rocks is identical with the original stratification of the clays, sands, mud, and silt, out of which they have been elaborated. It is difficult to believe that the tremendous processes they must have undergone can have left untouched the slight seams, or variations of colour and texture, which constitute the marks of stratification in ordinary sedimentary strata. Moreover in the clay-slates, which lie generally at the outer limit of the metamorphic rocks, we observe the gradual and ultimately complete obliteration of these marks, and the assumption of a new divisional structure—cleavage—under the action of but a moderate degree of metamorphism.   Is it possible, then, to imagine that in the micaceous schists and gneissic rocks, which have undergone a much greater amount of metamorphic action, these marks of stratification should have been pre- served in all their original distinctness, or, rather, rendered far more distinct than they originally were?   It is surely more reasonable to believe that in them likewise (as in the clay- slates) the seams of stratification have been wholly obliterated and a new divisional structure introduced by the powerful chemical and mechanical agencies to which they have been exposed.   The mode in which this laminated or foliated struc- ture may have been impressed on a granitic magma I have explained by the analogy of the laminar trachytic lavas, to which, beyond all question, that identical and peculiar struc- ture was given at the same time with, or immediately previous to, their crumpling, crushing, and zigzag folding.   And, indeed (as I have already observed), it is difficult to conceive any other result from the tremendous squeeze which the upper layers of a semi-solid granitoidal mass must have sustained during its pressure against and upheaval through an axial fracture in the overlying rocks, by the expansive force of some still lower matter *.   Moreover, the crumpling effect resulting

* M. Delesse distinguishes two kinds of granite—one eruptive, the other metamorphic; the latter taking often a *gneissoid* structure. Gneiss is known often to show polished and striated surfaces, proving a sliding and slipping of its parts (Schlagintweit on the Bavarian Alps).   In mica-schists these marks of friction are yet more general and unmis- takeable.

This is not the place in which the question of the origin of the folia- tion of the gneissoid rocks can be fully argued.   In a paper read before the Geological Society 12th May, 1858, I entered upon it to some ex- tent (see The Geologist, no. ix.)   The late Mr. Sharpe and Mr. Darwin, as is well known, concurred in the opinion here given, that, at least as respects the oldest or fundamental gneiss, its foliated structure is not due from the intense friction of such a wedge of rock, forcing along with it the laminated layers of semicrystalline matter, seems to explain satisfactorily the otherwise enigmatical fact of the frequent '*fan-shaped*' disposition of the gneissoid schists, and their general dip *towards* the axis of the mountain-range on the flank of which they appear, as in the Alps, the Ural, the Himalayas, and other known instances.

§ 18. A word remains to be said as to the circumstances which may *immediately* occasion any development of subterranean energy, whether on a large or a small scale. If the reasoning pursued in the foregoing pages has led to the conclusion that the latent force of subterranean expansion under very large areas of the earth's surface, as well as smaller ones, is often pressing with great and increasing energy against the resistances opposed to it from above, we may be sure that occasions will occur when, these resistances being almost overcome and on the point of giving way, the slightest casual diminution in any of their elements will give the signal for their yielding, and bring about the paroxysm which had, in fact, been long imminent, and needed but this slight change (like the last drop added to the cup) to burst into action.

The facts observed on already, in regard to the eruptions of Stromboli and other permanently active volcanos being most violent in stormy weather and with a low barometric pressure (p. 41), accord with this reasonable supposition; and we should anticipate that a similar change in the atmospheric pressure upon a wider area may give the signal for the occurrence of earthquake-shocks on a larger scale. This view is supported by some of the results obtained by Mr. Mallet and M. Perrey as to these phenomena, which are certainly found to be more frequent in the winter than in the summer, and most

to original sedimentary deposition, but to the movement of the particles under great pressure while the mass was in a condition of imperfect igneous fluidity. Professor Naumann has still more recently advocated the same view, which is, however, resisted by Messrs. Lyell, Murchison, Geikie, and others.

frequent at the autumnal equinox, when sudden falls of the barometer are common, and also at times of new and full moon, and when the moon is nearest the earth. M. Perrey concludes from these results that a tidal action operates on the fluid matter (I should rather say, on the elastic steam or gases contained in the matter) beneath the earth's crust.

§ 19. But if this be so, the question arises, whether such tidal action, suddenly and violently called forth at any time, may have determined the particular position of the principal dislocations of the globe's surface? Will it, for example, account for the very remarkable preponderance of elevated land in the northern hemisphere, especially between the 40th and 70th parallels of latitude ;—or for the peculiar angular projections of nearly all the great masses of land towards the south, while towards the north they exhibit a great longitudinal extension parallel to the equator? This configuration is so general as to give to each of the continents an outline composed of a series of unequal and irregular but nearly equilateral triangles, the apex of each uniformly pointing southwards, and the sides of course ranging obliquely to both the equator and meridian, and varying little from N.E.-S.W. and N.W.-S.E. Again, the two longest rectilinear stretches of continuous (or all but continuous) land in the Old World have almost precisely these same opposite directions of N.E.-S.W. and N.W.-S.E., the one extending from Behring's Straits to the Cape of Good Hope, the other from Donegal to Van Diemen's Land*. At the intersection of these two lines we find the Himalayas and Thibetan platform—the highest and most massive elevation on the surface of the globe; and through nearly every part of their remaining course they either coincide with or run parallel to, and at no great distance from, the principal mountain-ranges of the continents they respectively traverse.

In the western hemisphere a similar law prevails. The

* See the Map appended to this volume.

great western axial and eruptive ranges traverse the whole extent of both Americas from Behring's Straits to Cape Horn, with a general N.W.–S.E. direction, and are met, or nearly met, on the eastern side by two transverse N.E.–S.W. ranges; one stretching along the whole north-eastern coast of Greenland (but broken at Baffin's Bay), through Labrador, Nova Scotia, and the Alleghanies, towards Mexico; the other from the coast north of Cape St. Roque, through Brazil, to the Andes about Potosi. This last line may be considered, perhaps, as continued to the east beyond the narrowest part of the Atlantic, from Cape Verde, through the Atlas range of Morocco, across the Straits of Gibraltar, Spain, Brittany, Wales, the north of Scotland, and Norway, to the North Cape.

Another remarkable fact of a similar character is, that the general N.W.–S.E. direction of the great eruptive fissure of Western America is almost *antipodal* to the similar eruptive fissure of Eastern Asia, of which, indeed, it is but the prolongation; the two together—continued (as we may presume them to be) through South Shetland on one side, and South Victoria on the other, of the South Pole—nearly bisecting the entire surface of the globe *.

Does not this singular chain of coincidences—which cannot be merely fortuitous—suggest the notion that the sudden and violent rise of a tidal wave of subterranean expansion, accompanied by its antipodal swell, may have fractured the earth's crust along lines diagonal to the direction of rotation? Is it possible that the attraction of some erratic planetary body passing rapidly near the earth north of the equator and in a meridional direction, by momentarily lessening the amount of pressure upon the superficial area beneath it, may have occasioned the rise of a tidal wave in the underlying elastic

---

* The two spots of the whole earth in which the volcanic forces are now most active, viz. the wreath of islands around Borneo in the Pacific, and the circuit of the Caribbean Sea, are exactly antipodal.

matter, powerful enough to split the crust in this series of oblique zigzag fissures and their antipodals? The removal of but a moderate amount of the atmospheric pressure may have sufficed to produce this effect; and if we must suppose the waters of the ocean to have necessarily partaken of the disturbance, there are not wanting appearances, in the sudden breaks of the succession of stratified rocks and their organic contents, indicating the possible occurrence in past time of more than one cataclysm of this character.

§ 20. However this may be, geologists are now generally agreed, from palæontological evidence, that oscillations of level have occurred frequently on many, if not on all parts of the earth's surface, each having been alternately elevated and depressed more than once—indeed, repeatedly. Where the continents now spread themselves, seas once rolled; and where the ocean now prevails, land formerly existed. This alternate rise and fall has unquestionably been going on more or less, though irregularly, for countless ages; but whether in a uniform or in a progressively diminishing ratio is a question upon which geologists are divided in opinion. As respects the volcanic forces, their activity certainly does not appear to have lessened in any degree from the earliest ages to which geological observation can look back, up to the present day. Analogy therefore might be held to support the opinion that the development of plutonic action has been equally uniform. And if we believe—as we not unreasonably may—that the subterranean granitic matter, which has been so generally thrust up through sedimentary strata, may have been derived from the fusion and crystallization of similar strata depressed to a level sufficiently within the influence of the internal heat, the successive processes of fusion, crystallization, upheaval, conversion into sediment by meteoric and organic agencies, depression, and refusion, may have been continuing in endless rotation from all eternity.

This supposition, however, involves the further assumption

that the outward transmission of heat from the interior of the globe—the *primum mobile* of the whole series—has likewise been going on without diminution through all past time. On the other hand, the opposite view, namely that there has been a progressive decrease in the energy of plutonic action, coincides with the popular notion that the globe is slowly cooling from a once wholly molten, perhaps gaseous, or nebular state. I will not attempt any argument for or against either of these theories, more especially since the evidence by which the question must be determined is mainly palæontological. I content myself with saying that the latter appears to me to offer the most probable solution of the question as to the source of the internal heat of the globe, which seems, moreover, to be supported by considerations derived from astronomical data.

General Conclusions on Telluric Phenomena.

1. The earliest ascertainable condition of the lowest-known matter forming the substance of the globe is that of a granitoidal triple mineral compound, consisting generally of felspar, quartz, and mica, in a crystalline or granular, but nevertheless at times, if not always, in a softened and semi-liquid state, owing apparently to the interstitial mechanical mixture of water or the vapour of water, holding probably more or less silex in solution, among its crystals; this 'magma' being at an intense, and occasionally an increasing temperature, and consequently in a state of violent elastic tension, by which it presses forcibly against the overlying and resisting solid masses.

2. The highest layers of this matter seem to have been by that upward pressure, acting on them while in a pasty or semi-solid state, so squeezed and set in motion whenever the giving way of the overlying rocks allowed of any upward movement, as to acquire a more or less laminar arrangement of their component crystals, and in this state to have been repeatedly split and penetrated by the intrusion of some of

x

the more liquid matter beneath, and often bodily forced up the axial fissure of dislocation in crumpled zigzag folds or upright walls of solid crystalline laminated rock, to the outer surface of the globe. In their rise they would necessarily shoulder off on either side huge masses of the overlying strata, which, shoved horizontally, or sliding down laterally by their own weight, have been in their turn likewise, when sufficiently soft or pliable, crumpled up into more or less irregular parallel folds, of which the flexures are necessarily deepest, most frequent and close near the axis of elevation, and gradually become lower and wider as they recede from it—subject, of course, to frequent irregular variations in the direction and amount of flexure, determined by their greater or less solidity and structural arrangement, as well as by the interference of pre-existing resistances, or subsequent change of position.

3. The fissures which by these disturbances are formed in any of the solid rocks in such positions as to open or gape downwards into the heated lava or granitic matter beneath (and such will be for the most part formed along the margin of the elevated areas, or the inferior bends of the rocky flexures, where the strain is most intense), will be injected by the instantaneous intumescence of this matter (owing to its comparative relief from pressure), which on consolidation will seal up the fissure and produce a plate or 'dyke' of crystalline igneous rock.

4. The snap and jar accompanying the rending of every such fissure and the violent injection of heated matter into it occasion an undulatory vibration in the adjoining masses of solid rock forming the sides of the rent, which, transmitted through the continuation of these beds, produces the effect of an earthquake-shock, more or less violent in proportion to the force and magnitude of the rent, the intensity of the preceding tension, the position of the point or line of fracture, and the nature of the rocks through

which the shock is transmitted. And the superficial fissuring of the strata above by the transmission of these earthquake-waves probably produces many of those minor cracks, or solutions of continuity, and faults (that is, irregular elevation or depression of the alternate sides of a crevice) which are so numerous in all elevated strata.

5. It is only when a crevice penetrates to some pocket or focus of liquefied igneous matter that it gives occasion to the formation of a dyke ; and should the upward projection of such matter force it so far up a fissure as to obtain free, or approximatively free, communication with the atmosphere or shallow water, it will enter into violent ebullition (*i. e.* volcanic eruption), more or less temporary, until the extravasation of heated matter and escape of vapour have cooled down the contents of the fissure, or the portion of underlying matter with which it communicates, so far as to give the predominance to the ever-powerful forces of repression.

6. The mineral matter (lava) so erupted is sometimes in a state of complete glassy fusion, but more usually in one of more or less imperfect crystallization, the mobility imparted to the component granules or crystals by interstitial heated water or steam in a great degree occasioning its fluidity, though this is often very imperfect. The escape of this vapour hastens the consolidation of the matter, and the resulting lava-rock is usually more porous and fine-grained than the plutonic lava consolidated under greater pressure, and is more varied in mineral character, owing, probably, to changes which it has undergone during repeated fusion (or liquefaction) and recrystallization under varying conditions of pressure and temperature previous to its eruption.

7. The erupted matters, both fragmentary and consolidated, generally accumulate over the vent in a conical mound, the orifice whence the eruptive explosions proceeded being marked by a saucer- or cup-shaped hollow or crater. By the accumulation of repeated ejecta a volcanic mountain is

x 2

formed, composed generally of alternating beds of fragmentary matter and consolidated lava. These are usually
penetrated by numerous dykes, by the injection of which
into its mass the bulk and height of the mountain are also
more or less augmented.

8. These outward eruptions of the internal heated matter are
sometimes accompanied or followed by the subsidence of
the surrounding or adjoining area (together with the superposed volcanic masses), sometimes by its elevation. And,
as a general fact, the upheaval by plutonic action of any
area of the globe's surface is usually attended by the subsidence of some other not very distant area, and by volcanic
eruption from some adjoining or intervening point or
series of points.

9. There is reason to believe that the originating cause of
these changes in the crust of the earth is the unequal transmission through it of heat from beneath upwards, owing to
variations in the covering surfaces from the deposition of
marine and other aqueous sediments at the bottom or on
the shores of the ocean, and the abrasion of the land,—heat
being thus driven to accumulate partially, increasing in
some parts, diminishing in others, according to the varying
weight and conducting powers of the overlying masses.
Where the temperature is increasing, and the subterranean
matter consequently swelling, the area above suffers elevation with all its accompanying phenomena : where it is decreasing, the overlying subaqueous or subaërial areas un-
dergo depression, from the shrinking of the matter below.

10. The source of the internal heat of the globe, which is the
primum mobile of the whole series of changes, is a question
the solution of which I will not attempt, further than by
saying that I do not believe it to be owing to the oxidation of any metallic nucleus by penetration of water or the
atmosphere (a theory given up by Davy, its inventor);
nor do I understand how it can be due to the generation

of electric currents within the globe, as has sometimes been suggested. Some writers still insist upon the fact of volcanic orifices being found generally in islands, or adjoining the sea, as proving that their phenomena are occasioned by the access of water from above to the volcanic focus beneath. Two leading objections to this view have always constrained me to reject it : viz., 1st, that a motive power is wanting to initiate the series of operations, by forming the fissures through which the water is to penetrate to the volcanic focus ; 2ndly, that, supposing these to be formed in some unknown manner (of which the theory does not afford a glimpse), the result might possibly be a sudden explosive outburst, but hardly the long-continued, sometimes even permanent and almost tranquil eruptions, which are among the ordinary phenomena of active volcanos. I incline rather to the supposition of a gradually cooling nucleus, still retaining much of the intense temperature possessed by it at the time of its original formation. The nature of heat, however, is as yet such an impenetrable mystery, that this is a region of conjecture into which I scruple to enter.

The theory suggested above as to the emanation of the central heat not only provides a reasonable origin for plutonic upheavals and the formation of fissures and faults, but also for the occasional extravasation and ebullition of some portion of the subterranean mineral matters (known, as far as we are acquainted with them, to contain water) which increased temperature or diminished pressure has liquefied and caused to effervesce. It moreover accounts for the relative geographical position of the elevated ranges and the eruptive ones. One hypothesis alone suffices to explain the whole series of terrestrial phenomena—elevations and subsidences in mass, earthquake-shocks, and volcanic eruptions, as well as their mutual relations—that hypothesis being, the shifting of the flow of heat (which we know to be continually rising out of the interior of the earth) from one subterranean mass of mineral matter to another. It has been shown that such a shifting is not only probable, but inevitable, through the ever-varying capacities for the conduction of heat of those areas of the globe which are respectively subaërial and subaqueous —variations that must necessarily arise from the varying influence of the oceanic, meteoric, and organic forces.

This theory seems to me to explain both the plutonic and the volcanic phenomena better than any other, and the harmony and general accordance of all its parts is the best test of its truth.

APPENDIX.

CATALOGUE

AND

DESCRIPTION OF VOLCANOS,

NOW, OR RECENTLY, IN ACTIVITY; WITH SOME ACCOUNT
OF EARLIER VOLCANIC FORMATIONS.

PREFATORY REMARKS.

In the following list of the known volcanos and volcanic formations, the arrangement pursued has been dictated rather by
reference to the main lines of volcanic disturbance and eruption on the surface of the globe than by purely geographical
considerations. I have availed myself to a certain extent of
the previous compilations of Von Buch, Humboldt, and Daubeny, as well as of original sources of information; and I have
endeavoured to condense the accounts of phenomena reported
by various observers at different times and places, avoiding as
much as possible the repetition of identical and unimportant
facts. And I have not thought it consistent with my purpose
to dwell upon the obscure and more or less fabulous statements of such events as have been transmitted from distant
ages by the early poets or historians.

The chief claim that I venture to advance in favour of my
compilation, as compared with those of the authors mentioned

above, rests upon the intelligible character of the phenomena described, arising from the simplicity and general uniformity of the laws of volcanic action to which, as indicated in the preceding pages, I refer them; whereas the descriptions of the same events or appearances given in the works referred to are in too many instances rendered vague and unintelligible (to my comprehension at least) by continual references to the purely imaginary theory of upheaval or elevation-craters. It is impossible for a disciple to possess a clear idea of the proceedings of any one volcano, or of the origin of any volcanic mountain, who is taught that it is an error to suppose such mountains to be formed by *accumulation of erupted matters*, and that they are, on the contrary, mere hollow blisters (*vessies*) formed by the inflation, at one stroke (*d'un seul coup*), of previously horizontal or nearly horizontal beds*. The observer sees accumulation rapidly going on before his eyes, by the ejection of fragmentary matters and the pouring forth of lava-streams, one layer heaped up over another, from every active volcano he examines. This fact is not denied. But he is told nevertheless that volcanic mountains are not built up in that manner, but by some imaginary, never-witnessed† process of sudden inflation; and no test or criterion is afforded him by which to distinguish masses admittedly the product of accumulation, from those which are declared to have swelled up "in a night, like the gourd of the prophet." Under such guidance, it is impossible that he should acquire any clear or definite idea of the volcanic phenomena, or of the origin or disposition of any volcanic formations; and

* Humboldt, Kosmos, iv. part 1. (Sabine's translation) p. 224; Von Buch, Canaries; Elie de Beaumont, Recherches sur le Mont Etna, p. 133; Daubeny, Volcanos, ed. 1848, pp. 634–6; Dufrénoy, Terrains de Naples, p. 360.

† The only two instances adduced by Humboldt, in which such an event is supposed to have been actually witnessed, are that of Methone, reported in Ovid's Metamorphoses (!), and that of Jorullo, now ascertained, by M. de Saussure's recent examination, to be a case of ordinary eruptive accumulation. (See pp. 81–83, *supra.*)

any description of such phenomena or formations drawn up under the influence of this strange theory must be confusing and perplexing to the student rather than instructive.

This has been my chief motive for framing the accompanying Catalogue, which at least avoids that source of vagueness and uncertainty in the description of phenomena and results, which in the works of the authors referred to could not but arise from their indefinite and erroneous views as to the character of volcanic action.

In the body of this work it was stated that a more or less distinct parallelism or coincidence is traceable between the leading mountain-chains of the two hemispheres, and the linear bands of volcanic vents, active or extinct, insular or continental, by which they are traversed.

In the New World this circumstance is perhaps most strikingly apparent, which I shall proceed to show when describing its volcanic formations. The Old World presents considerable irregularity in the direction of its elevated ranges. It is, however, crossed throughout its greatest breadth from west to east by one principal and almost continuous ridge, beginning in the N.W. angle of Spain, "passing along the Pyrenees, the Cevennes, the higher Alps, the Balkan, the Caucasus, and the mountains of Elburz, through the Hindu Koh, up to the great system of Asiatic mountains which enclose the plateau of Thibet and form the frontier of China *."

Of the physical geography of Africa little is known as yet; but there seems reason to believe that the direction of its principal ranges coincides with that of its coast-lines, with the exception of one great central equatorial platform.

Other lesser ranges branch off from these main ones, at various angles; such is that which runs nearly due north, from Brittany, along the western coasts of our own island, and is, perhaps, continued beyond the North Sea, in the Scandinavian granitic axis; another is seen in the more directly meri-

* Herschel, Phys. Geogr. p. 127.

dional chain of the Ural which divides Europe from Asia, and runs nearly at right angles to the central Asiatic range.

Now it is the fact that these several mountain-chains are bordered at moderate distances, on one or both sides, by linear bands of rocks of volcanic formation, along which, therefore, eruptions have at some time or other taken place. One of these (containing, however, no present active volcano) traverses the north of Germany from the west bank of the Rhine to Saxony and into Hungary, maintaining a general parallelism to the leading direction of the Alps and Carpathians; while another parallel band on the south of that chain may be traced along the whole length of the valley of the Mediterranean, from Portugal, through Sicily, the Greek islands and Asia Minor, into Persia and Northern India; and from these again other volcanic belts break off at considerable angles, parallel in turn to the transverse axial ranges already spoken of. Such is that which borders the Apennines to the west from Sicily to their inosculation with the Alps.

I propose to begin the Catalogue of known volcanos and volcanic formations with this last belt, since it contains those which have been most frequently and thoroughly explored by scientific observers; on which account the description of their phenomena can be better authenticated and given in greater detail than in the case of less well-known examples, and will therefore form the fittest introduction of the subject.

VOLCANIC FORMATIONS OF EUROPE.

SOUTHERN ITALY. VESUVIUS.

Vesuvius.—I commence with this, the best-known and most frequently visited and cited of all active volcanos—an advantage which it derives from its immediate proximity to the luxurious city of Naples, whence its phenomena may be hourly noted.

Its elegant outline, rising from the sea-shore in a sweeping curve, which gradually increases in steepness towards the double summit, forms a beautiful object in the celebrated panorama upon which the city of the Siren looks. The cone of Vesuvius proper occupies precisely the centre of the circular area covered by the whole mountain. Its base is marked towards the sea only by a slight terrace-like step, called the Pedamentina, corresponding to and exactly continuing the semicircular curve of the cliff-range of Somma, which half embraces the cone of Vesuvius on the landward side. From that curved ridge the outer slopes of Somma descend with the same sweeping incline on all sides towards the low grounds that lie between it and the foot of the Apennine. Together, indeed, Somma and Vesuvius may be said to represent a normal volcanic mountain, possessing all the usual most characteristic features of such formations :—the recent cone with its central crater, from time to time emptied by paroxysmal explosions, and refilled by succeeding eruptions of a less violent type; the encircling cliffs of a still larger crater, the result of some yet earlier and more violent paroxysm; lastly, several minor parasitic cones upon its lower flanks, marking the outburst of scoriæ and lavas from as many lateral vents. (See fig. 48, p. 196, and fig. 60, p. 232.)

The first recorded eruption of this volcano is that of the year A.D. 79, by which Herculaneum, Pompeii, and Stabiæ were buried under its fragmentary ejecta, and the great crater was formed, of which a segment, called the Atrio del Cavallo, still separates the cone of Vesuvius proper from the cliffs of Somma.

Before this epoch it is probable that Somma alone was in existence—a single conical mountain, of which the volcanic character was scarcely, if at all, suspected *. After that tremendous paroxysm, an interval of tranquillity seems to have ensued, lasting till the year 203, in the reign of Severus, when a second eruption is described by Dion Cassius and Galen. The third took place in 472, and is said by Procopius to have covered all Europe with ashes, and to have spread alarm even at Constantinople. It was doubtless, therefore, of a paroxysmal character. Other eruptions are recorded as having occurred in the years 512, 685, and 993. The next, in 1036, is said to have emitted a stream of lava, both from the summit and sides of the cone, which reached the sea. In 1138-9 the volcano was again in activity; but after that date it remained for nearly two centuries, viz. up to 1306, in complete repose.

In 1500 it was once more eruptive, but then became quiescent again for 130 years. The old crater of Somma (the Atrio) at this time contained woods and a few small lakes; and the cone of Vesuvius proper was only 350 feet high above the Pedamentina or terrace-flat from which it rises, and which marks the level of the truncation of the south-western half of the mountain by the eruption of 79. It also contained a deep lake within its crater.

The next recorded outbreak was in 1631—another paroxysm, which, by discharging the lakes just mentioned, let loose torrents of water upon the villages at the foot of the mountain, no less destructive than the lava-streams.

Further eruptions broke out in the years 1660, 1682, 1694, 1697, and 1698, since which last date there has rarely been an interval of tranquillity of more than four or five years together. In 1737, Torre del Greco was overwhelmed by a stream of lava of great magnitude, and met with the same fate in 1760, when eruptions burst out at once from fifteen different points of a fissure broken from the summit to the base

---

* Dr. Daubeny's second edition of his 'Description of Volcanos' contains an interesting account of the early condition of Vesuvius, as far as can be discovered from any notices of it in the Roman or Greek writers.

of the mountain, each of these openings vomiting lava as well as scoriæ *. The frequent eruptions from this mountain, and its consequent changes of form between this date and a great paroxysmal eruption which took place in 1794, are recorded with clearness by Sir William Hamilton, and represented in the illustrations of his admirable work on the 'Campi Phlegræi.' (See pp. 186–8, *supra*.) After 1813 there was almost continual activity of a moderate and persistent character, until inter-rupted in 1822 by a paroxysmal eruption, the principal fea-tures of which have been described in the body of this work. (See p. 25.) The wide and deep crater hollowed out by that eruption remained tranquil, although emitting copious vapours, for four or five years. In 1827, eruptions recommenced from its bottom, and threw up there a small cone, which gradually increased until in 1830 it had risen 150 feet above the rim of the crater, and in the course of the following year poured its lava-streams over the lip down the outer side of the cone. Violent explosive discharges took place during the winter of 1831, which again emptied out the crater almost to its former depth. Then two fresh cones were formed within this cavity, and these increased until they had in turn, by their mingled lavas and ejected scoriæ, once more filled the crater, and per-mitted the outflow of lava over the external slopes, where it destroyed the village of Mauro, near Ottaiano, on the eastern side of the mountain. In 1839 another violent paroxysm occurred, which again gutted the cone, after producing two

* While these pages are passing through the press, accounts are arriving of a new and most formidable eruption having commenced close behind Torre del Greco, and threatening to overwhelm it once more beneath a lava-current nearly a mile in width. It has already thrown up two or three minor or parasitic cinder-cones. The height of the column of vapour, scoriæ, and ashes projected from these new openings is said to have reached 10,000 feet. Explosions also took place from the great central cone, as soon as those of the new vents ceased; and it is said that the shore along this part of the base of the mountain has been perma-nently elevated by some feet. The old lava-beds beneath and behind Torre del Greco were split up by radial fissures. In all these respects the phenomena seem to have been identical with those of previous eruptions from the same, or nearly the same spot, especially that of 1794.

currents of lava, one flowing to the east, the other to the west. In 1841 new cones were thrown up at the bottom of the new crater (see fig. 42, p. 188), which, by the continuance of minor eruptions, was at length filled entirely, but was emptied anew in 1850 by violent explosions.   Since that date two new cones have been formed in it, which have increased until they have raised the summit into a high and broken platform similar to that of 1821-2.   During the last five or six years the volcano has been in frequent eruption, the lava-streams generally bursting out at some point on the outer side of the cone, and often flowing into the Atrio, which is being fast filled up by their accumulation and the scoriæ which fall or roll down the slopes of the cone into it.

The recent lavas of Vesuvius are leucitic basalt; the crystals of leucite, when visible, being, as usual, dodecahedral, and sometimes as large as peas or nuts, but generally much smaller, and sometimes undistinguishable from the base, which is a confused mixture of granular leucite, magnetic iron, and augitic matter.   Its scoriæ and scoriform portions have usually an external semivitreous film, indicating the nearly complete fusion of their surfaces.   The earlier lava-currents of the volcano, of which sections are seen in the half-encircling crater-cliffs of Somma, rising above the Atrio, are also leucitic, as well as the numerous dykes which intersect them.   These leucitic lava-beds, as is generally the case with the strata composing a volcanic cone, dip everywhere away from the rim of the semi-crater at an angle of about 25°, and are interbedded with scoriæ and ash of the same mineral character.   But the external flanks of the mountain, up to nearly two-thirds of its height, are in great part formed of whitish or yellowish trachytic tuff, composed of pumice and lapillo, of the same character as that which forms the substance of the other volcanic hills east of Naples.   And since the mass of fragmentary matter which overwhelmed the buried cities at the foot of Vesuvius in the year 79 is also pumiceous, there is every reason to believe that the earliest products of the volcano—which that paroxysmal eruption broke up and ejected in such abundance—were trachytic, and probably coeval with

those of the neighbouring volcanic vents, to be presently described. Large accumulations of this pumiceous tuff occur
in places high up the outer slopes of the mountain, overlying
the leucitic strata. They were most likely thrown up there by
the explosions of 79. A signal example of the confused notions
respecting volcanic formations in general, which the unhappy
' Elevation-Crater theory' could not fail to disseminate, is
shown in the views of M. Dufrénoy on the origin of Somma—
the whole of which mountain, he maintains, must have been
upheaved from below the sea, because of the occurrence of
this pumice-tuff on its surface nearly up to the summit ! The
simple explanation, that these mantling tuffs are merely the
fragmentary ejections of the great paroxysmal eruption which
formed the crater of the Atrio, throwing out the very bowels
of the early trachytic volcano, is repudiated, and the sudden
blister-like elevation of the whole mountain from the bottom
of the sea—*" une véritable ampoule"*—taught as the only true
doctrine !

In this loose pumiceous tuff of Somma occur many fragments of non-volcanic rocks, especially of limestone, rendered
more or less crystalline and otherwise altered by the heat and
other influences to which they were long exposed, when forming probably the sides of the eruptive fissure at a great depth
below the mountain. In the cavities of such metamorphosed
masses are found those many rare minerals—some scarcely
occurring anywhere else—of which Dr. Daubeny has given a
reduced list, still containing not less than forty distinct minerals—all, I believe, in crystalline forms.

Volcanic district adjoining Vesuvius.—Although Vesuvius
is the only volcano of Italy at present in full activity, there
are many other points on the western side of the Apennines
at which eruptions have occurred within a very recent geological period. In the immediate vicinity, to the north-west
of Naples, lies the district called by the Romans the Phlegræan fields, from the abundant traces of volcanic action over
all its surface. Within an area of about twelve miles by ten,
there is to be seen a series of hills composed of pumice-tuff,
which in their curving and, often, regularly circular forms

exhibit the traces of from twenty to thirty distinct cones and craters : Naples itself lies within one of these. (See fig. 60, p. 232.) Behind it, upon the highest point of the group, the monastery of the Camaldoli, 1643 feet above the sea, overlooks a wide crater-shaped basin named Pianura, at the bottom of which are quarries of a remarkable mottled and highly porous cellular greystone, called Piperno, generally employed in Naples for building. It is remarkable for being interspersed with lenticular concretions of a darker colour than the base, less porous but more cellular, and far more augitic : these portions have evidently separated by segregation in the same manner as flints in chalk ; they are elongated and flattened out in the direction of the flow of the current. Continuous with this eminence is the promontory of Posilipo, a ridgy hill of erupted tuff, the product probably of explosions from several orifices. And at its S.W. extremity lies the little island of Nisida, which is a very regular cone with a deep circular crater, opened by a breach towards the sea. Here, as in many other of the like hills around, the sea-worn cliffs show the internal structure of the cone to be a pumice-tuff, indurated by violent mixture with sea-water at the time of its eruption, and covered with loose arenaceous beds of the same fragmentary matter, which, no doubt, fell in a drier state from the air, after the mouth of the vent had been raised some height above the reach of the waves. The beds of both characters are arranged mantlewise, with that double quaquaversal external and internal dip, which has been shown to be the normal structure of a volcanic cone the product of a single eruption.

North of Nisida, proceeding westwards along the coast, is the hill immediately overhanging Pozzuoli to the north, which contains the remarkable crater called the Solfatara, still showing unmistakeable signs of the intense temperature of the mass of subterranean lava with which it is in communication, probably through some fissure, in the exhalations of hot sulphureous vapours (sulphuretted hydrogen) which continually rise from orifices in the small plain that forms its bottom, and in the cliffs of trachytic lava-rock that surround it. This rock is decomposed and whitened by the vapours, whence the name of ' Leucogæi Colles ' given to the hill from a very early date. Among the products of this decomposition are sulphur, the sulphates of iron, lime, soda, magnesia, and above all, alumina, muriate of ammonia, and sometimes sulphuret of arsenic. The decomposed matter carried down by the rains as a white mud and spread over the floor of the crater-plain hardens into beds resembling pipe-clay, which, probably through the vaporization of the water contained in it by the heat rising from below, is filled with small hollow vesicles ; and consequently when stamped upon this floor returns a hollow sound (*rimbombo*)— a circumstance which has led some writers to believe it a mere vault over a vast abyss. This notion was many years since contested by me in a paper upon the Phlegræan fields (Transactions of the Geological Society, 2nd ser. vol. v.) ; and though I have failed to convince Dr. Daubeny and Professor Forbes, I still retain the opinion that my explanation of the cause of the hollow sound is correct. All porous ground returns such a reverberation ; and the stratified tuff at the bottom of the Solfatara, being extremely cellular throughout, possesses just the structure fitted to send back these multiplied echoes with the greatest effect. An eruption broke out from the Solfatara in the year 1198. It was, however, probably by an earlier one that the bulky bed of trachytic lava was erupted which clothes the entire outer slope of the cone on the southeast side, from the upper rim of the crater down to the sea, into which it projects as a rocky promontory, called Olibano, much quarried for building-stone. It may be seen to rest conformably on the sloping beds of tuff which compose the substance of the hill, and therefore certainly flowed down the exterior of the cone, but spread to no distance, adhering like a bulky buttress, at least 50 feet thick, to the side of the hill. The mineral character of this very recent lava is remarkable, and accounts for its extremely imperfect fluidity. It is a highly crystalline or granitoidal ash-grey trachyte, or rather greystone, made up of large crystals of glassy felspar, imbedded in a felspathic and granular base, with a small quantity of augite. It is cellular towards the surface, but in general compact and very hard. It is identical with much of the greystone or

Y

trachyte of Ischia and the Ponza Isles, to be described presently. The upper part overhangs the crater of the Solfatara (which could not therefore have existed in its present form at the time this lava was emitted, or it would have been filled up by it), and was probably hollowed out, after the flow of this lava had ceased, by the subsequent aëriform explosions of the same or a still later eruption. The lava-bed is covered by layers of loose scoriaceous conglomerate, then, no doubt, thrown out of this crater. Another massive current of very similar character flowed, perhaps at the same time, possibly at an earlier date, from the same vent towards the north-east, descending almost as low as the town of Puzzuoli. Sir W. Hamilton gives a very faithful view of this last current in his great work.

Immediately behind and attached to the hill of the Solfatara is that which includes the royal preserve of Astroni—a very perfect circular crater, the inner walls of which are so precipitous as to form a natural barrier for the confinement of the wild animals (boars, &c.), which are very numerous in it; the only entrance being by an artificial cut through the ridge. Here too the component beds of the hill may be observed to have the quaquaversal mantling dip characteristic of a cone of eruption. From the plain at the bottom of the crater rise one or two small hills composed of a trachytic rock—no doubt a lava protruded in a semi-solid state at the close of the eruption which formed the crater. This central boss of trachyte is frequently quoted by the advocates of the Elevation-crater theory (Astroni being one of their favourite types of such a crater) as having *evidently* upheaved at the time of its production the sloping beds of tuff all round. But, in truth, there is no evidence of the kind beyond the existence of this mass of lava-rock in the bottom of the crater, than which nothing can be more in accordance with the ordinary laws of volcanic action. (See p. 134, *supra*.) The inclination of the beds forming the cone in all directions is just the same as in the neighbouring tuff-cones, viz. from 15° to 35°, but with the same irregularity in the character of the layers as results in all such cases from accidents of form, size, disposition, &c., necessarily affecting such fragmentary ejections.

Adjoining the cone of Astroni is that of Monte Barbaro (the 'Gaurus inanis' of Horace), having a well-defined cup-shaped crater, surrounded by steep banks of tuff.

Next to this, and on the bank of the Lake of Avernus, rises the Monte Nuovo, a tuff-cone 430 feet high, with a crater 370 feet deep. This entire hill was thrown up in the course of two days, in September 1538, by aëriform explosions from the valley or low plain at the foot of Monte Barbaro, separating the Lake of Avernus from the sea-shore, and scarcely above the level of the sea. Several contemporary narratives are extant of the visible phenomena of this eruption. All unite in declaring the spot to have been low and level—the bottom of a valley, whence violent explosions took place, throwing up stones and ashes in such prodigious abundance as to have formed the mountainous hill we now see there in the course of forty-eight hours. But because one of these writers, Francisco del Nero, speaks of the earth "swelling up" till it formed a hill, the Elevation-crater theorists have seized on it as a clear case of upheaval,—although the same writer in the same sentence says, "It *threw up* for a long time earth and stones, which fell back all around the gulf till they formed a hill of vast size, and even covered the ground and trees for seventy miles round with ashes." It was evidently then an eruption of fragmentary matters that produced the cone; and the expression previously used by the writer, of "the earth swelling up into a hill," is quite consistent with this, and only meant that a hill of earth had accumulated there at the time he mentions, viz. noon on the first day of the eruption. But a still better narrative of the event is to be found in a volume (now in the British Museum, the gift of Sir W. Hamilton) printed at Naples in the very year of the eruption. In this publication, Signor Marco Antonio Falconi, an eye-witness of it, thus describes its phenomena :—" Stones and ashes were thrown up, with a noise like the discharge of great artillery, in quantities which seemed as if they would cover the whole earth; and in four days their fall had formed a mountain *in the valley between Monte Barbaro and the Lake Averno*, of not less than three miles in circumference, and almost as high as Monte Barbaro itself,—a thing incredible to those who have not seen it, that in so short a time so considerable a mountain should have been formed." Another account in the same volume, by Pietro Jacobeo di Toledo, describing the same fact, adds,— "Some of the stones were larger than an ox. They were thrown up, the larger ones, about a cross-bow's shot in height from the opening, and then fell down, some on the edge of the mouth, some back into it. The mud ejected (ashes mixed with water) *was at first very liquid, then less so,* and in such quantities that, with the help of the aforementioned stones, a mountain was raised a thousand paces in height on the third day. I went to the top of it and looked down into its mouth, in the middle of the bottom of which stones that had fallen there were boiling up just as in a great caldron of water that boils on the fire."

The composition and structure of the hill correspond with the origin here attributed to it; the nucleus being, as in the case of all the other volcanic hills around it (the date of whose production is not known), of more or less compacted tuff, composed, no doubt, of the hardened mud whose eruption was witnessed by the writer just quoted, with overmantling layers of loose pumice and scoriæ (*i. e.* arenaceous tuff) resting upon it. The solid nucleus, in M. Dufrénoy's opinion, was upheaved bodily, not in a semiliquid, but in a solid form; and when the objection was made to him that the wall and columns of the Roman Temple of Apollo, standing close to its base, remained perfectly vertical, and its cornices perfectly horizontal, he solved the difficulty by the wild supposition that the entire nucleus of the hill (probably three-fourths of its bulk) existed before the eruption of 1538, and was only sprinkled over by the upper loose pumice-beds on that occasion—notwithstanding the unanimous testimony of all contemporary observers to its having been wholly formed on what was previously a low flat (in fact, the sea-shore), by eruptions of stones, scoriæ, ashes, and mud, in the last three days of September of that year.

That some slight elevation of the general level of the shore of the Bay of Puzzuoli accompanied the formation of Monte

Nuovo is probable—indeed is both mentioned by the writers already quoted, and visible in a raised beach, called ' La Starza,' at the base of the old cliff that borders the sea from the new cone to the town of Puzzuoli (see woodcut below). Some further oscillatory movements of the same shore, both up and down, are shown to have subsequently occurred, by certain facts respecting the columns and pavement of the Temple of Jupiter Serapis, for the details of which I may refer to Sir C. Lyell's ' Principles'; only remarking that such a general elevation or depression of some miles of coast (a not uncommon circumstance in a volcanic district) in no degree supports the idea of the sudden upheaval of a steep and lofty volcanic cone with a central crater.

Monte Nuovo, as seen from Puzzuoli.

The scoriæ of Monte Nuovo are vitreous, but darker and heavier than the usual type of pumice. The ejected fragments of lava approach to clinkstone, being laminated, and sometimes veined with pitchstone. The eruption produced no *current* of lava.

At the foot of the Monte Nuovo and Monte Barbaro are the nearly circular basins of the Lakes Averno and Agnano, both surrounded by tuff-hills of moderate elevation, whose strata, as usual, slope away from the hollows, each of which was, no doubt, a crater of considerable dimensions. From several fissures at the foot of these hills there issue exceedingly hot sulphureous vapours; and from one (the Grotta del Cane) volumes of carbonic acid gas. These mephitic exhalations account for the noxious properties attributed of old to the air above the Avernean lake. The ridgy tuff-hills are continued to the south-west, behind Baiæ, as far as the promontory of Misenum—a tuff-cone with distinct traces of a small crater, and the characteristic double (internal and external) dip of the layers that compose it (see fig. 6, p. 61). The Monte di Procida and the hill of Cumæ form portions of other eruption-cones which have been largely worn away by the sea. The former hill rests on a very black slaggy and scoriaceous trachyte, seen in the sea-cliff, and doubtless marking one of the eruptive orifices, the waves having destroyed the remainder of the cone. The island of *Procida,* now separated from this hill by about two miles of sea, may possibly have been once united to it. It is of precisely the same composition —tuff, interbedded with some very scoriaceous trachytic lavas; and some of its curving bays indicate more than one crateriform hollow, now so far cut up by the erosion of the sea as to be with difficulty traceable.

Ischia.—A short way off, in the same south-west direction, the much larger island of Ischia rises in all the dignity of a true volcanic mountain. The summit-heights of the island, called Monte Epomeo, attain 2600 feet above the sea. They are peaked and precipitous, chiefly composed, as is the greater part of the island, of a trachytic tuff, of a greenish tint, from containing much augitic matter, and in parts apparently consolidated by heat, and passing into earthy trachyte. On several points of the lower flanks of the mountain eruptions have taken place in recent times. Indeed its early history shows that the inhabitants were more than once destroyed or driven away by the violence of the earthquakes and eruptions to which it was then subject. Ten or twelve minor cones are the product of these later eruptions; one of them, the Monte Rotaro, was thrown up in the year 1302, and gave issue to a bulky stream of highly porphyritic trachyte, which flowed into the sea. The other recent lava-rocks of Ischia are of very felspathic greystone; in some instances remarkably ribboned and laminated, and sometimes having the scaly texture of clinkstone. Some, as in Monte Vico, near Foria, are entirely composed of compacted crystals of glassy felspar. Some parts

of these currents are brecciated ; different varieties of rock
having seemingly been broken up, and reunited by fresh lava
as the stream flowed on.

Several hot sulphureous springs break out in the island,
justifying the belief that its internal activity is not yet quite
over. In truth, however, all the eruptive vents of this volcanic district may be considered to belong to the same subterranean focus, and, to some extent, to relieve one another.
The Ischian eruption of 1302 occurred at a time when Vesuvius had been inactive for nearly two centuries; and, on the
other hand, since the latter mountain has been frequently in
eruption, Ischia and the Phlegræan fields have remained dormant, with the single exception of the eruption of Monte
Nuovo in 1538, which also took place during a quiescent interval of Vesuvius that lasted a century and a half.

The earlier trachytic rocks of Ischia rest upon strata of clay
and marl containing many shells still existing in the Mediterranean—probably, therefore, of post-pliocene date. From
this it is evident that the mountain has sustained a great
amount of elevation in mass since the deposition of these
beds. They are said to be found even at heights of 2000 feet
above the sea. Indeed the entire eastern coast of this portion
of Italy shows similar signs of having gained considerably in
elevation since the commencement of its volcanic activity.
The whole of the fertile plain of the Campagna Felice, and the
bottoms of the valleys running up to the eastern foot of the
Apennines, consist of stratified pumice-tuff, evidently formed of
the fragmentary ejecta of the volcanic vents of the Phlegræan
fields, strewn over the shore of the sea at a time when it
covered all this district, and now raised some hundred feet
above its surface. Natural sections of this formation may be
seen in the cliffs of Sorrento, more than 200 feet high, and
on many other spots along the coast.

Eastward of Ischia several other small volcanic islands rise
from the Mediterranean. The two nearest, *Ventotiene* and
San Stefano, are evidently the remnants only of a much larger
island, preserved from the wearing action of the sea-waves,
which has destroyed the greater part, through the resistance

offered in both by massive beds of greystone which form the substratum of the islands. This rock rises at the eastern extremity of Ventotiene into a beetling cliff above 500 feet high, and is covered throughout the length of the island by a stratified conglomerate of ash and fragmentary lava, averaging 100 feet thick. The surface is encrusted with a concretionary calcareous sandstone,—the lime having evidently been derived from the decay of innumerable land-shells, and conveyed by rain-water into the arenaceous beds below. This last is a similar formation to that which Mr. Darwin describes as common in nearly all the volcanic islands of the tropical seas. It is almost equally so in those of the Mediterranean. (See the outline of Ventotiene, fig. 55, p. 209.)

At about the distance of twenty miles still further eastward rises the remarkable group of the *Ponza Isles*. These also have suffered an enormous amount of degradation since their formation. The highest rock of the principal island, *Ponza*, is a mass of greystone lava, extremely crystalline and porphyritic, and similar in every respect to that of Ventotiene, as well as to that of Olibano, near Puzzuoli. It forms a bed perhaps 200 feet thick, overlying unconformably at the southern extremity an earlier volcanic formation which composes the bulk of this island, and of the two adjoining ones, *Palmarola* and *Zannone*. All these islands may be considered as forming parts of one ancient submarine volcano. The rocks of this earlier system appear like so many more or less vertical and extremely irregular dykes of trachyte thrust up through masses of a very vitreous pumice-agglomerate. The principal rock is a whitish-yellow, sometimes pink and brown trachyte, very similar to domite, but generally harder, and divided into small irregular prisms from an inch to a foot in diameter. It contains glassy felspar-crystals, mica-plates, and titaniferous iron, and is remarkable for a very generally streaked or ribboned texture which penetrates the entire mass. The different layers seem to owe their several tints to a greater or less proportion of the colouring minerals, the whiter being wholly felspathic or siliceous, the darker stripes containing more of augite or mica. Many parts of this rock, indeed, are so highly siliceous as to have the conchoidal fracture, with sharp cutting edges, of flint. In this state their general white colour is stained with brilliant tints of scarlet, orange, blue, and brown. Others are carious, resembling buhr-stone (or the siliceous millstone-trachyte of Hungary), and penetrated by veins of quartz, the cavities being lined with quartz-crystals, sometimes amethystine. The white vitreous pumice-conglomerate through which this prismatic trachyte has been protruded is *uniformly* altered along the planes of contact to the depth of several yards. The nearest portion is completely vitrified, and changed into a compact green obsidian containing many white felspar-crystals. The unaltered white conglomerate-rock graduates into this through a yellowish waxy-looking pitchstone, enveloping nodules of the green and more glassy variety. The green obsidian has sometimes a large globular concretionary and lamellar structure ; sometimes it passes into pearlstone by the formation of minute spherulites throughout its substance. And on some points there is a clear passage from this pearlstone into the ribboned prismatic trachyte; the spherulites being dragged out—evidently by the motion of the matter when in a softened state—into planes, whose varying mineral character gives a laminated structure and ribboned aspect to the rock. These laminæ are generally bent, folded, and crumpled up in a remarkable manner, so as to resemble the structure of the most highly corrugated slate-rocks and the contortions of gneiss and mica-schist. (See p. 139, *supra*.)

Zannone, the last island of the Ponza group towards the coast of Italy, is entirely composed of highly siliceous trachyte, with the exception of its eastern extremity, which is a hard stratified limestone resembling that of the promontory of Circello just opposite (Jurassic), and becoming crystalline and dolomitic in the vicinity of the trachyte. The latter appears to have burst up through the limestone which it thus metamorphosed. (See, for further details, my paper on the Ponza Isles, Trans. Geol. Soc. ser. 2. vol. ii. 1827.)

Mount Vultur.—On the eastern slope of the Apennine ridge, about midway between the two seas, and nearly due east of Naples, rises this insulated volcanic mountain, of an imposing and regularly conical figure, reaching a height of 4156 feet above the sea, and measuring above twenty miles round its base. It is, in fact, a larger and loftier mountain than Vesuvius itself. It has been visited and described by Brocchi, Abich, and Dr. Daubeny. On its northern flank it displays a wide circular crater, surrounded by an amphitheatre of rocky hills, the highest point of which reaches 1000 feet above the two small lakes which occupy the bottom of the hollow. There are emanations of carbonic acid gas upon their borders. The mass of the mountain seems to consist of pumice-tuff of various degrees of compactness, enveloping beds of a lava-rock composed of crystals and granules of leucite and augite, with (according to Brocchi) pseudo-nepheline and melilite, and a still rarer mineral, haüyne, in clear blue or green crystals. No record exists of any eruption from Mount Vultur; nor is there any appearance on the spot of such an event having occurred within recent times. Dr. Daubeny justly remarks that it rises on the prolongation of a line drawn through Ischia and Vesuvius. And since at an intermediate point on the same line the Pool of Amsanctus is found, celebrated of old for its mephitic exhalations of carbonic acid and sulphuretted hydrogen gases, it is probable that some deep-seated fissure transverse to the general direction of the Apennine axis exists, or once existed, along this line through the rocky substratum of the Peninsula. The eruptions of Mount Vultur were probably all subaërial.

Rocca Monfina.—To return to the western slope of the Apennines. Near the town of Sessa, on the road between Naples and Rome, we find another insulated volcanic mountain, likewise extinct. There are, at least, no well-authenticated accounts of its activity, although some of its products have an appearance of freshness, leading to the belief that they are of recent date. It consists of a nearly circular hill-range, sloping upwards from the base all round with the usual gradually increasing talus-like inclination, to a height at some points of 2000 feet, where the slope terminates in a steep circular and precipitous escarpment, surrounding (except at one point to the east, where there is a depression)

a wide basin, evidently a great crater, formed by some paroxysmal eruption which destroyed the upper part of what
must have been once a much loftier mountain. The diameter
of this basin is two and a half miles, and its circumference
seven and a half. In the centre rises a conical hill, called
the Monte della Croce, to the height of 3200 feet from the sea
—considerably above the highest point of the surrounding
crater-rim. The rock of which this hill is composed is a
trachyte approaching to greystone, fine-grained and compact,
but not of close texture, containing a confused mixture of
small felspathic granules, glassy and having a fused aspect,
with much green augite and brown mica in hexagonal tables*.
The encircling hill-range almost wholly consists of mantling
layers of leucitic tuff, composed principally of scoriæ, sand,
and lapilli, with many loose blocks of *leucitic* lava ; the leucites
being often of great size, as big as large walnuts, with some
crystals of augite and a felspathic base. Some continuous
layers of this lava-rock rest upon and are interbedded in the
tuff. There is a second large crater closely adjoining this
great one on the north side, called La Concha, above a mile
in diameter, and a third smaller one about two miles further.
There are also several minor eruption-cones, which have produced very recent-looking felspathic lava-streams on the outer
flanks of the mountain.

The advocates of the Elevation-crater theory rely upon
Rocca Monfina as a typical example in support of their views.
They consider the protrusion of the central boss of trachyte to
have caused the upheaval, in an annular bank, of the surrounding tuff-beds. There is however, in truth, no good
reason to be given for this opinion. There is nothing to rebut
the supposition—which would be in accordance with the
normal action of volcanos—that, after the formation of the
great crater by an eruptive paroxysm, a body of very imperfectly liquid trachytic lava should have been emitted from
this central vent, and have piled itself up into the massive
conical form of the Monte della Croce. All the examples of
the greystone lavas of this volcanic district, in Ponza, Vento-

* Daubeny, ' Volcanos,' p. 180.

tiene, San Stefano, Monte Olibano, &c., show that they possessed a very low degree of fluidity when emitted. The circumstance of a volcano which had previously thrown up
leucitic lavas changing the mineral character of its products,
and giving issue to a trachytic rock, is by no means extraordinary; and the change, at all events, gives no support to
the theory of the upheaval of the earlier tuff-beds. Moreover,
Abich describes the latest lavas produced by this volcano *on
the outskirts* of the circular hill-range as trachytic, and resembling the central rock of Croce. I lament that the observations of this last-named writer, illustrated as they are by admirable drawings of this group, should be rendered less
valuable than they otherwise would be by his adherence to the
unhappy doctrine of upheavalism. Rocca Monfina is situated
at no great distance from the volcanic hills of Latium, which
will be presently described. But I will first proceed southwards towards the Lipari group and Etna.

The Lipari Islands.—The group of seven islands, called
from the largest the Lipari Isles, that lie between Naples and
Sicily, together with several insular rocks rising above the
sea-level about them, all of volcanic character, may be considered as belonging to the same system, if not as separate
mouths of a single submarine volcano.

Two at least of these may be reckoned as at present in a
state of activity, viz. Stromboli, which has been already dwelt
upon in an earlier page as one of those rare examples of a volcanic vent in permanent eruption—its constant activity being
attested by writers antecedent to the Christian era, as well as
by numerous subsequent authorities,—and Volcano, which has
been frequently eruptive within that period, and exists at
present in the condition of a solfatara.

Stromboli is in figure a very regular cone. The remarkable
circumstance in this small but most interesting volcano is that
the column of lava within its chimney is shown, by the constant
explosions that take place from its surface at intervals of from
five to fifteen minutes, casting up fragments of scoriform lava,
to remain permanently at the same height, level with the lip of
the orifice at the bottom of the crater, and therefore some 2000

feet above the sea-level.   It is evident from this that a nearly
perfect equilibrium is preserved between the expansive force
of the intumescent lava in and beneath the vent, and the
repressive force, consisting in the weight of this lofty column
of melted matter, together with that of the atmosphere, above
it ; consequently a very small addition to or subtraction from

View of Stromboli, from the North.

the latter, such for instance as any change in the pressure of
the atmosphere, must to *some* extent, however small, derange
the equilibrium.   It need not therefore surprise us that the
inhabitants of the island, chiefly fishermen, who ply their
perilous trade day and night, within sight of the summit of the
volcano, declare that it serves them in lieu of a weather-glass,
warning them by its increased activity of a lightening of the
atmospheric pressure on the volcano—equivalent to a fall of the
mercury—and by its sluggishness giving them assurance of the
reverse.   It is the tension of heated steam or water disseminated
through the lava in and beneath the vent which occasions its
eruptive action, and the boiling-point of every drop or bubble
must be sensibly affected by every barometric variation.   Even
in the time of Pliny the same observation was made by the
navigators of the Mediterranean.   " E cujus fumo," he says,
speaking of this volcano, " quinam futuri sunt venti in tri-
duum prædicere incolæ traduntur, unde ventos Æolo paruisse
existimatum."

In the foul weather of winter I was assured by the inhabitants that the eruptions are sometimes very violent, and that the whole flank of the mountain immediately below the crater is then occasionally rent by a fissure, which discharges lava into the sea, but must very soon be sealed up again, as the lava shortly after finds its way once more to the summit, and boils up there as before. Capt. Smyth found the sea in front of this steep talus unfathomable, which accounts for the remarkable fact that the constant eruptions of more than 2000 years have failed to fill up this deep sea-hollow. The scoriæ ejected by Stromboli are highly augitic; indeed, some of the outer slopes of the cone are covered with loose and perfect crystals of augite, often cruciform, and of titaniferous iron. Its recent lavas are augitic; but rocky masses of coarse-grained trachyte form the southern side of the island. The stratified tuff which chiefly composes it resembles that of Ischia, though in many parts it is more augitic and ferruginous.

South of Stromboli is a cluster of small islets and rocks so placed as to make it probable that they are the remains of one volcanic island, partly perhaps blown up by paroxysmal explosions, partly worn away by the waves. *Panaria* (the largest), *Basiluzzo*, and the others, all consist of a hard and highly crystalline granitoidal trachyte, in some cases columnar, composed of glassy felspar, mica, and much quartz. Much of this rock has a lamellar and even a slaty structure and a striated texture, evidently occasioned, like the ribboned trachyte of Ponza, by the dragging out of matter of different mineral character into distinct planes while in a softened state.

Lipari comes next—a considerable island, made up of three or four separate mountains, the product of as many eruptive vents. They are partly composed of pumice-tuff, but chiefly of alternate layers of pumice and obsidian, both fragmentary and disposed in continuous sheets or currents of glassy lava. In their vesicular or fibrous parts the fibres and vesicles are uniformly elongated in the direction of the downward flow of the current. In the Monte Guardia these beds form a mass several hundred feet in thickness. They have flowed from its summit in successive streams, which have consolidated in the manner of ordinary lavas on the slopes of the mountain at an angle of from 20° to 30°. The lower beds are less vitreous than the upper. This glassy lava passes at some points into a compact stony trachyte resembling claystone, at others into pearlstone. Both are laminated, and divide readily into tabular masses, as indeed does the pumice and obsidian-rock likewise. Another of the mountains in Lipari, called Campo Bianco, has a perfect crater on its summit, and the cone which contains it is itself encircled by another older crater-wall. All the rocks are of pumice or glassy lava, either in loose blocks or continuous streams. Campo Bianco supplies all Europe with the fibrous pumice of commerce, and may continue to do so for thousands of years. These enormous accumulations of highly vitreous lava-currents, composing entire mountains, are well worthy the attention of those geologists who consider all the crystalline and stony lavas to have been vomited in a state of complete vitreous fusion, and to owe their lithoidal character to slow subsequent cooling.

There can be no reason for supposing that these obsidian and pumice streams were more quickly cooled than the augitic lavas of the neighbouring volcano of Salina, or the highly crystalline trachyte of Basiluzzo and Panaria, or the greystone lavas of Etna, or the leucitic lavas of Vesuvius. How then came the former to retain a vitreous texture throughout, and the latter to be everywhere stony and crystalline, if both were erupted in a similar condition of molecular fusion ? The only possible explanation of the different texture exhibited by the two classes of rock seems to be, that the latter lavas were not thoroughly fused at the time of their emission, but consisted of a granulated magma, the imperfectly formed crystals being lubricated by a more liquid base, and by heated intercalated steam, composing a red-hot granulated paste. (See pp. 116–122, *supra.*)

Some of the rocks of Lipari towards the east are siliceous, white in colour, and cellular, crystals of quartz lining the cells. They resemble the millstone-porphyry of Hungary and the siliceous trachyte of Ponza described above.

There are several hot-springs in the island, traversed by

emanations of sulphuretted hydrogen gas. One very copious source of nearly boiling water gushes from the side of a hill at a height of 300 feet above its base. The escarpment whence it issues is composed of numerous *perfectly horizontal* and *parallel* strata of incoherent volcanic ashes, alternating regularly with others of a solid lithoidal and siliceous nature. The first are from 2 to 3 feet thick, the latter not above 4 or 5 inches. Both contain vegetable remains, leaves, twigs, &c. The stony layers are grey, veined with red, and resemble agate and jasper. They often split into small prisms. Their surface is porous and cellular. These cells puzzled greatly M. Faujas de St. Fond. He supposed such air-cells indubitably indicated fusion by fire; yet this idea could not be reconciled with the leaves they enveloped, or the thin strata of the rock. The fact is, that this is a very analogous formation to that still going on upon the surface of the plain within the crater of the Solfatara of Puzzuoli. The aluminous earth washed down from the decomposed trachytes and tufas that form the sides of this crater is collected within flat pits surrounded by embankments for the purpose, and then slowly deposited from the water that holds it in suspension,—the coarsest parts first, and the finest last, in horizontal layers. Of these layers the upper, being composed of the finest sediment, before it is dry forms a fine white mud or clay; as it is gradually dried up by the rays of the sun, it hardens into a crust as fine and compact as the purest porcelain biscuit, and in many parts of a semi-vitreous or glassy lustre. This crust is so compact and so suddenly hardened by the rays of the sun, that the moisture of the lower strata cannot escape in vapour through it, and consequently collects beneath the crust in globular air-bubbles. During its desiccation the mass also often contracts and splits into rude columnar prisms. In the strata of Lipari it appears that the sediment was still *finer* and probably more *siliceous* than at Puzzuoli, consequently the upper crust was still more glassy and jaspiform, being coloured by oxide of iron; whereas that of the Solfatara is almost pure alumine. In the Lipari rocks the intermediate beds of loose ash probably fell in that state from the air (ejected by the neighbouring craters) during the intervals of the storms of rain which carried down and deposited the siliceous and hardened sedimentary layers.

The plants of which the remains are found in the tuff are both dicotyledonous and monocotyledonous, and prove the eruptions which buried them to have been subaërial. Indeed, the hot-springs, together with some vague notices that are to be met with in many early writers, seem to show that eruptions may have occasionally occurred from some of the craters of this island within the historic period.

Volcano.—That certainly has been the case with the principal crater of the adjoining island of *Volcano*, in the interior of which copious emanations are emitted of sulphurous acid gas (Daubeny, p. 258); these, by decomposing the rock around, give rise to deposits of boracic acid and sal ammoniac, both largely collected for commercial uses. A dyke of compact scaly trachyte (clinkstone) traverses one side of the crater. There was an eruption from this volcanic mouth in 1775, and again in 1786. On the last occasion the present crater was, no doubt, formed, since the mountain is said to have vomited scoriæ and ashes for fifteen days without ceasing. In 1775 a stream of glassy lava flowed from the summit of the cone; at which time, of course, the crater could not have been in existence. The cone which contains the present crater is itself almost surrounded by the precipitous walls of another much larger exterior crater-ring or hollow cone; and on the opposite or breached side is a third, smaller cone, with a still smoking crateriform depression at its summit, called *Volcanello.* (See figs. 46 and 47, p. 193.) The bulk of these three mountains consists of beds of earthy trachyte, pumice, obsidian, and their conglomerates, resembling those of Lipari, and disposed with a quaquaversal dip from the ridges of the several craters.

To the N.W. of Lipari, a very regular volcanic mountain rises from the sea to the height of 3500 feet, called *Salina.* It has a crater, and is composed of beds of peperino or augitic tuff, and of lava equally abounding in that mineral; all

z

sloping outwardly, as usual, from the circular rim of the crater.

Felicuda and *Alicuda* are two other volcanic insular mountains. The first emits from its crevices sulphureous vapours. Both possess craters on their summits. Some of their lavas resemble clinkstone; some are large-grained trachyte, and regularly columnar. In their cliffs ten or twelve beds of lava may be seen alternating with conglomerate-layers.

Ustica, another island in the continuation of the same east and west line, at a distance of forty miles north-west of Palermo, contains three large breached craters, and is composed of a brown tuff, or peperino, alternating with thick beds of greystone lava, full of felspar and augite, with some olivine, all sloping from the summit to the sea around. Pumice also occurs in it, as well as much calcareous matter cementing the volcanic conglomerate, and containing marine shells, attesting the recent emergence of part of this island from beneath the sea.

I visited the Lipari Isles in 1820, and was greatly struck by their remarkable features, which convey a more vivid notion of the conduct of volcanic action, especially when productive of vitreous felspathic lavas, than perhaps any other equally accessible localities. I strongly recommend them to the study of those geologists who desire to form their own opinions on volcanic phenomena.

Sicily and Etna.—The southern angle of Sicily, forming the province of Val di Noto, is partly composed of calcareous strata of newer Pliocene age, interbedded with volcanic tuff and basalt. It appears, therefore, that this district was the site of eruptive igneous action when probably the whole surface of the present island lay beneath the waters of the Mediterranean.

Etna itself is based upon similar beds of marine origin (marls and clays of the age of the Norwich Crag, in the opinion of Sir Charles Lyell*). They are to be seen in sea-cliffs 500 and 600 feet high along its eastern base, asso-

---

* Phil. Trans. 1858, " On Etna," &c., p. 78.

ciated with basaltic and other igneous products. They even
reach, at the distance of a few miles inland, on the eastern
side of the mountain, a height of more than 1200 feet above
the sea-level, where likewise they are capped with lava-beds.
It is therefore certain that, within a comparatively recent
geological period, the area forming the base of the mountain
must have been gradually upheaved by at least that amount
of vertical rise, while its superior mass was being as gradually erupted in open air and distributed into its present
position. There is every reason to believe that the entire
conical mountain, at least from that level up to the central
apex, now nearly 11,000 feet above the sea, is wholly composed of alternate layers of lava-rock and its conglomerates.
This internal structure is disclosed more or less in all the
water-courses that drain its slopes, but best of all, in one
deep and wide valley pierced through its eastern flank and
opening down to the sea—the well-known Val del Bove. The
beds, wherever observable, have the usual quaquaversal dip
from the central heights to the plain around, except in the
upper part of the Val, where they dip *towards* the present
central axis of the mountain, or, rather, mantle round a point
within the higher basin called Trifoglietto, which is thence
justly supposed by Sir C. Lyell to have been at some former
period the site of the principal eruptive chimney,—a shift
having since taken place to the actual centre of eruption, and
highest cone, called Mon Gibello. The evidence adduced in
proof of this by Sir C. Lyell is quite conclusive ; and indeed
nothing can be more accordant with the ordinary laws of
volcanic action than such a shifting of the axis of eruption.
(See p. 227.) The question as to the origin of the Val del
Bove itself might perhaps be more open to dispute ; at least,
it is one on which Sir Charles Lyell still expresses a doubt
whether subsidence or explosions had the chief share in its
production. To me, I own, it has always appeared to have
originated in a great fissure, enlarged into a crater by some
paroxysmal eruption which blew out of the heart of the mountain, and since widened by the abrasive violence of aqueous
debacles, caused by the sudden melting of snows on the heights

z 2

above by the heat communicated from erupted lavas, and showers of red-hot scoriæ falling over the surface. One such torrent, indeed, is recorded, which rushed down this same valley in 1755, in the month of March, the *volcano being at the time covered with snow*; on which occasion the flood is said by Recupero to have run at the rate of a mile and a half in a minute, for the distance of twelve miles,—a rate which would give an enormous abrading and carrying force to a great body of water. Accordingly its track, two miles in breadth, is now visibly strewn to the depth of from 30 to 40 feet with sand and fragments of rock. And that similar debacles had previously for many ages taken the same course is demonstrated by the accumulation of a vast alluvial formation at the opening of the valley to the sea near Giarre, more than 150 feet deep, measuring ten miles by three in area, and now resembling an upraised line of beach 400 feet high above the sea. Sir C. Lyell terminates his detailed reasoning on this subject by saying, " I do not hesitate to attribute the bulk of the Giarre alluvium to the excavation of the Val del Bove" by such floods. (See, too, p. 211, *supra.*)

Before the occurrence of the paroxysmal eruption which originated this great crater (or caldera, as Sir C. Lyell calls it), the summit of the volcano situated somewhere within its limits, and probably in the axis of Trifoglietto, rose, no doubt, far higher than the present Etnean apex, perhaps by several thousand feet. Such, indeed, must also have been the case with the latter cone, Mon Gibello, before it was truncated by the formation of a great elliptical crater, 2500 feet wide according to Von Waltershausen, the upper rim of which is still visible, encircling the Piano del Lago—a sort of platform which supports the recent and now active upper cone. This platform is not flat, but rather dome-shaped, and has been formed, no doubt, from the up-filling of the vast crater, of which its ridge marks the original cliff-range, by subsequent eruptions, in the same manner as that which formed the summit of Vesuvius before it was blown off by the explosions of 1822. (See pp. 187–190.)

The present active cone of Etna rises about 1200 feet above

its base on the Piano del Lago. It contains a crater which has varied considerably in size and form since the beginning of the century. During this period it has been frequently in eruption, and abundant lava-streams have flowed out of it through a wide breach to the north, particularly in the years

Outline of Etna as seen from Catania.

1803 and 1838. But, generally speaking, the Etnean eruptions of lava take place from lateral openings; and more than two hundred small scoriæ-cones, marking the site of such mouths, are visible on the flanks and towards the base of the mountain. Most of them are breached, and have evidently poured forth lava. Many are more than half buried by subsequent flows of lava around them from higher points, or the accumulation of ejected scoriæ and ashes.

Some of these lateral mouths have opened within the Val del Bove. This was the case on the occurrence of the last great eruption of 21st August, 1852, of which Sir C. Lyell gives many interesting details. It appears that clouds of scoriæ were first thrown up from the central crater, and on the next day there were formed many openings—some say as many as seventeen—on a line of fissure from the summit to the base of the great precipice which forms the head of the Val. The two lowest threw up there two cones of considerable size, one about 500 feet high. From this a very great lava-stream took its course, and ran in the first day about two and a half miles, advancing more slowly afterwards, and stopping on the eighth day, having covered a surface of six miles in length, with an average width of two. The lava broke forth afresh in October, and again in November; the stream on both occasions precipitating itself over a lofty cliff, making a sound in its descent "as if metallic and glassy substances were being broken." The eruption lasted on the whole nine months. The lava-streams varied in depth from 10 to 16 feet; but where several accumulated one above the other, the mass reached 150 feet in thickness. The central vent of Mon Gibello sympathized throughout with these lateral eruptions, its explosions increasing in intensity when the latter recommenced their eruptive discharges. The surface of the lava-stream was marked with great longitudinal crested ridges with parallel intervening hollows,—slaggy masses topping these crests, and rising perhaps from 70 to 80 feet above the bottom of the adjoining depressions. These great wrinkles appeared to be formed of concentric layers of lava, as if they had been folded or crumpled up, while yet in a softened state, by lateral compression—which was perhaps the case. Or some may have been the summits of so many arched channels or gutters, through which lava so often runs, and which fresh outbursts tend to split, thrusting up fragments in rough slaggy masses at the crest of each. The sheet of lava which had cascaded over a precipice above 400 feet high formed a continuous bed of stony rock between two and three feet thick, with an upper scoriform portion somewhat thicker, the whole inclined at an angle of from 35° to 50°; demonstrating, even if no other evidence was forthcoming, the fallacy of the doctrine of MM. De Beaumont and Dufrénoy as to the impossibility of lava congealing into solid stone at angles steeper than 3° or, at most, 5°.

The steep slopes nearly surrounding the upper part of the Val del Bove have been similarly encrusted with lava-currents of earlier dates which have flowed down from the neighbourhood of the highest cone. Some of these are seen in the sketch in the next page, copied from Von Waltershausen.

Etna seems to have been in frequent activity during the four centuries preceding the Christian era, but was comparatively tranquil for 1000 years after, since which time, for more than eight centuries, it has continued in the phase of intermittent paroxysms—eruptions of considerable violence

ETNA—ITS GRADUAL FORMATION. 343

having succeeded each other at short intervals. Von Waltershausen's admirable map of the mountain represents the
chief scoriæ-cones that have been thrown up during this long
period on the slopes of the mountain, and the lava-streams
that have flowed from them, radiating in all directions outwards from the central heights to the sea or the plains around.
It seems evident from the mere inspection of such a map,
but still more of the mountain itself, that if in the course of

View of the truncated cone of Etna (Mon Gibello), coated over with lava-
streams, as seen from the Val del Bove.

only eight centuries such an immense amount of matter has
been added to the former surface, the multiplication of similar
eruptions through an indefinite number of ages—say from the
beginning of the Post-pliocene era to the present day—would
necessarily have accumulated a proportionately greater mass,
equal probably in bulk to the entire mountain, and it becomes a matter of wonder how any geologists could have
thought it necessary to adopt the notion of a sudden upheaval
of the whole, or the greater part, " like a bladder," in one day,
as the mode of its formation.

It is, no doubt, true that to a certain extent the bulk of the
mountain will have been increased through internal swelling
from the injection of fissures broken across its framework by
the lava rising in the vent at the epochs of its several eruptions. The amount of this internal accretion will be measured
by that of the more or less vertical dykes which intersect its
horizontal or sloping beds. But though these are certainly
numerous towards the central parts of the mountain, their
aggregate bulk cannot at the utmost be estimated at more

than one-sixth of that of the beds they traverse—the accumulated products of outward eruptions.   Nor can they, even in
this reduced proportion, be admitted to afford the least countenance to the Elevation-crater hypothesis, which assumes the
existence of a hollow vault beneath the crust of the volcano,
and attributes its upheaval, not to successive heavings accompanying each of its innumerable eruptions, but to one single
expansive effort of subterranean gases.   There is no need, however, to dwell further on this exploded and wholly untenable
theory. (See p. 167.)

The lavas produced by Etna are generally of that intermediate mineral character between trachyte and basalt which
I have called greystone.   Neither pumice nor obsidian has
been found on any part of the mountain—a remarkable fact,
considering the proximity of the Lipari Isles, which abound
in the vitreous and highly felspathic lavas.   Professor G. Rose
has determined the Etnean lavas to consist of an intimate
mixture of Labrador felspar and augite.   There is a very
general uniformity in the character of its lavas of all ages
—the more recent, however, being more ferruginous than
the earlier.   The scoriæ of Etna are consequently heavy and
harsh.   Nor is there such a variety in the ejected extraneous
fragments as at Vesuvius.   Some of granite have been occasionally, but rarely, found.

There appears to be no correspondence in point of time
between the eruptions of Etna and Vesuvius, or of the intermediate volcanic vents, from a comparison of their dates, so far
as historical records of such phenomena allow of a judgment
being formed upon the subject.

Pantellaria.—The island of this name, thirty-six miles in
circumference, lying about sixty miles off the coast of Sicily
on the south-west, and about half-way towards the nearest
point of Africa, is wholly composed of volcanic rocks.   Some
of them appear, from the statement of Hoffman, to be highly
felspathic, and to resemble those of Lipari in the abundance
of pumice and obsidian, which, as in the latter island, occur in
currents, and often contain large cracked crystals of felspar.
With these are associated many ferruginous lavas resembling

those of Etna; so that in this case the two varieties of volcanic product peculiar to the two last-described localities are combined. There seem to be traces of a vast crater encircling a space twelve miles in diameter (!), the enclosing walls being composed of beds of trachytic lava-rock and pumice-conglomerate, all sloping towards the sea. Though there is no record of any eruptions from this volcano, it was probably in activity at no very remote period, since steam issues upon many points, as well as several hot-springs.

Graham's Isle.—In the interval between Pantellaria and the nearest part of Sicily, at about an equal distance from both, the submarine eruption took place in 1831 which produced the small island so named by Capt. Smyth, and by the French the Ile Julie, already noticed in the body of this work (see pp. 61 & 237). It very shortly after disappeared, under the wearing action of the waves, probably to reappear on the occurrence of some future eruption, and in the end perhaps become a permanent isle.

Linosa and Lampedusa.—Two other wholly volcanic islands, about five miles in diameter, rise from the Mediterranean on the south-west of Sicily, and nearly half-way between Pantellaria and Malta, but rather to the south of the direct line. The first is surrounded, according to Capt. Smyth, by unfathomable water. It possesses four distinct craters. The highest peak measures 850 feet from the sea.

From what has gone before it appears worthy of notice that the triangular island of Sicily is almost surrounded by sites or lines of volcanic eruption. The range of the Lipari Isles and Ustica runs nearly parallel with its northern shore, Etna and the volcanic rocks of the Val di Noto border or compose its eastern shore, while the three last-mentioned volcanic isles front its southern flank. I mention this as one of the examples of the general law of the parallelism of volcanic chains of vents with the adjoining coast-lines noted in the body of this work (see p. 275, *supra*).

North Africa.—The island of Pantellaria lies within fifty miles of the coast of Tunis, about Cape Bon, and, in conjunction with the last-mentioned sites of eruption, appears as a link connecting Sicily with Africa, especially as the intervening tract is known to be very shallow compared with the average depth of the Mediterranean. We should hence be justified in the expectation of finding traces of volcanic action upon some part of the African coast in that direction. I am not aware, however, of any authentic information upon this point. Further south, and immediately at the back of Tripoli, we hear of basaltic hills. And, again, in the interior, on the route from Tripoli to Fezzan, considerable portions of the Atlas range, called Djebel Soudan, and the Black Haroutch, are described by Ritchie and Hornemann as wholly basaltic. They seem to belong to the tertiary age, being mixed with calcareous sedimentary strata of that period, so as to form a very close parallel to the traps of the Vicentine, petrified fish even being found in a fissile limestone in the same abundance as at Monte Bolca. This range is presumed to be the *Mons Ater* of Pliny. And as Solinus speaks of the snowy summits of Atlas shining with nightly flames, it has been thought not improbable that some of these mountains may have been eruptive within the historical period.

Sardinia.—It appears, from the full and able description of the geology of this island by General della Marmora (Voyage en Sardaigne : Turin, 1857), that it exhibits on many points, scattered over a large part of its surface, the results of volcanic eruptions, repeated from the earliest period to that which immediately preceded the introduction of man upon the globe. The basal rocks of the island are gneissose and Silurian, penetrated and upheaved by granite, porphyry, and greenstone (diorite) with steatitic portions. This axial elevation ranges due N. and S., and is continuous and probably coeval with that of Corsica. During the upheaval of this central axis, and from a fissure parallel to it on the west, eruptions took place, and produced several groups of hills composed of trachytic porphyry and tuff, with much stratified pumice. The trachyte is seen to rise through granite and the other basic rocks, but generally rests upon its own tuffs in current-like beds (*coulées*). Its eruption was subsequent to the deposition of the Nummulitic (Eocene) strata, which

appear at that time to have been forcibly folded into parallel
wrinkles. There are also indications of fractures and dislocations about this period in an E. and W. direction at right
angles to the main axial ridge, and it is generally at the intersection of these transverse systems of rents that the eruptions are found to have burst forth. Some of the trachytic
tuffs contain Miocene freshwater shells and vegetable remains.
A subsequent eruptive era threw out other trachytes containing hornblende, and passing into clinkstone. These are
intermingled with the marine Subapennine (Pliocene) strata;
and both these classes of rocks are seen in turn to have been
pierced by dykes of basalt, connected with a later and very
copious emission of basaltic lavas, which cover large areas of
the island with plateaux capping hills of Subapennine limestone and the earlier rocks. These plateaux have suffered great
denudation and dislocation since their flowing as lava into
their present position. One district alone of this character
occupies an area of fifty geographical miles in circumference—
which consists, in fact, of one great volcanic mountain (the
Monte Ferru), rising 3200 feet above the sea. Its central
eminences, forming separate peaks, surround what may have
once been a great crater, filled up by subsequent eruptions
and the effects of atmospheric degradation. From these
peaks currents of basalt slope downwards on all sides, having
their origin generally in scoriform knolls, which mark the
issue of the lava, though the lighter and looser scoriæ have
disappeared. These basaltic currents reach to a great distance, with a gradually diminishing slope, and have been
cut up by waterworn ravines, and perhaps by earthquake-
shocks likewise, into isolated plateaux capping hills of considerable elevation above the intervening valleys. The deeper
ravines that furrow this mountain's flank show that its nucleus
is trachytic. We have here therefore, as in Somma, in the
Mont Dore and Cantal of Central France, and many other
parallel cases, a volcanic mountain whose earlier eruptions
were chiefly felspathic, the later augitic. Indeed, the analogy
of the volcanic rocks of Sardinia of different ages with those
of Central France is remarkably close. The trachytic por-

phyry is described by General della Marmora as resembling domite. It differs, however, in one respect from the trachyte of Auvergne, by being occasionally vitreous in texture, and containing bands of obsidian and perlite, ribboned and brecciated,—in this corresponding to some of the trachytes of Ponza, of Lipari, of Pantellaria, and Hungary. The newer lavas of Monte Ferru contain olivine. Some of them are rather greystone than basalt; some amygdaloidal, and occasionally columnar.

Another volcanic mountain, called Monte Arci, of about the same date, but not quite on so large a scale, offers precisely the same constitution—a central nucleus of trachytic rocks, supporting currents of basaltic lava with quaquaversal dip from the borders of a summit-crater half-encircled by precipitous cliffs, whence opens a great valley or baranco, recalling the Etnean Val del Bove. The traces of a still larger crater seem to exist in the circuit of the neighbouring Bay of Oristano, nearly surrounded by basaltic platforms with an outward slope.

Other similar eruptive vents of this period, such as the Monte di Bari, &c., occur on several other points of the island; their more or less early date being indicated in the greater or less amount of aqueous erosion which the lavas emitted from them have undergone, and the state of preservation of their cones of scoriæ or craters.

The age of these augitic eruptions was followed, according to Della Marmora (probably rather accompanied), by the deposition of a considerable marine formation of Post-pliocene sand and sandstone (*Grès quaternaire* ; *Panchina* of Leghorn, Genoa, &c.) which encloses rolled pebbles of basalt. This formation, however, never reached any considerable elevation above the sea-level ; and it is probable that the eruptive vents just described were subaërial. At a still later period there occurred eruptions from several points in one special district in the centre of the island, between Cagliari and Sassari, where a train of some twenty or thirty very fresh-looking cinder-cones, each productive of a lava-stream, rises on a N. and S. line. Large isolated crystals of augite and specular iron occur in the puzzolana and ash scattered around. The lava usually contains the same crystals, with large knots of olivine, and is in parts columnar. The scoriæ decompose into a red earth. These eruptions have burst through both the earlier trachytes and basalts and the tertiary marls. Notwithstanding their very recent aspect, and that the lava-currents have occupied the lowest levels of the existing surface, it would seem that these evidently subaërial eruptions preceded the occupation of the island by man, since some very ancient remains of the earliest presumed inhabitants are found to have been built with, and upon, their lavas. In connexion with this it may be mentioned that General della Marmora describes a very recent stratum of marine shells, of the species which now inhabit the sea around, as gradually sloping from the sea-level up to the extreme height of 100 metres, in which stratum he found a piece of very early *pottery* and some other fabricated articles among the sea-shells—unopened Ostreæ, Patellæ, Mytili, &c. Hence it would appear that parts, at least, of the island have suffered a very considerable upheaval during the human period, perhaps synchronously with the latest volcanic eruptions just described, and probably with the similar elevation of the opposite (western) coasts of Italy.

Corsica.—The north and south direction of the leading lines of volcanic action through the island of Sardinia may be considered to be prolonged through this island, in which erupted rocks of volcanic character have likewise pierced the tertiary strata towards its central parts. And the same meridian line, if continued further north, would strike the group of serpentine and trap-rocks which form the coast-range to the west of Genoa near Savona.

But I now leave this presumed line of volcanic development, and return to the nearly parallel one of the western coast of Italy north of the Neapolitan group.

CENTRAL AND NORTHERN ITALY.

Alban Hills.—On the road from Naples to Rome, and at no great distance north of the volcanic mountain of Rocca Monfina already described (p. 331), is found a group of volcanic hills containing several circular lake-basins, which were evidently craters. The largest of these, that of Ariccia, is eight miles in circumference; that of Albano six. The lakes of Nemi, Juturna, Gabii, and Cornufelle near Frascati, are others. They may all, as well as several cones of scoriæ in their vicinity, be considered as lateral vents of the principal volcano, of which the existing summit is now the Alban Mount. This rises nearly 3000 feet above the sea. It has within its central heights a very regular cone and crater, the Camp of Hannibal, or Monte Cavo, broken away on the side of Rome, towards which it has poured out two very abundant and well-marked streams of leucitic lava. One of these terminates in a lumpy hillock near the tomb of Cæcilia Metella, on the Appian Way; the other near Ardea. The upper cone of Monte Cavo is nearly encircled by the walls of a far larger and earlier crater, ten miles in diameter. They consist partly of loose scoriæ, partly of peperino, or compact basaltic tuff, of which latter rock the slopes of several of the volcanic hills of this district are for the most part composed. Its elements are sufficiently compacted together to make it serviceable as a building-stone—a quality which, like the tuff of Naples, it owes beyond question to the forcible mixture of the materials with water at the time of their ejection in a fragmentary form from within crater-lakes, such as the existing Lake of Albano, or marshes, which are shown to have covered a considerable surface of this district by the abundant deposit throughout it of calcareous travertin full of terrestrial shells, interbedded with volcanic lapillo and ash.

There seems some reason to believe the Alban Mount to have been in eruption since the historical era commenced. Its lavas are greystone, composed of leucite and augite. The imbedded crystals, of the former mineral especially, are often

very large. Some parts, however, to the eye, are homogeneous, compact, and possess a conchoidal fracture. One remarkable variety of this lava, called *Sperone*, appears near
the summit of the Alban Mount, and in great part of the
contiguous hill-range about Tusculum, both as dykes and
currents interbedded with the loose scoriaceous conglomerate
which composes the mass of the mountain. It is sometimes
waxy, at others light, coarse-grained, porous, and rough in
texture, resembling some trachytes in these characters, and in
parts has a globular concretionary structure, appearing to
consist of nodular concretions with interstices between them.
These nodules are sometimes arranged in layers, as if stratified; this is, however, probably the result of a flattening
or dragging out of the matter by movement during the concretionary process. This lava passes into the more ordinary
leucitic greystone. At Rocca di Papa it is charged with small
garnets. The *sperone* is quarried largely for building-stone
near Tusculum. It is not very dissimilar from the Piperno
used for this purpose at Naples. The beds of peperino
have a general slope from the hills towards the plain of the
Campagna. They are often separated by layers of lapillo, or
loose volcanic sand. The matter has evidently often flowed as
a torrent of mud down the slopes of the volcano, vegetable
matter being found under its beds, and even surfaces of turf,
the grass-stalks depressed in the direction of the descent.
(Ponzi, Storia Nat. del Lazio. Roma, 1859.)

Sir R. Murchison, in an interesting paper on the earlier
volcanic rocks of Italy (Proc. of Geol. Soc. 1850, p. 298),
expresses the opinion that all the Peperino of this district is
submarine, and (as I understand him) that its compactness is
owing to the pressure of a deep sea above. This is clearly an
error. The subaërial mud-eruptions of the South American
volcanos, of Java, and many other localities, produce a rock as
tough, solid, and massive as peperino. The tuff which covers
Herculaneum, and which we know to be of subaërial origin, is
equally compacted. And I myself saw strata of tuff formed
in 1822, high up on the flank of Vesuvius, from the ash washed
down by rains, so hard and tough as to require a severe blow

of the hammer to break them.   It is not to pressure or deposition of the materials *under* water that the peculiar compactness of the solid trachytic tuffs or augitic peperinos is attributable, but to their tumultuous admixture *with* water, whether
finally deposited beneath the water-level or in the open air.
The loose arenaceous tuffs or scoriæ-conglomerates, on the
other hand, owe their want of compactness to their having
either fallen, after ejection into the air by the volcano, at the
spots they now occupy, or been strewed upon the surfaces they
overlie by gentle currents.   The two sorts, in truth, often pass
into and alternate with one another, as would be the case
supposing their distinctive character to be owing simply to
these different circumstances of deposition (see p. 242, *supra*).
The peperino-beds that slope from the edge of the crater of the
Alban Lake were certainly subaërial, for they alternate with
layers of loose lapillo and the remains of terrestrial vegetation.

The famous seven hills of Rome itself are partly volcanic
and of the same mixed character.   The tuff, however, is generally more felspathic, and contains a less proportion of augite
and mica than the peperino.   It is also more friable and of a
darker colour.   It overlies the fossiliferous strata of Subapennine (older Pliocene) age.

North of Rome are numerous indications of volcanic action
on a large scale, which probably took place at a time when the
low lands west of the Apennines were but partially elevated
above the sea-level.   The great circular lake of Bracciano,
twenty-two miles in circumference, is surrounded with volcanic
sand and lapillo, some pumice, and fragmentary augite, leucite,
and titaniferous iron, and must be regarded as a crater.   Two
other smaller basins of similar character adjoin it towards
Baccano ; and beds of leucitic lava cap some of the hills
around.   At La Tolfa, westward of this, occur trachytic rocks
penetrated by sulphureous vapours, which decompose them
into aluminite.

The Monti Cimini.—Near Ronciglione, an eminence called
the Monte Rossi shows a very distinct crater on its summit,
now a lake, whence a current of black basaltic lava has issued
and followed the course of the contiguous valley.   The town

of Ronciglione itself stands upon stratified tuff forming the southern and lowest lip of a lofty circular hill-range called the Monti Cimini, which encloses another large crater-lake, from the midst of which rises a small secondary cone. This is, in fact, a very regular and considerable volcanic mountain, of which the outer slopes and the beds that compose them on all sides are inclined with much uniformity from the upper rim of the great crater to the plain around, on every side but the north, where two considerable bosses of trachyte project from the principal mass, called Monte di Viterbo and Monte Soriano. The lavas produced by this volcano are chiefly trachytic. They are accompanied by their proper tuffs.

Bolsena.—At a short distance further north, the oval lake-basin of Bolsena is surrounded by low hills composed of a stratified conglomerate of scoriæ and lapillo, alternating with beds of basalt. Upon one of these, which forms the steep northern ridge of the basin, stands the post-town of Acquapendente. The basalt, which is leucitic, rests here upon tuff, and seems to have flowed down the slopes of the hill towards the river Paglia. At Civita Castelletto and Borghetto are other beds of very cellular leucitic lava, the vesicles and the leucite-crystals likewise being elongated in the direction of the current,—a clear proof that the latter were in course of formation before the lava had ceased to flow, and were broken up and dragged out into stripes by the movement. The lake of Bolsena is so large, measuring twelve miles in diameter, that it is with some hesitation I venture to class it among the wide saucer-like craters of eruption, of which the western slopes of the Apennines have already afforded us several examples on a smaller scale.

Some miles further north, the town of Radicofani (another post on the high-road from Florence to Rome) stands upon a massive bed of lava, at an elevation of 2470 feet above the sea. The rock is a dark greystone or basalt, containing crystals of quartz. It is columnar, heavy, dense, and crystalline in the lower part, but in the upper scoriaceous—indeed, so extremely cellular as to cut with a hatchet almost like a honeycomb, the partings of the cells being very thin and brittle, and the mass so light as to float, like pumice, upon water. The eruption of Radicofani broke out upon the slope of the adjoining still higher eminence, the *Monte Amiata,* of which it terminates one of the embranchments. It appeared to me that the leucitic lavas and tuffs of Acquapendente had once a connexion with the heights of Radicofani. Though now separated by subsequent denudation, the interval is strewed over with blocks of basalt.

The lavas of Monte Amiata itself are trachytic. It constitutes indeed a considerable trachytic volcano, of an early date, having been greatly eaten into by aqueous erosion since these lavas were produced by it. The upper portion of the mountain is thickly clothed with forest; but a range of abrupt cliffs, almost encircling it, marks the level at which the currents of trachytic lava terminate, and the subjacent Subapennine clays and marls crop out from under them. This cliff-range passes immediately above the villages of Pian Castagnaio and Santa Fiora.

The Monte Amiata is the most northern point at which any volcanic rocks show themselves to the west of the Apennines. There are, however, several hot-springs on the continuation of the same line, both at San Filippo in Tuscany, and in the Lagune of Volterra, which may owe their phenomena to the same principal subterranean N.N.E.-S.S.W. fissure from which the various eruptive vents just described were produced. And, indeed, the large formation of travertin (calcareous tufa) which covers a very considerable area of Western Tuscany in the Maremma may be considered as originating in the same volcanic source, as no doubt was the case with the equally large similar deposits of the Roman Campagna.

With regard to the age of the volcanic products north of Rome, there can be no doubt of their being more recent than the latest Subapennine clays and marls of the Pliocene period. Some of the tuffs, which appear to have been distributed under water, may yet have been formed, like the contemporary travertins, with which they are often associated, in freshwater lakes or marshes, or spread over subaërial surfaces by floods or mud-eruptions. But as on the western coast, north of Civita Vecchia, hills 300 feet in height are composed of Post-pliocene shells mingled with volcanic debris, it is evident that the whole district must have partaken to a considerable degree in the general elevatory movements which raised the more southern shores of the Peninsula above the sea-level; and eruptions, both subaërial and subaqueous, probably accompanied that elevation, through a large portion of the Post-pliocene, and, as in the Neapolitan district, of the more recent period during which man has inhabited the country.

Volcanic rocks belonging to a much earlier geological age are associated on several points with the Apennine limestone. Between Bologna and Florence, for example, as well as in the Duchy of Parma, its strata have been pierced, over-flowed, and altered by eruptions of greenstone, passing into serpentine, jasper-breccia, and diallage-rock near the junction with the limestone, which for some distance is also darkened in colour, rendered crystalline and dolomitic. This probably occurred contemporaneously with the deposit of the limestone (Jura).

Again, the limestone of the Gulf of Spezia and Carrara Mountains and that of the Bocchetta (Val del Polcevera) have been invaded by protrusions of steatitic rock, diallage, serpentine, jade, and greenstone (granitone), all passing into each other as well as into the limestone, which to great distances is often rendered crystalline and occasionally micaceous, or changed to a calcareous slate.

NORTHERN ITALY.

A considerable interval separates the most northern volcanic rocks of Tuscany and the Genoese Alps from those found beyond the Po at the southern foot of the Alps.

The first of these met with in journeying from the South are the group of the *Euganean Hills*, near Padua. They are isolated from the Alpine slopes, and consist of several more or less connected eminences of no great elevation, the product, it would seem, of numerous submarine eruptions from as many vents during the tertiary period. The rocks that compose them are both trachytic and basaltic; the former, however, being most abundant. The characteristic variety is a rock (locally called *Masegna*) of an ash-grey colour and uneven fracture, very like the rock of the Puy de Sancy (Mont Dore), or that of the Drachenfels. It contains numerous crystals of glassy felspar, sometimes decomposed, sometimes fresh, and occasionally plates or 'nests' of black mica, or of augite-crystals. Other rocks are compact, with a waxy lustre and glassy aspect, resembling hornstone-porphyry. Some varieties are cellular, and contain infiltrations of quartz and chalcedony, like the millstone-trachyte of Hungary and Ponza. Others approach to pearlstone. These lava-rocks are accompanied, as usual, by their conglomerates, more or less stratified. Where the trachyte comes in contact with the calcareous strata (of the chalk or tertiary ages) through which they have been erupted, these latter beds are hardened and rendered semi-crystalline, the flinty nodules they contain being reddened. Hot-springs gush out from several points of the hills, seeming to indicate the high temperature of the subjacent matter even in the present day.

The *Vicentine Hills*, forming the neighbouring lower embranchments of the Alps, are likewise volcanic, and clearly of Pliocene age,—trap-dykes penetrating the secondary and tertiary (Subapennine) strata, and the latter alternating with volcanic conglomerates, showing the eruptive character of the district during the deposition of these strata. The lava-rocks are here chiefly basaltic, but sometimes have a base of claystone, which assumes occasionally a vitreous texture. Near Schio it is metalliferous, being penetrated by veins containing lead-ore and arsenical pyrites, associated with manganese, quartz-crystals, calcareous spar, and sulphate of barytes. It is in parts cellular or amygdaloidal, the cells being filled with a variety of zeolites. In some cases the calcareous matter is so mixed up with the basaltic, that it is difficult to say whether the rock should be classed as a basalt, a limestone, or a peperino. Shells of the Pliocene age are often enclosed in this impure calcareo-basaltic tuff; as well as fish in vast numbers at Monte Bolca and some other localities, where thin strata of ichthyolitic limestone alternate with volcanic tuff and basalt. The evidently sudden destruction and remarkable preservation of these fish were probably due to the heating of the sea-water in which they lived by a submarine eruption of lava, followed by their immediate entombment in the layers of calcareous ash, which, upon the water becoming tranquil, sank successively to the bottom. No vestiges occur of craters, nor other signs of subaërial volcanic action, in this district, with the sole exception of one hill, called Montebello, between Vicenza and Verona, which has produced a current of recent-looking lava.

Other rocks, of both augitic and felspathic porphyry, some approaching to pitchstone-porphyry, break through the secondary limestones on the borders of both the lakes Lugano and Como, as well as on that of Maggiore, near Intra; and also at the foot of the Alps of Piedmont, west of Arona. These, however, belong to an earlier age than the tertiary lavas of the Vicentine and Veronese.

The same must be said of the black augitic rocks of the Val di Fassa, in Tyrol, which are remarkable for their eruption having apparently accompanied the proximate protrusion of syenite and granite, by which enormous masses of limestone seem to have been elevated in colossal pinnacles, and at the same time impregnated by some process of sublimation with magnesia, so as to be converted into crystalline dolomite. The granite on some points certainly overlaps the limestone (of cretaceous age), and is therefore of more recent origin. It would seem, from the disposition of these rocky masses, that the protrusion of the syenite and granite upheaved the limestone in anticlinal ridges, while the augitic lavas were simultaneously erupted through fissures broken across the intervening synclinal axes. This would accord with the views advocated in the body of this work, of the usual mode of production of these two classes of igneous rocks. (See p. 276.)

I return to the coasts of the Mediterranean.

The Balearic Islands.—These islands form so evident a

continuation of an elevated range connecting Sardinia with the eastern extremity of the Spanish Sierra Morena, that we might expect to find traces of hypogene rocks penetrating them in that direction; and, in fact, an axial dyke of diorite (greenstone), occasionally amygdaloidal, is to be seen in the cliff-sections of the northern coast of the largest island, Majorca, penetrating the oolitic and cretaceous strata of which this part of the island is chiefly composed.

The group of islands between Majorca and the Spanish coast, called the *Columbretes*, are volcanic. Captain Smyth describes the largest of them as having a breached crater, and beds of trachytic lava, obsidian, and scoriæ*.

Spain and Portugal.—Passing into Spain, we find this volcanic band continued along the coast-range of Valencia, Murcia, and Andalusia, from Cape St. Martin, through the district of Cartagena, to Cape de Gata. There has here been a great development of trachyte with its conglomerates, from which alum is largely extracted near Cartagena. This town, and others in the neighbourhood, suffered very severely from earthquakes in the year 1829. Several very recent-looking cinder-cones with craters are also observable on parts of this coast, and to the distance of some miles inland. One very large and remarkable, with a breached crater, is described near Orihuela. Lava has flowed from these vents into the existing valleys. The promontory called Cape de Gata is a vast mass of both trachyte and basalt, with their conglomerates, and would seem to be the ruin of a great volcanic mountain. Considerable volcanic formations are also said to exist at some distance inland, on the northern slope of the Sierra Morena, in the province of Ciudad Real. Still further westward, in the basin of the Guadiana and the province of Badajos, occur rocks of diallage and compact felspar, penetrating secondary and even tertiary strata. Between Malaga and Gibraltar some eruptive rocks of recent aspect are also observable.

Beyond the Straits of Gibraltar, the extreme western coast of the Peninsula exhibits volcanic rocks, hitherto not well

* Geogr. Journ. vol. i.

described, at Cape St. Vincent, and the Sierra Calderona, or Caldron Mountains, believed to be so called from the number of craters still visible there. The province of Beira, according to Dolomieu, possesses a lofty conical volcanic mountain, with a crater at the summit, called Sierra de l'Estrella; and at the mouth of the Tagus, as also for a considerable distance along its northern shore, vast platforms of basalt occur, which, however, from their position on the summit of the hills, must have been produced by eruptions of an early date. It may be that the permanent closing of these vents of early eruptive activity has given occasion to the fearful earthquakes to which this coast has been subjected in later periods.

The mountainous coast-range of Northern Spain from Corunna to Bayonne is, in fact, but the western prolongation of the Pyrenees, and, like that elevated chain of secondary and tertiary strata, appears to have been penetrated on many points by massive dykes of greenstone, porphyry, and other early varieties of trap. In the province of Biscay, north of Bilbao, trachytic and augitic lavas have been erupted on a large scale. The trachyte is described by M. Colletta as often cellular, white, and resembling domite, or sometimes clink-stone, occasionally taking a vitreous texture.

The most recently active volcanic district of the whole Spanish Peninsula is probably to be found in the basin of the Ebro in Catalonia, where, at little distance from the southern foot of the Pyrenees, near the town of Olot, rise some fourteen or fifteen cinder-cones of very fresh aspect, though there is no historical record of their activity. Each of them has given birth to as many currents of basaltic lava, by which the valleys were evidently filled to a certain height. The rivers have since excavated new channels through these masses, to the depth of from 40 to 100 feet, exposing the structure of their interior. The scoriæ are red, and as fresh in appearance as those of Etna. The stratified rocks through which the eruptions burst are of the nummulitic limestone (Eocene) and saliferous red sandstone. Although no eruption has been recorded in this district, a local earthquake laid the neighbouring town of Olot in ruins in the year 1421, whence it may be presumed that the volcanic focus beneath is not yet wholly extinct. (See Lyell, Principles, vol. iii. p. 185.)

The northern as well as the southern flank of the Pyrenees present on many points instances of trap-dykes piercing the secondary and tertiary strata, and probably belonging to an eruptive era coincident with the elevation of this massive mountain-range.

Further eastwards, along the *South coast of France,* some scattered points of volcanic eruption are met with, of a somewhat recent date, as between Agde and Béziers, where a well-preserved cinder-cone has emitted currents of basaltic lava in several directions. Another instance is to be seen in the extinct volcano of Beaulieu near Aix; and no less than seven different points of eruption are found in the department of the Var, to the north of Antibes, on the French slope of the Maritime Alps. Trachytes accompanied by their conglomerates appear to have penetrated the nummulitic strata, and are covered in turn by conglomerates of the tertiary age (Molasse). At Rougiers, Ollioules, La Motte, and one or two other insulated points, basaltic lavas have been erupted. At the last-named place there is to be seen a regular breached crater-cone, which must have been subaërial, and comparatively recent. The lava of the volcano of Beaulieu, in the neighbouring department of the Bouches du Rhone, is basaltic, and is interbedded with strata of Miocene freshwater gypseous marls, and is itself covered by the Molasse.

At no great distance to the north, and almost in connexion with those last described, we come upon scattered links of the remarkable chain of (extinct) volcanic vents that have broken out in the Cevennes and across the elevated granitic platform of Central France, extending northwards nearly to the parallel of Moulins.

Central France.—This remarkable district, perhaps from its facility of access, has been often spoken of as a type of volcanic formations. And, in truth, it does present admirable examples of the varieties of position, structure, and mineral character assumed by volcanic rocks in other parts of the globe, as well as of the effects of time and exposure to meteoric influences upon them. I have so fully described these in another work (Volcanos of Central France, 2nd ed., Murray, 1858), that I will not attempt to do more in this place than offer to my readers a very brief summary of the general facts.

The eruptions which gave birth to these volcanic rocks are remarkable as having broken out from an elevated boss of granite-gneiss, and other hypogene crystalline rocks, which was not only at the time above the sea-level, but had probably been so from a very early period, since no traces occur within its area (nearly as large as Ireland) of any later marine strata than the carboniferous,—and those, indeed, confined to certain troughs, which seem to have been at that time *fiords* in the granitic island. Several depressions in this area were also occupied at one period of the tertiary era—not earlier than the Miocene—by freshwater lakes, from which was deposited a vast amount of arenaceous and calcareous matter—mostly finely stratified marls. And these strata are in part inter-bedded with volcanic ash and basalt, showing that some erup-tions had taken place long before the lakes were dried.

The points of eruption are distributed along two lines—one directed north and south, and ranging across the entire gra-nitic dome; another branching off from this in a direction N.N.W.-S.S.E., which is identical with that of the axis of the central granite, the range of La Margeride, as shown in the main strike of its laminæ, and those of the associated gneiss and crystalline schists.

The volcanic formations consist of (I.) four principal and separate groups, each of which may be called the skeleton of a great volcano, or habitual vent of volcanic matter on a large scale; and (II.) the products of one long train of isolated vents traversing the whole district; each vent having been, it would seem, but once in eruption, so that its products are scarcely at all mingled with those of other orifices.

In point of age, some of the latter class of scattered or in-dependent vents appear to have been in eruption as early as the great habitual volcanos; some, on the contrary, at a much more recent period, long after the latter had become extinct.

I. The great habitual volcanos severally bear the names of the Mont Dore, the Cantal, the Canton d'Aubrac, and the Mezen.

1. The *Mont Dore* is a mountain mass rising in its highest peaks more than 6200 feet above the sea. These nearly encircle two wide crater-like chasms—the upper gorges of the principal river-channels by which it is drained; and from this ridgy crest the flanks of the mountain slope away with considerable regularity on all sides to the plains beneath. The ravines that intersect the slopes afford sections showing that its mass is composed of beds of trachytic and basaltic lavas alternating with their respective conglomerates, and dipping outwardly with the usual quaquaversal inclination parallel to the outer slopes. There are also several projecting bosses of trachyte, especially on the north flank of the mountain, interrupting at that point the regularity of its surface. There can be no doubt of the alternation of the trachytic and basaltic lavas (see p. 126, *supra*); but the first are uniformly bulky and thick, and have not flowed far from the central heights. The basaltic currents, on the contrary, are less thick, and have flowed nearly in all directions in wide sheets to distances of twenty miles or more. The trachyte is very porphyritic, the felspar-crystals often large and glassy. Its usual type resembles that of the Drachenfels, or of the Euganean Hills. There are also some bulky beds and detached pyramidal masses of a schistose greystone (clinkstone), some of it so fissile as to be used for roofing-slate. The basaltic beds vary much in mineral character and aspect: some are dense, black, heavy, and fine-grained; others iron-shot and cellular; others highly crystalline and coarse-grained. Many should rather, perhaps, be classed as greystone. The basalt, though usually prismatic, is rarely very columnar, but more often tabular; the trachyte sometimes columnar on the large scale. The conglomerates are various in character—some loose, and containing blocks of all sizes and sorts of lava-rock. Others are fine arenaceous tuffs, often laminated, and at times as white as chalk; frequently stained with iron, and containing vegetable remains, trees more or less carbonized, bones of

animals, &c. These conglomerates seem to have filled up hollows in the mountain-side, and extend to distances of twenty miles or more in vast irregular accumulations. The basaltic conglomerates are often mingled with the trachytic fragmentary matter in a compact ferruginous peperino. There are no cones of scoriæ left belonging to the old volcano, nor other indications of the site of its eruptive vents, except some superficial heaps of volcanic bombs, and heavy scoriform masses, towards the upper heights, whence the basaltic lavas were no doubt poured, but from which all the more friable scoriæ have long since disappeared under the degrading meteoric influences to which they have been exposed. Isolated eruptions of a comparatively recent character have, however, taken place within the limits of the area of the Mont Dore, which properly belong to the second class of scattered, or independent eruptive vents, shortly to be described.

2. The *Cantal* is another skeleton-volcano, very similar in character and composition to the Mont Dore, and probably of about the same date, since it has been equally cut up by meteoric denudation. It covers about four times the area, although none of its central peaks rise quite as high as those of the Mont Dore. The difference, however, is not great, the Plomb du Cantal being 6100 feet above the sea. As in Mont Dore, the highest peaks embrace a vast crater-like hollow, from the middle of which rises a pyramidal eminence of clinkstone. This was probably shot up as a dyke in a semi-solid condition within banks of loose conglomerate, since worn away. The environing crests are in part trachyte, in part basalt. The highest point, the Plomb du Cantal, consists of the latter. Thence, as in the Mont Dore, thick beds of both kinds of lava-rock descend with the usual inclination, steep at first, and gradually diminishing towards the base of the mountain, to distances of twenty and thirty miles. These lava-beds are accompanied and enveloped throughout by massive conglomerates, which form even a larger portion of the bulk of this mountain than in the case of the Mont Dore. Their sections are seen on both sides of the many deep, and, in some cases, wide, valleys which radiate from the central

heights. The conglomerate is of very mixed character: in some places a felspathic or pumiceous tuff, in others a ferruginous peperino; sometimes compact enough for a building-stone; at others loose and sandy, enclosing bouldered blocks of trachyte and basalt, or pumice and scoriæ. Within the limits of the freshwater basin there has been an intermixture of calcareous matter with the ash and basalt, showing that the lake still existed and deposited sediment when some at least of the eruptions of the volcano were taking place. On many points the tuff contains lignite-beds in sufficient abundance to be employed as fuel. The emission of trachyte appears in this mountain to have preceded that of basaltic lavas, since currents of the latter surmount the former throughout the extended area it occupies. And this order is, I believe, here never reversed. Vast high platforms of basalt, covered with a coarse herbage, are spread over the lower slopes of the mountain in dreary sameness, especially towards the east. Wherever they are cut through by torrents, they exhibit repeated and bulky beds of columnar basalt, with scoriæ-conglomerates separating them. The columnar configuration is sometimes (as at Murat) wonderfully regular, the columns reaching a length of 150 feet; some measuring 50 feet without a joint, though but 8 or 10 inches in diameter. Except within the limits of the freshwater basin, which is not extensive, the granite crops out from beneath the volcanic rocks wherever the valleys are excavated deeply enough to pierce these through. But the mass of volcanic rocks towards the centre of the mountain must be from 1000 to 1500 feet in thickness.

3. *Canton d'Aubrac.*—South of the Cantal, at a distance of about thirty miles from its summit, rises another but smaller group of basaltic plateaux near La Guiolle, between the valleys of the Lot and the Truyère. These lavas are the product of another separate habitual vent or cluster of vents. They rest uniformly on granite, capping the highest eminences, and belong therefore to an early period. I have not visited them; nor do I believe they have ever been thoroughly examined.

4. *The Mezen.*—The mountain region called the Mezen, in which the river Loire has its source, constitutes the fourth

and most southern great volcanic mass of Central France. Its highest point is 5820 feet above the sea, and is composed of clinkstone, as are all the other principal eminences about the central one, which has specially the name of Mezen. There is, indeed, scarcely any trachyte in this district which is not more or less schistose, and therefore to be ranked as clinkstone, though some of the varieties are almost white, and contain very little augite. In the immediate vicinity of the Mezen, a semicircular basin with scoriform and cellular banks is probably a late-formed crater. Thence radiate several elevated embranchments or strings of conoidal hills of clinkstone, more or less degraded. These are either the remnants of several massive currents which have been worn down at intervals where the substance was most friable, or the products of as many separate outbursts of this peculiar kind of lava from different openings on the line of a fissure of eruption. (See p. 165, *supra*.) These pyramidal masses of clinkstone rise from 500 to 1000 feet above their bases; some appear to rest upon basalt, some directly on the granite platform, and at one or two points the clinkstone certainly overlaps the freshwater sands and marls of the basin of Le Puy. The clinkstone is usually variolitic—that is, spotted with small greenish globular concretions, of a matter more augitic than the base, which is almost wholly felspathic, scaly in texture, and semi-crystalline. The grain is often compact, close, and fine; in other places coarse, and resembling domite. In many parts it is considerably decomposed, and passes into a greyish-white powdery earth, almost a kaolin. The heights of the Mezen have also given birth to vast currents of basaltic lava, which have spread in different directions around, chiefly to the east and north. One massive embranchment is directed to the S.E., and reaches nearly to the Rhone at Rochemaure. It is not, however, quite clear whether this vast basaltic plateau, called the Coiron, has flowed as a current from the Mezen, or has been erupted from different vents upon a line of fissure in that direction. The fact that several dykes are seen piercing the gneiss and Jura limestone (which abuts against it at this the south-eastern limit of the granitic platform), and in connexion with the overlying basalt, seems to favour the latter hypothesis, as well as the direction of this basaltic embranchment, which is N.W.-S.E., and therefore coincident with that of the train of independent volcanic vents I am about to describe. The Mezen group, on the whole, is much flatter in form than either the Cantal or the Mont Dore, and therefore has less of the character of a normal volcanic mountain. In point of age, it was probably in eruption at the same or nearly the same time with them, if we may judge from the similar amount of degradation which its products have sustained, and their partly overlying the freshwater strata. In this respect the great mural cliffs of columnar basalt, called Palais des Géans, which edge the eastern terminations of the great basaltic platform of the Coiron, and overhang the valley of the Rhone opposite to Monte-limart, at heights of 700 or 800 feet, are worthy of notice, as showing that the whole of that deep and wide valley must have been excavated since the flowing of these lava-streams. There is a comparative absence of conglomerates in the Mezen; though probably they once filled several vast hollow troughs in the freshwater basin of Le Puy with thick beds of peperino, of which the breccias of Denise, Roche Corneille, &c., are the remaining fragments. I am inclined to believe that the eruption of clinkstone usually takes place in an intumescent, but semi-solid, highly pasty mass, unaccompanied by much of aëriform explosions, or the ejection of scoriæ in any abundance. The scoriæ of lavas so very felspathic as the clinkstone of the Mezen should be pumice; but this substance is rarely found throughout the entire district.

II. *Independent vents of eruption.*—Besides the great volcanic mountains already described, there have been eruptions from time to time throughout this entire district from *several hundred* separate points upon a broad linear band directed nearly north and south. The earliest of these were probably coeval with the activity of the greater vents, since their cinder-cones have equally disappeared for the most part, and the basaltic lavas they poured out form elevated plateaux far above the level of the existing valleys. Some certainly broke out before the freshwater lakes were drained, as is shown by the occasional intimate mixture of their lavas and lapillo with the marly sediment, evidently at the time in a state of soft mud, into a calcareous peperino, as well as by the alternation, in some cases, of basalt with the limestone beds.   Other lavas have broken out on the high granite platform, and after flooding its slopes, continued their course over the freshwater formation, at what must then have been the lowest levels of its surface, but which now, from the subsequent excavation of valleys in the unprotected portions of the formation on either side, form high and flat lava-topped hills.   A still more recent eruptive period threw up groups and strings of very regular and numerous cinder-cones, which appear almost as fresh as some of the most recent parasitic cones of Etna ; many of them breached by the outburst of lava-streams, which have flowed down the existing valleys, filling them to heights often of a hundred feet or more, for distances of several miles., and sometimes giving rise to lakes by damming up their water-channels.

The Puys Noir, Solas, and La Vache, Monts Dôme (Auvergne).
Breached volcanic cones.

The lower portions of some of the currents, especially those of the Vivarais (the most southern eruptive points of the band), show columnar ranges, of almost architectural regularity. The lavas of these independent and scattered vents are chiefly basaltic.   Some, however, as near Volvic, are of very cellular greystone ; and among the range of cones of the neighbourhood of Clermont, four or five bell- or dome-shaped hills of a very porous trachyte (called Domite, after the largest and highest of the whole range, the celebrated Puy de Dôme) rise

out of, or close to, decided craters within as many unquestionable eruptive cones, composed of pumice, scoriæ, ejected blocks,
and ash. These trachytic lavas must have been produced in
a state of such imperfect fluidity as *not to run,* but to accumulate in heaps over and around the orifice whence they were propelled. (See p. 132 *et seq., supra.*)   Several lake-craters occur
among the vents of this range—some bored through granite,
others through basalt.   The banks of scoriæ and lapillo which
surround them, though not heaped up into any considerable
cone, are yet in sufficient abundance to prove the hollow to
have been produced by explosions from a body of melted and
ebullient lava bursting through and shattering the overlying
rocks.   On the whole, this region offers an admirable field of
study of the varied mineral character and disposition of the
products of volcanic action, within a journey of a day or two
from Paris or London, and in the immediate neighbourhood
of towns affording all the conveniences of civilized life.

Perhaps the most valuable lesson that its physical geography or geological features teach, is the enormous amount
of slow and gradual change effected in the relief of the country
during the time its volcanos (all unquestionably subaërial)
were in activity—changes which must have been due to ordinary meteoric agencies (aided, probably, by earthquakes), not
to any extraordinary diluvian floods,—the varying state of preservation of the several cinder-cones, and the varying heights
of the lava-currents that flowed from them, above the bottoms
of adjoining valleys, attesting their relative ages with the
fidelity of a natural chronometer.   For a more detailed description of this most interesting volcanic region, I refer my
readers to the volume already mentioned.

To the north and east of the volcanic district last described,
the mountain-range of gneiss-granite which divides the basins
of the Upper Loire and Rhone, between Vichy and Lyons,
and again between Nevers and Dijon, has been broken through
by considerable eruptive masses of red and black porphyry.
These are rather to be considered of plutonic than of volcanic
character, especially since they have dislocated and thrown up
to the surface strata of Devonian sandstone.

Rhine Volcanos.—Eastward of this occur several groups of volcanic rocks, chiefly of an early (tertiary) age. The most important of them is the insulated hill called the *Kaiserstuhl*, in the valley of the Rhine, near Freyburg. It consists of basaltic masses, often having a scoriform structure on the surface, but without traces of craters. They appear to have been erupted on the spot in a condition of imperfect liquidity, accompanied by the explosive ejection of fragmentary matters, since considerable deposits of tuff are associated with them. Probably this was once a considerable volcano, but has been largely denuded in the long series of ages which have elapsed since it formed an island in the tertiary ocean.

The Odenwald, another group of hills near Heidelberg, offers the remains of another site of volcanic eruption of the same age and character, productive of augitic lava-rocks, with their conglomerates.

North of the *Lake of Constance* occurs a further series of similar formations, the basalt being accompanied by and passing into clinkstone ; and in Würtemberg volcanic rocks of the same general characters are met with in a large number of separate localities.

Volcanic band of North and Central Germany.— A still more remarkable band of eruptive sites, at present to all appearance extinct, crosses the whole of Germany in an east and west direction, from the Prussian provinces on the left bank of the Rhine, through the Siebengebirge, the High Westerwald, the Vogelsgebirge, the Rhöngebirge, the Meisner, and the Habichtswald. It runs parallel with the main ridge of the Alps, about four degrees to the northward of it.

Upper and Lower Eifel.—The eruptive points which occur on the left bank of the Rhine are generally known as those of the Upper and Lower Eifel. They have broken through the slate of the Ardennes, as well as some limestones associated with this, of Devonian age. No more recent strata showing themselves in this district, there is no further indication of the period of the development of volcanic energy, other than the very fresh appearance of the cones of ash and lava-streams produced, which would lead to the impression that they are of very recent tertiary, probably of Post-pliocene date.

I visited and examined these districts in 1825, and described them in a paper printed in the 'Edinburgh Journal of Science' for June 1826, from which I take the bulk of the following statements.

1. *District of Andernach, Mayen, and the Lower Eifel.—* Upon reaching the summit of the steep and richly-cultivated slope which, near Andernach, forms the left bank of the Rhine, you suddenly find yourself in a rude and barren country, presenting a strong contrast to the soft and luxuriant scenery you have left behind, and consisting of an elevated mountain plateau of greywacke-slate, across which the deep valley of the Rhine appears but as a narrow trough-shaped channel which the eye overlooks entirely, the plateau being continued at the same level immediately on the eastern side of that river. On the westward the general level rises gradually to the rugged heights of the Upper Eifel, and it is also partially broken by the narrow and sinuous gorges through which a few tributary streamlets find their way into the Rhine, and still more so by a number of isolated hills of volcanic formation, mostly of a subconical form, with which the surface of the plateau is irregularly studded. Some of these hills are very complete volcanic cones, with or without a central funnel or crater; as the Hirschenberg, near Burg-brühl; the Bousenberg, between that village and Olburg; the Poter, Pellenberg; and lastly, the Camillenberg, perhaps the highest and largest of these hills, which appears to rise about 1000 feet above the level of the surrounding slate-plateau. Others are less regular, seeming to owe their want of symmetry to their being thrown up on an uneven surface, as the steep side of a valley. Others form elongated ridges, composed of the mingled products of three or four neighbouring volcanic orifices : such are the hills above Nieder-nich.

Many have regularly funnel-shaped craters; others are breached on one side by the subsequent emission of a lava-stream; and some are still more irregular, and appear to have suffered more or less destruction from the mechanical action of some denuding force since their production. All these cones of every kind are composed wholly of loose conglomerate, or volcanic ash, containing numerous fragments of pumice, of a phonolitic lava, of slate partly calcined, &c. Thin beds of these fragmentary matters also occasionally cover the flat parts of the slate-platform in the vicinity of the cones, or occupy a few bosoming hollows in the slopes of its valleys.

Many of these valleys are also filled to a considerable height, often to more than half their total depth, with indurated tuff, called in the dialect of the country *Dukstein*, or Trass, of which an immense quantity is quarried on numerous points, and carried down the Rhine into Holland, where it is employed to form a cement which will *set* under water. The lower part is the most compact, and hence is preferred by the quarrymen. It becomes gradually arenaceous towards the upper part of the deposit. This tuff resembles extremely that of Capo di Monte and Posilipo, near Naples. When freshly quarried, it is thoroughly saturated with water, which is driven out by every blow of a hammer upon it. In this state it is of a dull bluish-black colour; but on drying it assumes a shade of light grey. It appears to be almost wholly composed of fragmentary pumice, and is evidently a conglomerate. It contains also fragments of a slaty, or phonolitic, and of amorphous basalt, of burnt clay-slate, and a great quantity of carbonized wood, not in fragments or beds, but consisting of whole trunks or branches, which penetrate the rock in all directions. The condition of this wood is very nearly that of common charcoal; but it pulverizes more readily, and often of its own accord, on exposure. In the valley of Burg-brühl the trass rests sometimes immediately on the slate; but on other points a bed of calc-tuff intervenes —the deposit of some mineral spring prior to the formation of the trass. A similar incrustation occasionally overlies it, and has enveloped fragments of pumice, producing a calcareo-volcanic tuff. The indurated portion is sometimes divided into massive beds by intervening layers of loose pumice or lapillo and fragmentary slate.

On ascending the valley of the Brühl, I found this trass

deposit occupying it to a great depth the whole way from its embouchure in the valley of the Rhine, up to the foot of the Feitsberg, one of the hills which form the circumference of the lake of Laach; from whence this, as well as many other streams (if they may be so called) of tuff are derived.

The basin of the lake of Laach is nearly circular and crateriform, encircled by a ridge of gently-sloping hills of no great elevation. They are composed of irregular beds of loose tuff, containing numerous fragments and some very large blocks of a variety of lava-rocks. Those which are most abundant are of a basalt with very large and regular crystals of black augite, and of olivine. Fragments also occur of trachyte, sometimes of a whitish-yellow colour and conchoidal fracture; at others of a coarse grain, consisting solely of crystals of glassy felspar and hornblende. Some fragments are also found similar to those which are common in the conglomerates of Somma, composed of an agglomeration of crystals of mica, nepheline, meionite, vesuvian, and other rare minerals. No lava-rock appears in place within the interior of the basin; and on its exterior, the only rock of this nature which shows itself on the surface in the form of a regular current of lava is that in which the millstone-quarries of Niedermennig are worked. This stream certainly flowed from the crater of Laach, the ridge of which suffers a depression on that side. The eruption which produced it was probably the last, not only of this particular vent, but perhaps of the whole district, as its surface has an air of great freshness, and is not yet entirely clothed with vegetation. This may have been the eruption recorded by Tacitus (Annal. lib. xiii.) as having ravaged the country of the Jutiones, near Cologne, in the reign of Nero. The rock of which the current consists is greystone verging on trachyte, with very few visible crystals of felspar and augite, and extremely cellular, the cavities being very small and irregular. It is divided into rude columns at the lower part of the current, which is much more compact than the upper, but still cellular. It is here so hard as to be in great request for millstones, which are exported to Holland in great numbers, and from thence find their way to England. It envelopes numerous fragments of quartz (always more or less vitrified and cracked), of granite, and other problematical rocks like those described above as occurring in the conglomerate, crystals of lazulite, &c.

The origin of the trass has been variously accounted for, but appears to me to be derived simply from an ordinary modification of the volcanic phenomena. The pulverulent matter, of which it was principally composed, mixes into a retentive paste or clay with water,—so much so, indeed, as to be used for making pottery where it is found in a loose state. In this state it was ejected by the volcano, and thrown up as usual into a circular or elliptical ridge around the orifice. The rain, which falls generally in great abundance after the termination of an eruption, mixed with these trachytic ashes, must often have formed an impermeable crust at the bottom and upon the sides of this cavity. Hence the water that drains down these slopes would accumulate into a lake continually increasing in depth, until either the pressure of its waters broke down the banks on some one side, or a fresh eruption from below displaced it. In either case, a breach being made in the circumference of the crater, the contents of the lake must have rushed out in a violent *debacle*, carrying off great quantities of the fragmentary matter of the hills through which the water burst, and filling with these alluvial deposits the valleys by which it escaped on the plains at the foot of the volcano.

This process may have been many times repeated from the same volcanic orifice, and is without doubt the real history of the trass of the left bank of the Rhine. Whether the mass hardened afterwards, or remained incoherent, appears to have depended chiefly on the quality of the ashes, and their intimate commixture with the water. This induration is evidently a chemical process, analogous to the *setting* of cements and mortars. The mud-eruptions (*tepetate*) of Quito and the tuffs of Iceland are produced by the same train of circumstances in the present day. With regard to the trass of Laach and its vicinity, this explanation is peculiarly applicable; and the lake would, even at this day, be subject to rise until it burst its bank, but for an artificial channel, or emissary, cut for its drainage by the monks of the abbey of Laach, a picturesque ruin which stands on its western side. Currents of liquid tuff appear to have been discharged in this manner from many points of the circumference of the lake. Those that issued on the eastern side occupied the valleys of the Brühl and other streams which empty themselves into the Rhine ; the remainder inundated the slate-plateau in the direction of Niedermennig, Bell, Olburg, and Kruft, and covered it more or less with beds of compact tuff, which alternate with others of similar composition, but loose and incoherent, probably derived from the fragmentary ejections of the neighbouring vents.

A cavern within the basin of the lake of Laach gives out a considerable volume of carbonic acid gas, presenting all the phenomena of the Grotta del Cane. There are also many mineral springs in the vicinity, as at Tonigstein, and near the Brühl, strongly impregnated with the same gas, which is usually the latest product of an otherwise extinct volcano.

At some distance from Laach, towards the south-west, and between the villages of Bell and Mayen, rises another group of cones, containing two or three irregular crateriform basins, from which different mud-streams appear to have flowed, covering the slate-plateau in their neighbourhood with their deposits. These volcanic vents differ, however, from that of Laach in having produced leucitic lavas, and consequently their conglomerates are of a different character, resembling exactly the peperino of Monte Albano. Such is the rock quarried near Bell, and called *Bakofenstein*. It is in request for lining ovens, from its capacity of resistance to fire, which it owes to its being almost wholly composed of leucite in a fragmentary state. It encloses many small white farinaceous leucites, fragments and blocks of leucitic lava, of burnt clay-slate, and large broken plates of mica.

The leucitic phonolite spoken of by Keferstein as existing in massive beds near Reiden and Meyr, is, I presume, derived from this system of vents.

Further to the south, and near the village of Kruft, rise

three other smaller cones, covered with vegetation, and with faint traces only of craters. Other cones, and some of a large size, are visible to the westward of Olburg, but my time did not permit me to examine them in detail. On the whole, the volcanic products of Andernach and the Upper Eifel seemed to me to bear the greatest analogy to those of Italy, particularly of the Campagna di Roma. The points on which they differ are the result of the former volcanos having broke forth on a high and dry slate-plateau, the latter from a submarine alluvial shore. In both these districts, as well as in the Campi Phlegræi, it is remarkable that the same, or at least very neighbouring vents have produced *trachytic, greystone, leucitic,* and *basaltic* lavas, with their appropriate tuffs.

2. *District of the Upper Eifel.*—The group of volcanic vents which occupies this district is in immediate contact with that of Laach and the Lower Eifel, though the points on which eruptions have taken place are rather more thickly sown towards the western limit, particularly along the course of the river Kyll, than at its eastern extremity. The epoch of their activity appears also to be equally recent, dating at least since the formation of all the valleys of the country, into which their lava-streams have invariably flowed, usurping the beds of the rivulets, which in but a very few instances seem to have had force or time enough to excavate a new channel to any depth below the level of their former one. Indeed, such is the freshness of aspect which many of the volcanic rocks of this district exhibit, that it requires the silence of all historical records on the subject to persuade us they have not been produced within the last 2000 years. Nor is such evidence, indeed, at all conclusive. It is probable that accounts of phenomena of this kind would rarely reach the meridian of Rome from distant and barbarous districts, unless when they were of a most destructive and terrific character, such perhaps as that spoken of by Tacitus, and mentioned in a former page; and if any such occurred during the middle ages, all traditionary account of them may well be supposed to have perished with so much of other and more valuable information.

The volcanic eruptions of the Upper Eifel have burst

through the exposed surface of the slate-formation on many points, and on others through masses of limestone strata which overlie the slate, throughout a considerable part of this district. Some of the vents have emitted currents of augitic lava (basalt); others have confined themselves to the discharge of fragmentary matters. The latter principally, and in some instances almost entirely, consist of broken greywacke-slate and sandstone, more or less affected by heat, and pulverized. It is probably owing to the clayey nature of these fragments, when reduced to great fineness, that the craters of this country have nearly, without exception, become reservoirs of water, or *Maare*, as they are called by the natives. Most of them still have small lakes or peat-marshes at their bottom. Some have been drained for the sake of cultivation; a few appear to have undergone the same process by natural means, either from the lake rising till its weight burst through the banks encircling the crater, or from the slow erosion of the stream by which it discharged itself. In the last case, the sides of the basin are cut through by this natural emissary, as is seen in the Meerfelder and the Drieser Maare, as well as in those near Strohn and Waldsdorf. In the other case, the regularity of the basin has been more or less destroyed by the bursting of its banks, and considerable deposits of trass, or rather of peperino, have been formed, evidently aggregated by means of water. Examples of this are met with in the remains of craters near Steffler, Schalkenmehren, and Rockeskill. On those points where lava has been emitted in a liquid form, a regular crater is rarely to be seen—at least at the source of the lava-current. There are, however, always one or more such craters in the vicinity, which appear to have produced violent aëriform explosions, ejecting scoriæ and ashes, while the lava was flowing from the neighbouring orifice. The force of these explosive discharges of confined vapour is attested by the great size and depth of the cavities they have hollowed out of the solid greywacke strata. That of Meerfeld, for instance (one of the largest), measures above 500 feet from the surface of the lake (which is itself 150 feet in depth) to the average height of the ridge which encircles it, and its diameter can fall very little short of a mile. The quantity of fragmentary ejections heaped round these basins is not at all proportionate to their extent. The greater part consists of slate and sandstone, in pieces of every size, and appearing half-burnt, probably from having fallen repeatedly upon the surface of lava within the vent whence the explosions of steam were discharged.

The most westerly point on which any traces of volcanic eruption are met with is Ormont, where, upon the wild and elevated plateau of alternating slate and quartz-rock, two small cones are seen to rest. They are in contact at their bases, and have neither craters nor visible lava-currents. The scoriæ and fragments of which they are composed are basaltic, with much augite and large plates of brown mica. Isolated crystals and pieces of augite also occur, some nearly as large as the fist.

At no great distance to the east of Ormont, the slate-rocks are overlapped by strata of sandstone, inclined at a high angle, with an easterly dip. Resting upon these, to the south of the village of Steffler, rises a volcanic cone, composed of scoriæ and puzzolana, partly incoherent, partly compacted into a peperino. Steffler is built on strata of this latter kind, which, however, by their inclination are proved to have been deposited by an eluvial torrent descending from another hill N.E. of the village, which still exhibits a large circular crater on its summit. At a short distance on the S.E. lies a small *maar*, or crater-lake.

The village of Roth is built on a current of basalt derived from the cone which rises above it, and which has also emitted a considerable mass of lava towards the north and west. A small cavern, the mouth of a deep fissure in one of these lava-currents, half-way up the side of the cone, is noted for exhibiting a phenomenon met with frequently amongst volcanic formations. The floor of this grotto was paved with a thick crust of ice when I visited it, at noon, on a very hot day at the end of August. During the summer, the peasants of the neighbourhood say it is always found there, while in the winter there is none, but, on the contrary, that the shepherds creep into the cavern for warmth. The cave is probably the mouth of one of those arched galleries which are so often found under currents of lava in Iceland, Bourbon, and elsewhere. If the other extremity of the gallery communicates with the open air at a much lower level, for instance at the foot of the cone, or where the lava-stream terminates in the plain below, a current of air must be continually driven through this passage from the lower to the upper extremity. In its passage it would be thoroughly *dried*, from the absorbent nature of the rock (which is perhaps partly owing to the sulphuric or muriatic acids it contains) ; and the evaporating effect of this current on the wet floor of the grotto from which it issues, moistened by some superficial rill, will be sufficient to coat it with ice in summer,—since the more rarefied by heat the external air, the more rapid will be the current of cool dry air, and, consequently, the evaporation. In winter, a similar draught of air, though less rapid, will be produced, and taking the temperature of the rocks through which it passes (which, from the depth of the gallery, will be about the mean annual temperature of that climate), must appear warm, compared with the external air, to the shepherds who seek a shelter at the mouth of the fissure.

The cone of Roth connects itself with a smaller ridgy hill prolonged towards the Kyll, which has given rise to three or four small distinct streams of basaltic lava.

On approaching the Kyll towards Gerolstein, the traveller is struck by the appearance of an elevated plateau formed of limestone in horizontal strata, resting on the sandstone, and bounded by a range of picturesque and craggy cliffs, with a talus of massive debris at their base. From the surface of the plateau rise four large volcanic cones, besides small eminences of a similar nature. One has emitted a current of basalt, which descends the steep cliffs of limestone in a sort of cascade, on the western side, occupies a small bottom, and, winding round the base of the range of rocks, reaches the channel of the Kyll at Sarsdorf.

The two largest cones of this plateau lie N.W. of Casselburg, a romantic ruin about two miles N. of Gerolstein. The limestone of Gerolstein is crystalline and dolomitic, and is considered by Von Buch to have acquired these peculiar characters from the metamorphic influences of the volcanic action by which it has been evidently penetrated on several points, as seen in the eruptions of lava and scoriæ that have taken place through it.

Round Rockeskill there are traces of another formation of peperino similar to that of Steffler, and appearing to have originated in the hill immediately behind that village. Further north, the Waldsdorfer Kopf is a very regular cone, and at its foot lies a crater-basin, once a lake, but now reduced to a peat-moss. The cone has emitted one of the largest currents of lava of this district. It has flowed towards the west, and reaches nearly to Hillesheim.

Arnsberg is a large and complete cone, which has also produced much lava. Eastward of Waldsdorf lies the Drieser Maar—a wide crater, which has been artificially drained. Masses of olivine, often of three or four pounds' weight, and as large as a man's head, are found in the fragmentary strata which form the sides of this basin. Part of this encircling ridge rises into a high cone on the south-west; and this is again connected with a third hill above Dochweiler, which exhibits a well-characterized crater at its summit, and has sent forth powerful streams of basaltic lava. The road from hence to Daun leaves on the right three or four considerable cones near Nerod and Steinborn. They consist in great part of lava which has burst from their summits or flanks, and flooded the lowest levels of the surrounding plain.

On the east of Daun, a massive and elevated bed of basalt, bordered by abrupt cliff-sections, in which a rudely columnar configuration is visible, descends towards the town from a higher eminence at its eastern extremity, which is composed of scoriæ, and exhibits vestiges of a crater. This bears the appearance of being the least recent of all the volcanic formations of the neighbourhood.

South of Daun rises a group of hills, which appear, as one mounts them, to be solely composed of greywacke-slate, and in which, consequently, no volcanic appearance could be anticipated, when, on reaching the summit, the traveller suddenly finds himself on the edge of a deep circular lake-basin, evidently drilled through the greywacke by repeated and powerful discharges of subterranean vapour. There are three of these *maare* strung together on a line, in a N. and S. direction, and in immediate contact, the same ridge forming the barrier of two neighbouring craters. The fragments of which the surrounding slopes are formed consist chiefly of slate partially calcined, the remainder of augitic scoriæ. A large rock of greywacke-slate, evidently *in situ*, projects from the bottom of one of these basins. The water in the three lakes appears to stand at the same level, and they probably communicate by means of some fissures in the intervening rocks. One only, the Schalkenmehrener Maar, has any visible outlet, and there are traces of trass-streams in that direction.

A few miles farther to the south, the Pulvermaar of Gillenfeld is met with—a magnificent oval basin, presenting exactly the same general characters as those just described, but remarkable for its large dimensions and extreme regularity. The ridge of fragmentary matters which girds it in is without a break, and nearly everywhere preserves a uniform level at about 150 feet above the water surface. The depth of the lake is above 300 feet; the sides slope in the interior at an angle of about 45°, on the exterior of 35°. Immediately at the foot of the cone of the Pulvermaar, on the south side, rises a hill containing a much smaller crater, with a peat-bog at its bottom.

Still farther south, between the villages of Strohn and Trittscheid, is a double cone of large dimensions. It has two considerable craters, both broken down towards the N.W. The southernmost is large and circular, and bottomed by a morass. The other has produced a current of basaltic lava, which, after forming some considerable hummocks in a N.W. direction, turns its course along the bed of the neighbouring rivulet to the S.W., and occupies its channel to a distance of two miles or more, crossing the great Coblentz road.

But, unquestionably, the group of volcanic vents which presents the greatest interest of all in the Eifel district is the

Moseberg near Bettenfeld, with the neighbouring Meerfelder Maar. The Moseberg is one of the highest hills of the whole country. Its base, up to a considerable elevation above the level of the plain around, consists of greywacke-slate and sandstone. Its summit is formed by a triple volcanic cone, the accumulated ejections of three small craters, which remain very distinct. The two most northerly ones are entire, and reduced to the state of peat-marshes. The third has been broken down on its south-east side by a current of lava, of very recent aspect, which, issuing from the breach, descends the slope of the mountain in a stony flood, until it reaches the bed of a small river below.

The lava and scoriæ of these cones have enveloped a great quantity of half-fused fragments of sandstone and slate. The adjoining circular crater, called the Meerfelder Maar, is remarkable for its size and depth. It measures nearly a mile in diameter, and is said to be more than a hundred fathoms deep. It has been hollowed out of both the Devonian slate and sandstone, forming the north base of the Moseberg; and the steep walls which encircle it exhibit on many points the abrupt sections of these rocks, which are only partially covered by a sprinkling of ashes, puzzolana, pulverized slate, and other fragmentary matter. The bottom of this cavity is occupied by water to about a third of its superficial extent; the remainder is a plain, on which the village of Meerfeld is seated.

The most southerly point of this district on which volcanic products have been met with is the vicinity of the baths of Bertrich, a village seated at the bottom of the deep and narrow mountain-gorge of the river Isbach, which flows at the distance of a few miles into the Moselle.

Here, a lava, which has congealed into an exceedingly hard, tough, and compact basalt, full of crystals of olivine and augite, appears to have been emitted from clefts in the grey-wacke, on three or four neighbouring points, upon the very brink of the steep slope, or rather precipice, which forms the northern flank of the valley. Very few aëriform explosions seem to have taken place, since scarcely any scoriæ were ejected, and the few that occur lie in beds *upon* the surface of the lava, around its three principal sources, and were therefore thrown up *after* its emission. At each of these points is a very small cone. The most easterly, called the Facherhöhe, has an evident crater encircled by rocks of basalt covered by scoriæ. From hence a stream of basalt may be traced uninterruptedly into the bottom of the valley (which is here about 600 feet in depth), falling in a sort of indurated cascade over the almost perpendicular cliffs of slate.

The next cone, called Falkenley, consists of a mass of basalt covered by a deep bed of scoriæ, and also gives rise to a copious basaltic current, which descends into, and has usurped, the channel of the Isbach to some distance, both up and down the stream. The third point of eruption presents two very low and small cones, formed entirely of scoriform basalt, and appears to have produced a current of no great magnitude, which may be traced at least part of the way down the nearest ravine into the main valley below.

The exceeding crispness of the scoriæ of this locality, particularly of the Falkenley, is remarkable. Fragments of both the pebbly and slaty beds of the Devonian strata, partly fused, and graduating on these parts into the basalt, are enclosed in great abundance by this scoriform lava-rock.

At the bottom of the valley it becomes evident that the mountain-torrent called the Isbach has cut through and carried off the greater part of the basalt-streams which once filled its channel to a considerable height, throughout an extent of more than a mile above, and rather less than this below the village of Bertrich. Patches only of basalt are left now on either side of the present bed of the river, and most usually in the concave elbows of the valley; but some of these present cliffs 50 feet in height. The lower part of these masses of basalt is regularly columnar, the columns being divided by frequent joints, from 6 inches to 2 feet apart. Where they have been long exposed to erosion from the torrent, the angles of these short prisms yielding sooner than the nucleus, the columns appear to be formed of rude and flattened spheroids piled one upon another. This is, in short, an example of the columnar divisionary structure passing into the globular, by the increase of the number of joints.   An arched passage, which goes by the borrowed name of Fingal's Cave, or the Cheese-Cellar, nearly a mile above Bertrich, exhibits this structure in the most perfect manner.   It has evidently once formed the channel of the little torrent which now runs on one side of it, and which has thus partly worn away the columns, till they are reduced to mere piles of balls.

The eruptions of these three or four contiguous vents were no doubt simultaneous, or very nearly so.   The lava-streams produced by them can with difficulty be distinguished from each other, all uniting in the valley below ; and the basalt of all is identical in mineral character.   It seems probable that the thermal springs of Bertrich-bad owe their moderate warmth to having percolated through some mass of lava not yet quite cooled in the interior of the schist-rocks, occupying perhaps the prolongation of the fissures through which the lava-streams were expelled.

There occur a few other vents in the vicinity of Ulmen, Kellberg, Adenau, and Boos, which form the connecting links between this district and that of Andernach.   Some of these I did not visit ; but from those which I saw, as well as from Steininger's account of the others, they appear to be mere repetitions of the least interesting of the cones and *maare* already mentioned.

Upon the whole, though the vestiges of volcanic pheno-mena to be observed in the Prussian provinces on this side of the Rhine offer, without doubt, a highly interesting field of study to the geologist, yet they cannot be recommended as types of volcanic formations to those who, without visiting other more distant vents of subterranean energy, either active or extinct, might seek, in the short tour between Spa and Coblentz, to acquire a general knowledge of the effects of this class of natural agents.   They are far less instructive than the analogous formations of Auvergne, the Velay, and Vivarais, where almost every possible modification of the volcanic phe-nomena is to be clearly traced, and on a much larger scale. In the Rhine districts there is a comparative littleness, and an appearance as if the volcanic energy had been damped and impeded by the mass of transition and secondary strata which it had to pierce, and perhaps by the fragile nature of the grey-wacke-slate, which, shattered and pulverized by the first few aëriform explosions of every eruption, would be likely to accumulate in great volume above and within the vent, and stifle its further activity. They are, however, well worth examination; and as I am not aware of any other equally complete sketch of them, I hope to be excused for having reprinted it here in such detail.

One of the small craters of the Upper Eifel, not yet mentioned, is that of Rodderberg, on the left bank of the Rhine, immediately above the well-known basaltic rock of the Rolandsek. The crater, like so many others in the neighbourhood, has been excavated through highly inclined strata of slate and quartz conglomerate, and its banks are strewed with half-fused fragments of these rocks, as well as scoriæ, lapillo, ash, and numerous volcanic bombs. The basalt of Rolandsek appears to be so intimately connected with this crater, that I believe it to be contemporaneous, and to have been emitted from the same vent as a current of very imperfectly liquid lava which adhered to the slope of the hill, much as a run of wax or tallow adheres to the side of a 'guttering' candle, rapidly coagulating there into a buttress-like rock. It is, however, not impossible that it may have extended some way across the channel of the Rhine, and dammed up its waters for a time, the impediment having been since removed by their erosive agency.

The Siebengebirge.—On the right bank of the Rhine rise these seven hills, all volcanic ; the Drachenfels being the nearest to the river and most conspicuous. It is a high dyke-like mass of coarse-grained trachyte, containing large crystals of glassy felspar, and very similar to the trachyte of the Mont Dore, and to domite. It is seen to overlie the Brown-coal formation, and is therefore probably of middle tertiary date. It is accompanied by a pumice-conglomerate not unlike the trass of the opposite side of the Rhine. In its partially decomposed state it resembles clay-stone, and is employed, under the name of *Bakofenstein,* as a fire-stone for the lining of ovens, &c. The other hills of the group are chiefly basaltic; but there are, in fact, spots where the one class of rock passes into the other. The lavas of the Siebengebirge appear to have been protruded in a condition of considerable consistency, so as to have accumulated into knolls or dome-like hills above their sources, instead of flowing as currents, or spreading widely as plateaux, after the manner of more liquid lavas. The surface of the basaltic lavas appears generally broken up by retreat-fissures into a rude chaos of prismatic blocks, more or less cellular. There are no traces of craters, nor any cinder-cones remaining to attest the site of the aëriform explosions which probably accompanied the expulsion of these bulky lava-masses. The contrast is indeed remarkable between the general character of the development of volcanic energy on the west and east sides of the Rhine. In the former, ash-cones and craters, the result of aëriform explosions, are, as has been shown, numerous; while massive lava-rocks, on the contrary, are but rarely visible. In the latter, lavas, both trachytic and basaltic, abound, but no cone or crater is to be seen. The difference is no doubt attributable chiefly, but perhaps not wholly, to the greater amount of denudation to which these earlier formations have been exposed.

The Westerwald.—The group of the Siebengebirge is prolonged, and its characteristic formations repeated, in this extensive range of volcanic hills, which also consist partly of trachyte and greystone, but chiefly of greenstone and basalt. These rocks form overlying masses, and dykes penetrating secondary strata. They are accompanied by pumice and scoriæ conglomerates and trass. At its eastern extremity, the range of the *Vogelsgebirge* continues the linear train of vents, in a massive group of basaltic knolls and plateaux, covering an area of at least fifty square miles, the result of numerous repeated eruptions through the variegated sandstone (Trias), as well as some freshwater (Miocene) beds, outliers of the Mayence formation. The erupted rocks have here partially disturbed and altered the strata. Among these volcanic masses, that of the *Münsterberg* is conspicuous for its magnitude. There are appearances of crater-lakes in this district, and con-

2 c

siderable deposits of scoriæ and ash, the latter occasionally assuming the character of trass.

Further to the north and east, on the borders of Bavaria and Hessia, occur numerous scattered eminences of volcanic rock, each probably the site of an independent eruptive vent. One of these, the *Meisner*, an imposing plateau of basalt, over-lies the brown-coal, which it has in some places converted into anthracite. At the *Blaue Koppe* the basalt is seen to rise through the Bunter sandstone in massive dykes. Near Cassel, a lofty mountain-ridge, called the *Habichtswald*, is composed of basalt, scoriform lava, and peperino, resting likewise on brown-coal and the calcareous beds of the freshwater (Miocene) formation.

Next in order towards the east, the *Rhöngebirge*, a chain of mountains near Fulda, exhibits considerable masses of basalt and clinkstone, on some points highly scoriform, in others passing into pearlstone, as near Coburg. The volcanic action appears to have been very recently developed here, according to Professor Leonhard. In the *Thüringerwald* much red and black porphyry has been thrust up. Again, on the north-east of Bohemia, a series of basaltic cones borders the *Fichtelgebirge* from Egra to Pashstein, which have pierced the Keuper sandstone. Some of these rocks are of early date; but there are also indications of very recent eruptions, as at the *Kammerberg*, near Egra. The extremely hot springs of Carls-bad in the vicinity indicate that volcanic influence is still active at no great depth beneath the surface.

From Egra to Töplitz, and thence to the *Riesengebirge* in Silesia, another chain of numerous basaltic and clinkstone hills extends in a direction nearly parallel to that of the primary range of the Saxon Erzgebirge. Near Töplitz, basalt and clinkstone compose a series of lofty hills, and are accom-panied by beds of tuff alternating with tertiary limestone. According to Ehrenberg, a crater-shaped valley is seen near Franzenbad four miles in diameter, containing a small central volcanic cone called *Kammerbuhl*. Scoriæ and cellular lavas appear abundantly in the *Mittelgebirge*, where they contain leucite. Eastward of the Erzgebirge occurs another range of basaltic mountains, described by Daubuisson, chiefly assuming the form of high plateaux, resting on granite, gneiss, or mica-schist. One of these, the *Stolpen*, has long been noted for the great regularity of its columnar structure. This is the district which Werner chiefly studied, and from which he imbibed his long and obstinately maintained doctrine of the precipitation of basalt from an aqueous menstruum.

Passing from Saxony into Lusatia, we find the range of basaltic eminences continued, with occasional intervals—on some points spreading widely in flat plateaux, on others rising in isolated cones. In the *Riesengebirge*, basalt based on granite reaches a height of 4660 feet above the sea. Still further eastward a series of basaltic cones extends from the Oder at Falkenburg to Troppau, and thence to Freudenthal in Moravia; near which place, to the north of Olmütz, there are indications of comparatively recent eruptions in a hill called the *Raudenberg*, where basalt and red and black scoriæ appear as fresh as those of Vesuvius, and several funnel-shaped cavities are considered to be craters. Further east still, and near the frontier of Hungary, at Banow, is a small volcanic formation, described by Dr. Boué as a cone of grey clinkstone containing hornblende, and having pores elongated in a vertical direction—probably therefore a dyke, or the nucleus of an eruptive vent*.

Thus it appears that, traversing the centre of Germany, in a line nearly parallel to the central ridge of the distant Alps, there occurs an almost continuous string of basaltic cones and plateaux, the product of eruptions mostly belonging to the Pliocene or later tertiary age, but on some points much more recent. And as this same band of territory is liable to frequent earthquake-shocks, and possesses numerous hot-springs, it is possible that there yet exists beneath it volcanic matter in a condition far removed from complete inactivity.

Hungary.—Another and perhaps still more remarkable series of volcanic rocks is met with along the southern foot of the granitic mountain-range that separates Hungary from

---

* I am indebted for much of the above to Dr. Daubeny's second edition. I have not myself visited the district.

Gallicia, upon the northern border of the extensive and marshy plain, once no doubt a vast lake, which is now drained by the Theiss and Danube, and their tributaries, through the gorge of the Iron-gate near Orsova. The great bulk of the erupted matters of this district is trachytic. It has been described in great detail in the voluminous work of M. Beudant, as containing five independent centres of eruption, or volcanic mountains, viz. that of Schemnitz, that on the northern shore of the Platten-See, and three others immediately beneath the south-western slopes of the Carpathians. There are also minor masses scattered between these points. The principal trachytic groups consist of several more or less distinct and chiefly dome-shaped hills of different varieties of rock. These varieties M. Beudant classifies as—1. Trachyte properly so called; 2. Trachytic porphyry; 3. Pearlstone; 4. Millstone-porphyry; and 5. Trachytic conglomerate. From his description of these rocks, and an examination of the rich collection of specimens brought away by him, now in the Ecole des Mines at Paris, I have no difficulty in recognizing the two first divisions as identical with the usual varieties of the well-known trachytes of the Puy de Dôme and Mont Dore. The third presents the characters of ordinary pearlstones, such as are found in Ponza, Lipari, Sardinia, the Cordilleras of America, and many other localities. Like them, it often exhibits a ribboned and laminar structure, and passes into laminated crystalline trachyte. The fourth class, a highly siliceous and carious trachyte, has also its parallel in Ponza and among several of the South American volcanos. The fifth, or conglomerate class, is composed of the fragmentary ejecta and eluvial matters, chiefly pumiceous tuffs, common to all trachytic volcanos, disposed in the usual modes about the lower slopes or around the base of the mountains, which are composed for the most part of erupted lavas and such ejected matter as had not been transported far from the central vents. Some of these conglomerates are interbedded with marine strata of Miocene age, thus affording a clue to the period at which the trachytic volcanos were in activity. (See p. 175 *et seq.*)

Unfortunately M. Beudant, perhaps from not having very clear ideas on the laws of volcanic action, or the normal arrangement of erupted matters, furnishes little information as to the structural characters of these great trachytic groups. I gather, however, from his description of them, that there must have existed there through a long period at least five volcanos habitually erupting trachytic lavas of very imperfect fluidity, which therefore accumulated for the most part in dome-shaped hummocks or bulky beds in the immediate vicinity of their source, surrounded by eluvial agglomerates carried down by aqueous torrents accompanying, or in the intervals of the eruptions,—the mountains thus formed having since suffered a great amount of ordinary denudation. Whether any craters may yet be traced among these skeleton volcanos does not appear; but I think it probable that it would not be difficult on the spot to recognize the habitual centres of eruptive action, and trace the greater number of trachytic masses to their respective sources. The Hungarian trachytes are distinguished from those of other countries by the fact that they afford gold (auriferous sulphuret of silver) on one or two points, as well as many beautiful varieties of opal. The diluvial drift (Loess) of the Danubian basin is full of trachytic ash.

These trachytic formations of Hungary are continued further to the south-east in Transylvania; and here at some points distinct craters are to be recognized, some of which are in the condition of active solfataras.

Styria.—Dr. Daubeny describes and gives a section of a lofty conical hill near Friedau in Styria, called the *Gleichenberg,* composed of beds of an augitic tuff, with silvery plates of mica, arranged mantlewise at a high angle around a central nucleus of trachyte. Basalt also shows itself beneath the tuff; and the whole rests on tertiary marls, without any volcanic matter. Dr. Daubeny, as a partisan of the Elevation-crater theory, naturally sees an example of it here. For myself, I presume it to have been an ordinary augitic volcano, of which the last-formed central crater was filled by the eruption of a great boss of imperfectly liquid or semi-solid trachytic lava.

The Levant (or Eastern Mediterranean District).—The

trans-European volcanic band which we have so far traced, may be considered as prolonged from Hungary across the Danube, through the provinces of Servia and Roumelia, along the southern skirts of the Balkan, to the borders of the Ægean Sea and the Bosphorus, whence it passes into Asia Minor. Throughout this course several considerable groups of trachytic eminences are met with, closely resembling in general character those of Hungary. They are described by Dr. Boué in his work on European Turkey. Of the igneous rocks of the Thracian Bosphorus, the best account we possess is that given by Mr. Strickland in the 'Transactions' of the Geological Society (vol. v. 2nd ser.). They traverse strata of Miocene age, as well as older rocks, and consist of trachyte and its conglomerates, intersected by dykes of clinkstone and basalt, the latter frequently columnar. Minor veins of chalcedony penetrate the conglomerate. In some spots the rocks are tinged blue or green with copper, whence the name of Cyaneæ given to the Symplegades. Since we find these volcanic masses forming either side of the narrow opening of the Bosphorus into the Black Sea, it is difficult to resist the impression that their eruption coincided in time with the formation of this cleft-like passage, by which the waters of the great inland sea were discharged that once certainly covered the vast depressed area of Central Asia and Eastern Russia, now drained into the low-level salt-lakes of the Caspian, Aral, &c. Such a prodigious revolution of the physical geography of this enormous area (three millions of square miles in extent) may possibly have been effected by a single earthquake or eruptive effort bursting open the isthmus-barrier of the Bosphorus, by which Europe and Asia seem to have been previously connected.

It is not, however, to the Bosphorus alone that this hypothesis applies. The islands of Lemnos, Imbros, Samothrace, Tenedos, and others on the Ægean Sea, off the mouth of the Dardanelles, are likewise of volcanic origin—pitchstone, porphyry, and pumice being reported as among their constituents. An extensive tract of augitic porphyry forms the shore of the Sea of Marmora up to Mount Olympus, near Broussa.

The Troad.—South of the Hellespont, the mountains surrounding the plains of Troy present many traces of volcanic action in hot-springs, and masses of trachyte, clinkstone, basalt, and their tuffs, especially near Æné, Asso, and Mantosia. Trachytic rocks also occur both north and south of the bay of Smyrna, and are more recent than a calcareous lacustrine formation which they broke through and overspread in broad horizontal sheets. Further south, at Budrum (the ancient Halicarnassus), are several lofty hills, composed of trachyte and pumice-conglomerate.

The Ægean Archipelago.—Nearly in the same parallel of latitude with these last sites of volcanic energy rise a string of volcanic islands from the Southern Ægean—Patmos, Cos, Nisyros, Santorini, Policandro, Milo, Argentiere, with several smaller islets; and near the coast of the Peloponnesus, Poros, the promontory of Methana, and the island of Ægina. Professor E. Forbes describes not only Patmos, but also several other islands off this coast, as of recent volcanic origin. Nisyros appears, from the description given of it by Dr. Ross, to be a circular island, possessing a vast central crater 2000 feet deep, environed by cliffs, still in the state of a solfatara, and containing hot-springs. Radiating streams of lava are traceable from the rim of the crater on all sides towards the sea, into which they project as headlands. This island furnished millstones of cellular lava (millstone-trachyte?) to the Greeks in the time of Strabo.

Of *Santorini* and its adjoining islets something has been already said in an earlier page (see p. 200). The formation of the great crater, now nearly environed by the three curving islands, Thera, or Santorini, Therasia, and Aspronisi, dates, no doubt, from a pre-historic age. But the rise of two of the smaller central islets, called the Great and Little Kaimenis, seems to have been more or less vaguely recorded by Pliny, Justin, and other early writers, especially as respects Hiera, or Great Kaimeni. The two others are known to have been produced by eruptions in 1573 and 1707 respectively. They are all, no doubt, the summits of as many minor cones, that have been from time to time thrown up in the ordinary manner from the main vent of the volcano in the centre of the great paroxysmally-formed crater.

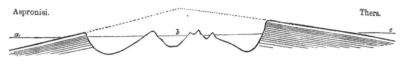

Aspronisi.   Thera.

a, b, c. Sea-line.

The dotted line shows the probable surface of the volcano before the eruption which formed the crater.

The encircling islands show steep cliffs towards the interior. On the other side, their surfaces, as well as the irregular beds of trachyte, tuff, and ash (chiefly the latter) of which they are composed, slope outwardly, with the usual gradually diminishing quaquaversal dip of from 20° to 10°. Although frequently instanced as an illustration of the Upheaval-crater doctrine, there is nothing to lead to the conclusion that these beds have been formed in any other than the normal mode of accumulation round an habitually eruptive vent. The trachyte consists of the ordinary varieties, contains crystals of glassy felspar, and occasionally passes into pitchstone and obsidian. Professor E. Forbes found patches of the muddy sea-bottom containing recent Mediterranean shells attached to the scoriform lava of some parts of the Lesser Kaimeni, which it had evidently carried up with it in its emergence (probably from no great depth) beneath the sea. In some parts of the old crater-basin, however, between these central islets and the surrounding cliffs, no soundings could be obtained. The inhabitants of the islands assert that subterranean noises are still heard, and sulphureous vapours rise occasionally in parts of the circuit, showing the continued activity of the volcanic focus beneath.

Argentiere and *Milo* are trachytic, with some tertiary strata, much altered by acid vapours. The trachyte is said to be in parts extremely siliceous, resembling probably that of Ponza, already described; and a white earth is found there, called Cimolite (from the ancient name of the island, Cimoli), containing 63 per cent. of silex, which forms an article of commerce, being employed to cleanse woollen stuffs. It is probably the residuum of trachytic lava decomposed by acid vapours, after the greater part of the alumine has been washed away by rains.

The island of *Poros,* on the coast of Argolis, is in part limestone, in part trachyte, alternating with beds of pumiceash. Near it is the promontory of *Methana,* which, on the strength of a passage in Ovid's ' Metamorphoses,' has been put forward by Humboldt and others as an example of sudden blister-like upheaval. It consists of a group of trachytic peaks and hummocks, the highest of which reaches 2200 feet above the sea, combining the ordinary varieties of the rock, porous or domitic, porphyritic, siliceous, and compact, partly altered by acid vapours, together with their usual conglomerates. Some of these trachytic lavas are of older date than the tertiary strata with which they are associated ; some are much more recent ; and at the western extremity of the Peninsula occurs much black slag and scoriform lava, resembling in its freshness of aspect that of the new Kaimeni, and probably therefore marking the site of the eruption to which the poet alludes *. There is nothing in this description indicating any departure from the ordinary mode of emission of trachytic lavas of very imperfect liquidity, and therefore accumulating in bulky heaps above the orifice whence they are expelled. (See p. 132 *et seq.*)

Asia Minor.—Returning to the Asiatic continent, we find, from the relation of Messrs. Hamilton and Strickland, a second range of volcanic rocks, running across Asia Minor nearly east and west, from the Gulf of Smyrna along the valley of the river Hermus. The most notable portion of this volcanic band is that which, from its generally recent character, was called of old the *Katakekaumene,* or Burnt-land. This district offers in many respects an exact repetition of Auvergne, the volcanos having broken out on numerous points through strata of tertiary lacustrine marls, the surface of which they have deluged with sheets of basaltic lava. As in Auvergne, the relative ages of these erupted matters are indi-

* Virlet, Expédition Scientifique, 1837.

cated by the relative heights at which the basaltic platforms overhang the existing valleys, which have been slowly excavated by "rain and rivers" alone since the dates when the lavas of these subaërial vents flowed into their present position. Mr. Hamilton recognizes three eruptive periods, the lavas produced by the most recent eruptions ranging up to 80 feet above the present valley-bottoms, those of the earliest 800 feet; those of intermediate age, at a medium height of from 200 to 300 feet. The most recent are traceable to the crater-lips of very fresh-looking cones of scoriæ; and the surface of these newest lava-currents is harsh, jagged, vitrified, and bare of vegetation. The number of cinder-cones remaining is between thirty and forty. Those belonging to the oldest lavas are worn down to mere stumps. The newest are as fresh as those lately thrown up on the flanks of Etna. As in Auvergne, numerous calcariferous, or petrifying, springs rise in the vicinity of these volcanic sites, and have produced much travertin.

Yet further in the interior of Asia Minor, the great high plain which stretches north of the Taurus mountain-range has been pierced by a long series of eruptions. At Asiom Kara-Hissar, and about Baiad, volcanic rocks abound, having the various characters of phonolite, basalt, trachyte, white pumice-tuff, and scoriaceous lava, with cinder-cones. The quarries of white crystalline marble at Eski Hissar, largely worked by the ancient Greeks, are described by Mr. Hamilton as opened in a patch of ordinary secondary limestone enveloped in trachyte, and altered by contact with the igneous rock.

Eastward from thence, an elevated table-land, 4000 feet above the sea, having a drainage-system of its own into some brackish lakes, and composed of an alternation of marine tertiary marls and volcanic rocks, chiefly tuff, in horizontal strata, extends over an area measuring 200 miles in its longest (E. and W.) diameter. From this plain rises another lofty volcanic mountain, *Hassan Dagh*, 8000 feet above the sea. It consists chiefly of trachyte and its conglomerates; but at its base are several cinder-cones, which have given vent to streams of black vesicular lava which has flowed into the plain, and is evidently of recent origin.

Adjoining this to the north-east is the still more imposing mass of *Mount Argæus*, now Erjish Dagh, an insulated cone not less than 13,000 feet in height, and wholly volcanic. Its summit, according to Mr. Hamilton, is composed of a scoriaceous breccia, containing fragments of basalt and porphyritic trachyte, and forming the wall of separation between two enormous breached craters, one open to the north-east, the other to the north-west, but covered with perpetual snow. Numerous cones of pumice and lapillo rise from the flanks of this mountain, and streams of black basaltic lava have flowed about its base. Other vast ridge-like currents of both trachyte and basalt radiate from this mountain on all sides. (Trans. Geol. Soc. 2nd ser. vol. vi.)

There are in this district the same convincing signs of varying antiquity in the volcanic formations as were observed in the Katakekaumene—some being anterior to, others contemporaneous with, the tertiary (Miocene) strata; and many eruptions having certainly occurred within a very recent period. Mr. Hamilton is of opinion that the elevation of the tertiary strata from beneath the sea must have been coeval with one of the earlier eruptive periods of this district. It is, however, as likely to have occurred in the interval between two or more periods of volcanic development, during which these escape-valves were closed—perhaps overweighted—and afforded no vent for the expanding matter beneath. The whole of Asia Minor is still liable to frequent and violent earthquakes, but for centuries has experienced no eruption.

Further to the south-east, at the bottom of the Gulf of Alexandretta, and north of Antioch and Aleppo, occurs another large development of volcanic matter, very similar in character to that last described.

Syria.—Proceeding thence southwards towards Palestine, many traces of volcanic action are observable. The coast of Syria is very subject to earthquakes, one of which, in 1759, is said to have destroyed 20,000 persons, and to have so terrified the inhabitants of Lebanon as to have induced them to abandon their houses and dwell for some time in tents. That remarkable long and narrow meridional hollow which is drained

by the river Jordan, and partly occupied by the Lakes of Tiberias and the Dead Sea (Lacus Asphaltitis), marks probably
the course of a deep subterranean volcanic fissure. Pumice,
bitumen, and sulphur are met with on the banks of the lakes,
as well as hot-springs. The traditional destruction of several
cities in this valley has been very plausibly attributed to volcanic
action. On the south-east side of the Dead Sea, cones of
scoriæ and several volcanic craters were observed by Mr. Legh.
Rüssegger speaks of the western side of the Valley of Jordan as
formed by rocks of Jura limestone traversed by numerous dykes
and streams of basalt; and other travellers describe the country
between the Jordan and Damascus as composed of scoriform
lava, with many cones and craters. The Lake of Tiberias is
partly encased in basalt; and on one side a stream of recent-
looking lava a league in breadth flows into it from the flank of
a mountain at a height of nearly 1000 feet. The river which
drains this lake has cut its channel through another lava-current, which therefore may possibly, by damming up the course
of the stream, have been the cause of the lake itself. On the
eastern border of the Valley of Akabah, which continues the
hollow of the Jordan to the Red Sea, are several volcanic cones.
If we suppose these to have choked up the valley by their lavas,
or that the bottom of the Wady was elevated, at some early
period, in their neighbourhood, sufficiently to cut off the former
continuation of the Gulf of Akabah along that hollow, the sea-
basin so insulated, not receiving in that dry desert region a
sufficient supply from its tributary streams to compensate the
evaporation to which it was subject, will have shrunk by
degrees to the extreme present depression of the level of the
Dead Sea, which is more than 1300 feet below that of the
Mediterranean and Red Sea.

Sinai.—In the peninsula of Sinai, according to Burckhardt,
there are several craters and cliffs of cellular basalt. I am
not aware of its having been explored by any competent
geologist.

Red Sea.—This north and south line of volcanic activity
may be considered as prolonged through the length of the
Red Sea itself, on the coasts of which (especially that of

Arabia) more than one still active volcano has been observed. Ancient chronicles report eruptions near Medina in the years 1254 and 1276 (Humboldt, Kosmos, vol. iv. p. 337). Von Hoff found porous lavas south of Mecca in various places down to Damar, 15° N. lat. Djebel Tier, an island-peak in the Red Sea (N. lat. 16°), sends out vapour continually, and is composed of volcanic rock, as well as an adjoining group of islands. The promontory of Aden, outside the Straits of Babelmandel, is entirely volcanic. Indeed, the town of Aden itself is said to occupy the bottom of a well-defined breached crater a mile and a half in diameter, encircled by precipitous walls from 1000 to 1800 feet high, and backed by still higher masses of volcanic rock. Tradition reports the occurrence of eruptions from this very point. Recent-looking streams of obsidian covered by pumice give some support to this story. If this be so, our settlement may be in some danger. On the African side of the Red Sea, so many sites of volcanic activity are mentioned by different travellers as to lead to the opinion that a nearly continuous N. and S. range of this character will be found there when the country is fully explored. In Egypt, between the Nile and the sea, smoking and sulphur mountains are reported to occur ; and in the province of Kordofan, in Nubia, a chain of hills composed of black and vesicular obsidian. Earthquakes are frequent, and subterranean noises occasionally heard. In Gondar and Shoa much basalt and trachyte are known to exist. On the coasts of Abyssinia similar indications of active or recent volcanic energy have also been noticed, as well as in the island of Socotra, opposite Cape Guardafui, from which rises a volcanic peak, called *Jebel Hajier*, 5000 feet in height.

So little is known of Eastern Africa, that it is impossible to say whether the line of volcanic spiracles is continued along its coast; but it seems not improbable, since the snow-covered mountain of Keenia (1° 20′ S. lat.) is reported to be a volcano (Humboldt, Kosmos, iv. p. 332) ; and the Cataracts of the Zambesi (16° S. lat.) cut their way through basalt according to Livingstone.

Returning to the north of Asia Minor, we find reason to

believe in an east and west line of early volcanic action (now, however, extinct), stretching from the Bosphorus through the interior to Samsoun, and thence along the coast to Trebizond, Poti, and Erzeroum. Trebizond derives its name from a flat-topped hill (Trapezus) of trachyte, covered with tuff and volcanic sand full of cruciform crystals of hornblende (Hamilton). From Erzeroum to Kars, and thence to Teflitz and Erivan —indeed through almost the entire space south of the Caucasus, separating the eastern coast of the Black Sea from the Caspian, as well as the country surrounding the lakes Van and Ourmia—volcanic formations predominate. Besides abundant trap-rocks of the age of the chalk, or early tertiary beds, numerous circular crater-lakes are to be seen there, with streams of lava and obsidian proceeding from them, and several conical volcanic mountains of great size, which have evidently been eruptive from a very early period down almost to the present day.

M. Du Bois de Mont-Pereux, in his elaborate Travels in these countries, describes the following six principal " volcanic amphitheatres," viz. :—

1. That of *Akaltsiké*, reaching from Poti on the Black Sea eastwards to the sources of the Kour River. 2. That surrounding the *Lake Sevan*. 3. That of *Armenia*, including the Great and Lesser Ararat. 4. That of *Lake Van*. 5. That of *Lake Ourmia*. 6. That of the ' volcanic valley ' of *Kapan*, near the embouchure of the Araxes in the Caspian.

1. The first of these ' systems,' that of *Akaltsiké*, exhibits much stratified volcanic tuff interbedded with tertiary marine marls and clays, together with basalt and trachyte of that age. The latter resembles domite. M. Du Bois describes also, and gives a drawing of, a vast crateral basin at the head of the valley of the Kour, having numerous minor cones with craters within it, which have poured out abundant augitic lava-streams. One encloses a small circular lake.

2. The *Lake Sevan*, to the north-east of Erivan, is surrounded by hills of volcanic origin, especially on its south-western side, where a massive eminence, with several craters, has given birth to numerous basaltic lava-streams.

3. To the south of the flat valley of the Araxes, on the borders of Armenia, rise the almost insulated twin cones of the *Great and Lesser Ararat*. The Great Ararat, whose peak is 17,250 feet above the sea, and 14,320 above the plain of the Araxes, presents on this side, according to Abich, an enormous horseshoe-shaped crater, called the Valley of St. James. The cliffs embracing this cavity show the internal structure of the mountain to be an alternation of massive and irregular beds of trachyte and its conglomerate. On this side no recent lavas appear. On the opposite or south-western side the mountain has a very regularly conical outline; but its flank is split, from near the summit down to a third of its height, by a remarkable cleft or niche between high ridges of black por-phyritic trachyte on either side. The bottom of the niche is occupied by a rapidly sloping current of trachyte having a convex cross-section, which widens downwards, until it termi-nates in ranges of abrupt cliffs halfway down the mountain's flank. These cliffs are bordered by still more recent flows of black vitreous lava, which have issued from cones of scoriæ above. Great numbers of similar cinder-cones stud the mountain-slopes from summit to base; some are arranged in strings, and seem of contemporaneous origin; all have craters, mostly breached, and have given issue to prodigious lava-streams, which have flowed to the base of the mountain and over the plains around. The lavas emitted on the southern side are mostly basaltic (*dolerite*, Abich). On the northern, towards Erivan, several bulky bosses and currents of trachyte are seen to issue from as many great crateriform hollows, sur-rounded with blistered and blackened cliffs. These trachytic lavas must have been very imperfectly liquid at the time of their eruption, since they generally stopped about halfway down the slope, terminating in steep escarpments.

Abich gives an interesting view of the Great Ararat, taken from the summit of the Lesser, in which these features are displayed (Bull. Soc. de Géogr. sér. 4. tom. i.). His descriptive account is, unhappily, obscured in some degree by a persist-ence in the Elevation-crater theory. It is astonishing that his observations on this great volcanic mountain alone should not have disabused him of this prepossession, since it would seem impossible to reconcile the idea of its nucleus having been suddenly upheaved like a blister (hollow, of course), with the subsequent expulsion from its summit and superficial slopes of those prodigious accumulations of erupted lavas (trachytic as well as basaltic) and fragmentary matters which he himself describes.

The Lesser Ararat closely adjoins the Greater, separated only by a flat plain or ' Col,' half a mile in width. It has the figure of a very regular pyramid or cone, truncated at the summit by a crater, which, however, appears not to have been eruptive in recent times. Its lava-rock is said by Abich to resemble the andesite of South America.

To the north-west of Ararat, towards Kars, Abich speaks of a vast volcanic system, called the *Tantoureck*, west of Bajazid; also of two great mountains (*magnifiques cratères de soulèvement*), called *Sordagh* and *Aslanlydagh,* and other "volcanic ranges, *Synak* and *Parlydagh,* surrounding the high lake Balykgoell." All of these are visible from the Little Ararat, as well as vast basaltic platforms beyond the Araxes and north of Erivan, probably those of the system of Lake Sevan.

The currents from Ararat seem to have spread to a distance of 110 miles, according to the statement of Mr. Loftus (Quart. Journ. Geol. Soc. vol. xi. p. 322). Some of its lavas are leucitic greystone; there is also much obsidian and pumice. A volcanic cone in this region, called *Suphan Dagh,* 10,000 feet high, is described by him as having a distinct crater on its summit, and its base is covered with basaltic lava-streams. Several other conical volcanic mountains are noticed by Mr. Loftus in the environs of the Lakes Van and Ourmia.

North of this great volcanic district rises the nearly east and west range of the *Caucasus,* chiefly composed of highly tilted Devonian rocks throwing off unconformable cretaceous strata on either side. From its axial heights rise several volcanic masses. *Mounts Kasbegh* and *Savalan,* both from 15,000 to 16,000 feet high, belong to this group. The loftiest peak of the Caucasus, *Elburz,* believed to be upwards of 18,000 feet in height, has a crater on the summit; its lavas are chiefly trachytic.   Many fill up the recent valleys, together with trass and tuff.

Although some of these volcanic vents seem to have been active within the most recent geological period, there exists no authentic record of any eruption from them.   Hot-springs are, however, numerous, and earthquakes not unfrequent.

Abich believes the principal eruptions of the Caucasus to have occurred about the dawn of the human period—its lavas and tuffs covering sedimentary deposits which contain the same Mytilus as that which now inhabits the Caspian Sea*.   Some of the volcanic rocks of this district are, however, interbedded with tertiary strata.   Others are still older, and, perhaps, coeval with the elevation of the Caucasian range.   It is evident that volcanic energy has repeatedly displayed itself here on a colossal scale, from an early period.   And it is by no means clear that it has ceased for ever.   In the region surrounding Ararat especially, earthquakes have been frequent of late years, and were, perhaps, accompanied by eruptions.   In A.D. 341, the mountains of Armenia are said to have split open and vomited clouds of flame and smoke.   Throughout that century similar phenomena were repeated.   And in the eighth total darkness is mentioned as having prevailed during forty days, which seems to indicate clouds of volcanic ash.   A tremendous earthquake in the year 1841 shook the two Ararats to their foundation, toppling down vast rocks from their heights, together with avalanches of ice and snow, into the valleys beneath.   The shock was felt with great intensity through the neighbouring provinces as far as Shusa and Tabriz on one side, and Teflis on the other.

At either extremity of the chain of the Caucasus, towards the Caspian on the east, and the Sea of Azof on the west, as also at the southern foot of its central portion, we find a low-lying district, over which are scattered vast numbers of mud-volcanos, as they are called, i. e. cones of a ductile unctuous clay, formed by the continued evolution of a sulphureous and inflammable gas, spurting up waves and lumps of liquid mud.   Some of them are 250 feet high.

* Abich, Geol. Sketches in Geol. Mem. vol. vi. part 3. p. 44.

The operations of these mud-volcanos have, apparently, been going on for countless ages, and have covered a great extent of territory with their products. The phenomena are certainly very distinct from those of true volcanos, since no scoriæ or lava, or heated matters of any kind are sent forth ; the mud being described as cold when emitted, notwithstanding that the gas, whose violent escape throws it up, is sometimes inflamed. The mud is frequently bituminous; and petroleum is obtained from wells sunk through a bituminous shale at Baku, on the west coast of the Caspian, in the vicinity of some of the sites of the mud-eruptions. The greater number of geologists consider these phenomena as entirely distinct from the volcanic, and to proceed from chemical action going on at no great depth beneath the surface among the constituents of certain stratified matters. Others (Sir R. Murchison especially, who has visited the locality) express the opinion that " they are as much connected with internal igneous agency as any other eruptive phenomena." Certainly their occurrence on the prolongation at either end of the volcanic train of the Caucasus affords much support to this view. And it will be remembered that the analogous phenomena of Macaluba in Sicily occur in a district not remote from the sites of vast volcanic disturbances. If we suppose the sulphuretted hydrogen which emanates so abundantly from volcanic spiracles to rise from great depths through a narrow tortuous fissure, and ultimately through beds of muddy sediment at the bottom of a shallow sea, it may reach the outer atmosphere at a low temperature, and produce effects similar to those described above.

The axial range of the Caucasus may be considered as continued on the west in the elevated southern coast of the Crimea, where the cretaceous plateau has been pierced on many points, and dislocated by eruptive greenstone, serpentine, and other trap-rocks. Tertiary strata rest unconformably upon these, and are mixed up with some beds of volcanic ash, peperino, &c., indicating the occurrence of, perhaps, still more recent volcanic action in that vicinity.

South of the Caspian, in the range of the Persian Elbourz,

occurs the lofty volcanic mountain called *Demavend*. It was ascended in 1837 by Mr. Taylor Thomson, who determined its height at 14,800 feet—the last 100 feet being a mass of pure sulphur, within which was a hollow sending out steam and sulphureous vapour. Pumice and scoriæ lie scattered about the mountain, and masses of augitic lava. The mountain towards its base is described as consisting on one side of strata of sandstone and limestone belonging to the coal-formation, highly inclined and dipping towards the axis of the volcano.

Volcanic formations are said to prevail between Teheran and Ispahan, as well as near Tabriz in Persia ; but no reliable account has yet been given of them.

Ural.—The great meridional mountain-range of the Ural, which divides Europe from Asia, sinks down to the level of the Kirghis Steppe between the Caspian and Aral Seas. The upheaval of the series of sedimentary strata which compose the western portion of this great range (as well as the plains of European Russia) seems to have been effected by repeated eruptions of numerous trappean masses—syenite, greenstone, serpentine, porphyry, and augite-rock, accompanied by ash and other fragmentary matters—along a great linear fissure, or series of fissures, upon the eastern side of the central axis. Sir R. Murchison " considers this eruptive agency to have been in activity from the remotest period," but to have chiefly operated at two distinct eras; viz. first, " after the formation of the carboniferous limestone," and, secondly, " after the deposition of the Permian strata on the dislocated and upturned edges of the earlier rocks." Since this last epoch he thinks there have been changes of level, but no outburst of igneous rocks in the Ural. Volcanic eruptions of the Tertiary period were apparently confined to the Caucasian range, which crosses the line of the Ural transversely at its southern extremity. (Russia and the Ural, p. 586 *et seq.*)

Hindostan and Central Asia.—Indications of volcanic action are said to exist in Affghanistan, as well as in the Valley of Cashmere; but detailed and trustworthy statements on this point are wanting. Along the southern flank of the Himalayas, east of the Jumna, the nummulitic (Eocene) strata, which here rise to heights of 7000 feet, have been pierced by repeated eruptions of trappean lavas—syenite, porphyry, green-stone, amygdaloidal basalt, and volcanic ash or 'grit,' passing (as is so often the case with the early traps) into one another, and altering the marls and sandstones with which they come into contact. These volcanic outbursts probably date from the early Tertiary period in which the great Himalayan range (as well as its prolongation in the Caucasus and the European Alps) was upheaved by many thousand feet. The fracture-lines through which the eruptions burst are parallel to the strike of the stratified and schistose rocks of the Himalayan range and their numerous flexures. Sulphuretted springs and deposits of travertin occur in the same localities.

The province of Cutch, south of the Delta of the Indus, is certainly in great part volcanic. According to Captain Grant, compact and cellular basalt, with scoriæ, abounds there; and in the Doura range is an extinct volcanic mountain with an irregular crater, which is even said to have been in eruption in 1819, while at the same time a formidable earthquake disturbed the neighbouring country. Other indications of recent erup-tions occur in the vicinity, and may help to account for the remarkable oscillations of surface-level which are known to have taken place in the extensive low tract adjoining, called the Runn of Cutch, especially during the above-mentioned earthquake of 1819. (See Lyell's ' Principles.')

In the central and western portion of the Great Indian Peninsula a vast extent of country has been deluged by basaltic lavas, which alternate with a freshwater deposit (probably of Eocene age), usually calcareous. The basalt forms elevated plateaux, hundreds of miles in extent, and seems to have flowed to vast horizontal distances, in repeated sheets, over the bottoms of shallow tertiary lakes; but from what particular vents does not appear, as no cinder-cones or craters are re-ported to exist there. The basalt is often amygdaloidal, con-tains much augite, and is occasionally nodular in structure rather than columnar. It has altered many of the sandy beds on which it reposes, or among which its dykes have pene-trated, into jasper and other metamorphic substances. (Hislop and Hunter on Nágpur district, Quart. Journ. Geol. Soc. vol. xi. p. 370, &c. Sykes, Geol. Trans. vol. ii. 2nd ser.)

M. de Férussac and Baron Humboldt have collected from various historical sources more or less obscure indications of the existence of volcanos in Central Tartary and China, between the Altai and Himalaya ranges. The great east and west range of Tiantschan, connecting the Altai with the Kuenlun, and through this with the elevated plateau of Persia, is said by the latter to be chiefly volcanic. One volcanic mountain, called *Peschan*, is especially mentioned as sending forth perpetually " fire and smoke, and melted stone which hardens as its cools "— the expression employed by a Chinese annalist. Flames are also described as rising from another mountain, called *Ho-tekeou*, near Tourfan, 420 miles further eastward. Sulphur and sal ammoniac are collected in the same range, and in that of the Altai. But so little is really known of the geology, or, indeed, the geography of this portion of Asia, that it is unnecessary to dwell upon the slight amount of information to be gleaned from these sources. There is, however, no reason to doubt the facts reported, which are chiefly remarkable from the great distance of these active volcanos from any sea. They belong, it will be seen, to the great inland-lake-basin of Central Asia.

ATLANTIC VOLCANOS.

An irregular string of volcanic vents, rising, at considerable intervals, out of the depths of the Atlantic Ocean, may be traced from the extreme north-east coast of Greenland, through the islands of Jan Mayen, Iceland, the Ferroe Isles, the Western Isles of Scotland, the North of Ireland, the Azores, Madeira, Canaries, Cape Verde Isles, Ascension, and St. Helena, to Tristan d'Acunha, in the latitude of the Cape of Good Hope; thus maintaining through 120 degrees of latitude a nearly meridional direction, or, rather, one approximatively parallel to the outline of the coasts of Europe and Africa bordering the Atlantic on the east.

I propose to notice these several sites of eruption in the order which they present from North to South.

The East coast of Greenland is said to be of volcanic formation through many degrees of latitude, massive alternating beds of basalt and basaltic conglomerate visibly composing its cliffs ; but no active volcano has, I believe, been noticed there.

On *Jan Mayen's Island,* in lat. 71° N., two volcanos were observed by Scoresby in actual eruption. Their lavas are described as basaltic.

Iceland.—This large island, six degrees south of Jan Mayen, is remarkable for the extensive scale on which its volcanic activity has been developed,—its whole area, which is considerably larger than that of Ireland, being exclusively composed of volcanic rocks, and more than twenty of its mountains having been witnessed in active eruption within the historical period. Indeed, no district in Europe or Africa can be compared to it in this respect. The only parallels must be sought upon the eastern or western coasts of the Pacific Ocean.

The lavas produced by the Icelandic volcanos are partly trachytic, partly basaltic. Probably the greater number should be placed in the intermediate class of greystones. Some are wholly vitreous, *i. e.* obsidian or pitchstone. As is very frequently the case among volcanic islands, a large portion of its area consists of horizontal platforms of basalt, showing repeated beds one above another separated by more or less of scoriæ-conglomerate wherever erosion has penetrated the surface, which, in the case of Iceland, the sea has effected in a multitude of deep fiords similar to those which pierce the neighbouring granitic coasts of Scotland and Norway. Probably these vast basaltic lava-streams were poured out beneath the sea, and have been since raised above it, and fissured, by slow action of the elevatory subterranean forces. They compose the north-western and south-eastern portions of the island. The central or intervening district is chiefly, but not exclusively, trachytic. It is in this that we find the greater number of volcanic mountains which have been eruptive within the ten centuries embraced by the records of the island. These active vents seem to be ranged upon two parallel lines traversing the island from north-east to south-west, and having a great longitudinal trough-shaped valley between them.

As might be expected in that northern latitude, where scarcely an eruption can ever occur without bringing molten lava and heated scoriæ into contact with vast masses of snow and ice, a large proportion of the formations of Iceland consists of conglomerates, formed by the tumultuous rush of floods from the eruptive heights carrying along enormous quantities of alluvial matter, which they spread in wild confusion over the lower levels, filling up some valleys and excavating others. Within the earlier conglomerates are interbedded layers of Surturbrand—a variety of lignite; leading to the supposition that vegetation was more abundant in former times than it is at present in any part of the island—unless it be supposed that the wood was drifted across the Atlantic, which is the case to a considerable amount in the present day.

The Icelandic eruptions that are on record have proceeded from some twenty or more different vents, all of them in distinct mountains in various parts of the island; chiefly, however, within the limits of its central trachytic portion. Hecla, perhaps, has been the most frequently eruptive, but is not the loftiest volcano, its summit being only 4800 feet above the sea. Its lavas are, for the most part, highly vitreous, sometimes consisting entirely of obsidian or pumice, according as the rock is vesicular or compact.

The recorded eruptions of Hecla occurred in A.D. 1004, 1137, 1222, 1300, 1341, 1362, 1389, 1538, 1619, 1636, 1693, 1766–8, and 1845. Since the last date it has been tranquil. Those from other volcanos are the following :—The Kötlugja Jokul was in eruption A.D. 900, 1245, 1262, 1416, 1580, 1625, 1660, 1721, and 1755, the date of its last paroxysm. In 1340 an eruption occurred near Reikianes, from a volcano which had also been active from 1222 to 1240. In 1563 an eruption was observed from the sea at a great distance off the west coast In 1716 one took place from the *Lake* Grimwatn; in 1720 and 1822 from Eyafialla Jokul. From 1724 to 1730

Mount Krabla was in vehement eruption; Orœfa in 1332 and 1362; Trölladyngia in 1150, 1188, 1359, and 1510; Thring-valla-hraun in 1510 and 1587. In 1783 Skaptar Jokul broke out with still greater violence. Others of less note are also mentioned by the early chroniclers of the island.

These several eruptions were usually characterized by the ejection, from a vent opened on the summit or flank of the mountain, of vast quantities of scoriæ or pumice and fine ash, during a term of many days, weeks, or months, by which the face of the country was covered, and even greater damage inflicted on the inhabitants than by the streams of lava emitted at the same time, though these have often been prodigious. The last great eruption of Hecla in 1845 was an example of this kind. It had previously been tranquil for nearly eighty years. The mountain lost on this occasion 500 feet in height—so much of the summit having been blown away by the explosions; and the lava-stream that flowed from it reached a distance of ten miles, with a thickness of from 50 to 80 feet. Its surface is described as fissured into rude blocks arranged in longitudinal ridges. The sides of its fumaroles were found to be coated with muriate of *ammonia* in large quantities, together with an abundance of muriate of iron *. The fumaroles of the crater still deposit much sulphur.

One of the most terrible eruptions that have occurred in modern times in Iceland was that of Skaptar Jokul in 1783. Its principal phenomena have been already mentioned in a former chapter (p. 82).

The latest great Icelandic eruption proceeded from the Kötlugja volcano in May 1860, and is described by Dr. Lauder Lindsay in the ' Edinburgh New Philosophical Journal' for January 1861 (see p. 213, *supra*). This volcano has been fifteen times in eruption since the year 900. The name implies " the great fissure of Kötl," and is derived from a vast cleft (or Baranco), resembling a deep valley, which penetrates the north-east shoulder of the mountain, whose proper name is Myrdals Jokul. The mountain has no other crater but this fissure—" which indeed," says Dr. L. Lindsay, " has

* Descloizeaux. quoted by Dufrénoy, 1846, Mém. de l'Acad.des Sci.

been seen by but few individuals, and that at a distance "—
so inapproachable, from the ruggedness of the country, are
the environs of some of these volcanic heights. Henderson
calls this chasm a tremendously yawning crater, distinctly
visible sixty-five miles off! Snow, ice, smoke, and steam
are said to prevent any near approach. During an eruption
of Myrdals Jokul in 1580, the mountain is reported to have
been rent asunder. This was probably the origin of the great
crater-chasm in its side, since called Kötlugja—a kind of
Baranco-crater.

The most characteristic phenomena of these Icelandic eruptions are the floods of ice, and bergs, and heated water bearing
along rocks and stones,

> . . . . " stirpesque raptas
> Et pecora et domos
> Secum revolventes,"

which they let loose from the mountains, and cover vast surfaces of the country with wreck. Dr. Lindsay explains with
clearness the character of these fire- and ice-floods :—" The
volcanic heat melts that part of the icy mantle of the Jokul
which is in immediate contact with the soil ; its adhesion is
loosened, and a stratum of water formed, which helps to break
it up and to float down the sides of the mountain the superincumbent ice." The devastating effect of such sudden floods
may well be conceived ; they not only heap up vast volumes of
conglomerate on the plains below, but also tear up and score
the mountain with ravines of proportionate size, groove,
striate, and polish its hardest rocks with the rolling ice and
stony flood, and add miles of new land to the coast-line. When
we add to this the dense showers of scoriæ and ash that fall
continuously for days together from the air, into which they
are ejected by the volcano, and the torrents of seething lava
that, issuing from its entrails, rush down the slopes in conjunction with the ice and water debacles, and cover many
square miles of surface with sheets of solid rock, it is evident
that few more powerful agents of superficial change can be
imagined among all the living forces of Nature.

At the commencement of the eruption from Kötlugja in

1755, the earth itself is said to have fluctuated like an agitated ocean, and the sea participated in the commotion, to the serious damage of the shipping. After the rocking had continued for some time, an exceedingly loud detonation was heard, and immediately fire and water (melted snow, or lake-water) were observed to be flung up from three apertures in the volcano,—the column of fire (red-hot scoriæ) reaching so high as to be seen nearly 200 miles distant, and the air darkened around to an equal extent with smoke and ashes. This eruption was remarkable as being contemporaneous with the great earthquake of Lisbon. It is said to have begun on the 19th of October, 1755, and to have continued until August of the next year. The earthquake of Lisbon occurred at half-past nine in the morning of the 1st of November, 1755, or eleven days subsequently,—an apparent difference, perhaps only owing to the one being reckoned by old, the other by new style. If this be so, we may infer that the Lisbon earthquake gave the signal for that which let loose the fires of the great Icelandic caldron, or, rather, that the same impulse occasioned both phenomena. Nor is the idea in the least incredible, since it is well known that oscillatory movements in the waters of the lakes of Switzerland and those of the northern Highlands of Scotland (half-way towards Iceland) evinced the sympathy of those distant portions of the earth's crust with the subterranean commotion, whose central point of disturbance seemed to be in the Atlantic, westward of the mouth of the Tagus.

The last eruption of this same volcano in 1860 began, as usual, with local earthquakes; then a dark column of vapour and ashes was seen to rise from it, and floods of water, ice, rocks, and lava poured down the valley leading from the great gorge. By night, a fountain of *balls* of fire (bombs) rose to a height calculated at 24,000 feet, since it was seen 180 miles off at sea. Many of these volcanic bombs, it is said, were seen, and heard bursting in the air with a loud detonation, from a distance of 100 miles. They must therefore have been of large size, as well as thrown up to a vast height. Their bursting there is rendered highly probable by the fact that fragments of volcanic bombs, evidently having this origin, are occasionally found near eruptive vents; and if we suppose the surface of the globular mass of liquid lava to be consolidated as it rises with a rotating movement to a great height in the air, the expansion of the contained gases in the rarefied atmosphere in which it finds itself at its extreme height is very likely to cause the bursting of the shell with a loud explosion.

The mountains of the central portion of Iceland are said to consist in great part of Palagonite-tuff—a peculiar trachytic conglomerate, stratified, and often fissile like shale (as many tuffs are in other countries), enclosing fragments of amygdaloidal and other trap-rocks, pumice, with infusoria and minute shelly fragments, proceeding probably from crater-lakes.

The lava-currents of Iceland are described as exceedingly rugged and uneven on their surface, the jagged edges cutting like a knife. They also frequently show parallel longitudinal rents, caused probably by the irregular subsidence of the surface as the liquid interior runs out to a lower level. The great rifts of Thringvalla are the most prominent examples (see p. 77). Notice has already been taken of the numerous boiling springs that are met with in several parts of the island, of which the Geysers are the best-known and most imposing, but by no means the sole examples. I have in a former chapter dwelt upon their phenomena and probable cause. It is known that many of the Icelandic lavas are blistered with cavities of large size; and Captain Forbes, one of the most recent visitors of the Great Geyser, states that the ascent and bursting of steam-bubbles within some subterranean cavity beneath the basin of the Geyser is heard in loud and frequent detonations, which even shake the ground above, and that this occurs while the water in the upper basin is tranquil. Probably, therefore, it is this internal ebullition which feeds the boiler beneath with fresh steam up to the point at which the elasticity of its contents overcomes the weight of the column of water in the pipe, and brings about an eruption in the manner indicated in page 149.

The volcanic mountain Krabla, in the north of the island, was first known to be in activity in 1724. Since that period it has had four noted eruptions, one of which produced a stream of vitreous lava (obsidian) nine miles long and four or five wide. Its crater is now in the condition of a solfatara.

In the vicinity, as well as in other parts of the island, there occur immense beds of sulphur, evidently the accumulated deposits of the vapours of sulphuretted hydrogen, which still issue in abundance from crevices in the rocks. The sulphur-beds alternate with others of white clay, proceeding from the aluminous matter of the decomposed trachyte.

Captain Forbes gives an interesting account of the great sulphur-deposits near Kriswick, on the southern coast,—a district twenty-five miles in length, covered with beds of earth and clay containing from fifteen to sixty per cent. of sulphur, besides numerous extensive crusts from 1 to 3 feet thick of pure sulphur. There is another similar district to the north, yielding almost as much of this mineral. These recently-formed sulphur-beds throw great light on the origin of the similar deposits of earlier age which are found in Sicily and elsewhere, associated with tertiary sedimentary strata.

Fiorite and siliceous sinter are also abundantly deposited by the hot-springs of Iceland—a fact which in its turn suggests the thermo-aqueous origin of the quartz-veins, metalliferous or not, so numerous in the metamorphic rocks of all ages.

Captain Forbes describes, at one spot which he visited, a sand and cinder cone crowned by a dark vitrified rampart of lava resembling an old embattled turret, about 600 feet in dia-meter, which is appropriately called *Elborg*, or "The Fortress of Fire." The origin of such 'chimney-pot'-summits to a volcanic cone has been spoken of (p. 66, *supra*).

Ferroe Isles.—All the islands of this group are composed exclusively of flat platforms of basaltic lava, superimposed one on another, similar to the north-western and eastern portions of Iceland already described. They belong apparently, like those, to a very early period. They were probably the product of submarine eruptions, and have since been raised above the water-level. No traces of scoriæ-cones or of craters have been mentioned by any visitors. Much of the rock is amygda-loidal, showing at least that these lavas were not produced at so great a depth beneath the sea as to prevent the expansion of the contained vapour into vesicles. In the upper surface of some of these rocks the vesicles are elongated in a perpendicular direction—a not very uncommon fact.

The British Islands.—I extract from the 'North British Review,' No. lxix., the following able sketch :—

" During the period of the [deposition of the] Lower Silurian rocks there were many *centres of eruption* in North Wales, as well as in Radnor, Montgomery, and Shropshire, from which enormous streams of felspathic lava and showers of ashes and scoriæ were ejected. There was also at least one focus of volcanic action north of the Tweed. Other points of eruption may probably be yet detected. Vast sheets of lava were then poured out, along with dense showers of dust and ashes. These materials consolidated into great ridges and hills, now forming some of the most conspicuous hill-ranges of the country, as the Sidlaws, the Ochils, the Campsies, the Pentlands, and the hills of Kilpatrick and Renfrew, which stretch away into Ayrshire. In Cumberland there are also traces of contemporaneous volcanic developments. During the Carboniferous period the subterranean forces continued in activity, but under a somewhat different aspect. Instead of wide-spread sheets of lava accumulating into long hill-ranges that swept across the country from sea to sea, the eruptions became smaller in extent and *more local* and *sporadic* in character. They seem to have resembled those of Auvergne and the Eifel, to have been in many cases nothing but monticules of loose ash, with sometimes a narrow column of lava closing up the crater. Such miniature volcanos dotted a large part of Central Scotland during the middle ages of the Carboniferous period. Professor Nicol ascribes the strange trappean conglomerates of Oban to the age of the Trias.

" In the oolitic group of the Inner Hebrides we encounter a vast succession of old lava-flows, now consolidated into mountain masses of greenstone and basalt. The area of eruption seems to have been confined to the district between Long Island and the western shores of Ross, Inverness, and Argyle, a district now occupied partly by the waters of the Atlantic, partly by the group of islands that extend from the Minch to the Linnhe Loch. Over the rest of Scotland, so far as we know, there were no volcanos at this period, unless in the case of Arthur's Seat at Edinburgh.

"And now, in a new region, the subterranean forces found for themselves a new vent, and poured out once more streams of molten rock—those great trappean hills which form so conspicuous a feature in the scenery of the West of Scotland. In Skye and Ramsay they are oolitic; in Mull, tertiary—a grand development of greenstones, associated with layers of shale. Rivers of molten rock, belched out from the craters in their neighbourhood, spread far and wide over the bed of the sea and its estuaries, and over tracts of land now wholly destroyed. Thick beds of lava were piled over each other to the height of several thousand feet.

"The massive dykes of the North of England belong to the same eras. They occupy long rents and fissures in the crust of the earth, through which molten lava welled upwards from the heated interior. Where visible now on the surface, they run over hill and dale as long irregular mounds, like the ruined ramparts of some primeval Hadrian or Antonine. The eroded chalk of Antrim is covered by the famous basalt of the Giant's Causeway. This igneous rock probably belongs to the same age as that which buried the leaf-beds of Mull, in the Tertiary period." (See Lyell's Manual, ed. 1855, p. 181; and A. Geikie on the Chronology of the Traps of Scotland, Trans. of Roy. Soc. Edinb. 1861.) It is also probably coeval with the trap-rocks of the Ferroe Isles last described.

I do not dwell further on these and other British examples, because their details are described in the works of so many distinguished authorities. It is scarcely necessary to say, that no traces of volcanic action of a recent character appear in any part of these islands.

Proceeding southward along the shores of the Atlantic, we meet with the volcanic formations of the west coast of Portugal and Spain, already noticed; and at no very distant interval the island group of the Azores rises from the depth of the ocean, all of them being entirely of volcanic origin.

Azores.—The largest island of this group, *San Miguel* (of which we have a very clear and full description by Dr. Webster, of Boston), is traversed east and west through its whole length by a chain of high volcanic cinder-cones, few of which are united at their base. They range from 1000 to 2000 feet in height. Many have craters, sometimes several miles in circuit, and with lakes at their bottom. Other smaller cones are scattered on either side of the central chain. There appears to be an alternation of trachyte and basalt among the products of these vents. Innumerable basaltic lava-currents may be traced to the craters whence they flowed in different parts of the island. At its north-western extremity is to be seen one remarkable crater fifteen miles in circumference at its rim. It has two large lakes at its bottom, as well as two or three minor cones. The surrounding heights rise 2000 feet above this base, at an angle of about 45°. These hills are composed of loose pumiceous tuff and coarse trachytic conglomerate—the ejecta, no doubt, of the explosive eruption which formed this vast caldron. Bituminized wood occurs abundantly. Probably the cone blown up by this paroxysm was covered with forests. Where the mountain has been degraded by the sea-waves, indurated tuff is seen on many points. When in such positions looser beds alternate with lava-streams, vast caverns have been worn out of the softer substance. There are hot-springs on several points; and near Villafranca is a solfatara, whose hot vapours deposit pure sulphur in abundance. A variety of blistered lava is described by Dr. Webster, containing several large caves, from the roofs of which hang, as in many similar instances, pseudo-stalactites of lava, and branching projections, having a glaze on the surface. No eruptions have taken place within the main island during the historical period; but in 1811, the small temporary islet was thrown up at a short distance off the coast, called ' *Sabrina* ' by the crew of the British frigate of that name, who witnessed the phenomenon. Like all such examples, it had a conical form with a crater on the summit, from which steam, scoriæ, and ashes were vomited for several days. Its extreme height was about 300 feet, and its area a mile in circumference. At the end of a few weeks after the explosions had ceased it was levelled again by the waves. (See p. 236, *supra.*)

Near the town of Villafranca, another scoriæ-cone is still to be seen rising from the sea. The crater is yet very perfect; but the external sloping beds of the cone, consisting no doubt of loose materials, have been carried away by the force of the waves beating against it; and the only parts that remain are the internal inward-sloping beds, which, being probably more compacted by the heat transmitted from the adjoining vent, have been able to resist for a longer time the erosive action of the ocean. (See p. 61, *supra.*)

The island of *Pico,* another of the group, consists of a single volcanic mountain, which rises to the height of 8000 feet above the sea. It is snow-capped, but supposed to be in permanent activity, as vapour is always issuing from its summit. This, however, may proceed from the fumarole of a very active solfatara. It broke out in a paroxysmal eruption in 1718. Its lavas are trachytic. So likewise are those of the island of *Fayal,* which possesses a great central ' *caldera.*'

The island of *San Jorge* was in violent eruption in 1812, when copious streams of lava flowed from the flank of the conical mountain, where a *crater* is still visible.

The central portions of *Terceira* consist of trachyte and trachytic tuff, overlaid by streams of basaltic lava. These are often traceable up to the craters whence they were poured forth. Two or three other islands of the group are non-volcanic, consisting of stratified schists.

Madeira.—Next in order to the Azores, proceeding southward, but rather nearer to the coast of Africa, the group of islands of which Madeira is the principal affords another example of a great volcanic mountain rising from the depths of the Atlantic Ocean.

The best account extant of the geology of Madeira is that given by Sir C. Lyell in the last edition of his ' Manual.' From this the island appears to be of submarine volcanic origin, of the Miocene tertiary epoch—tuffs and limestones with marine shells and corals occurring on several points up to the height of 1200 feet above the present level of the sea. Above these lie lavas and conglomerates alone, which have every appearance of having been produced by subaërial eruptions, principally from a central vent; the chief eminences of the interior of the island surrounding and marking the limits of the principal last-formed crater. In that central region, as is usual in the heart of a volcanic mountain, vertical dykes abound, penetrating the more or less horizontal beds of lava and scoriæ-conglomerate, which are partly trachytic, partly basaltic, or greystone. The highest points in the island, the Pico Ruivo and Pico Torres, 6000 feet above the sea, are composed of such ejected matters, strengthened by numerous dykes. The remains of many cones of eruption are still visible upon and about these central heights, more or less buried under flows of more recent lava, which seem to have united into nearly level platforms, just as happened on the summit of the cone of Vesuvius whenever its crater had been filled to the brim. (See p. 187.) But the lava-beds on the slopes of the mountain generally have an inclination of 10 to 15 degrees, the dip diminishing, as is usual, towards the base, where they reach the sea. In these lower positions there are also several parasitic eruptive cones ; many overwhelmed, more or less, by lava-streams from the higher levels. The aggregate thickness of the basaltic beds alternating with their tuffs, reaches to from 1500 to 3000 feet, as exposed to view on the sides of a deep valley called the Curral, which by some geologists has been considered to be a crater, from its encircling cliffs being composed of beds sloping outwardly on all sides, but by Sir C. Lyell is looked upon rather as the result of aqueous erosion. Probably, as in the case of the Val del Bove, &c., both actions (eruptive and erosive) have united to produce it. Sir C. Lyell states that on many points masses of trachytic lava have filled up valleys worn through earlier basaltic rocks, though some of the newest lava-currents are, again, basaltic. The greater mass of the upper heights is of an augitic rock with much olivine, which he calls felspathic trap. Its structure is largely spheroidal, especially shown in its exposed and partly decomposed portions. Some of the newest lavas are singularly fresh in aspect and rough on the

2 E

surface, so as to excite astonishment at no record existing of the eruptions which produced them. Everything in Madeira implies a long continuance of intermittent volcanic activity, chiefly subaërial, from the middle tertiary period down to a very recent date.

The adjoining island of *Porto Santo* is of very similar character, consisting of calcareo-volcanic tuff, rising to 1000 feet, and traversed by numerous dykes of a reddish-brown basalt. In the centre of the island is a shallow crater-basin, containing a sedimentary freshwater formation full of recent land and marsh shells. The tuff-strata are covered at the northeast part of the island by clinkstone, or rather, perhaps, a vitreous laminated trachyte with numerous crystals of glassy felspar. Dykes of the same traverse the underlying tuff-beds.

The neighbouring islet of *Basco* has very much the same composition. Its limestone-beds, traversed by dykes rising to 100 feet above the sea, contain numerous recent marine shells.

The *Canary Isles* come next in order towards the south.

Teneriffe.—This island is one great volcanic mountain, sloping upwards from the sea-coast on all sides, at an angle of 10° to 14°, to a height of from 7500 to 9000 feet, where the slope is cut down by a precipitous cliff-range, which nearly surrounds an extensive hollow circus—the "Great Crater." It is of an oval figure, measuring eight miles by six. From near the centre of this depression rises the Peak proper, a cone 12,200 feet high from its base, and 15,000 feet from the sea-level. Its surface, where not snow-covered, is composed of loose pumice and ash, together with streams of a glassy black lava which have flowed from the summit, where is still to be seen a small ash-cone and crater rising from a convex plain. Attached to the Peak on either side are two other cones of minor elevation (10,000 and 9000 feet respectively above the sea), called Chahorra and Montana Blanco. The former possesses a much larger and deeper crater than the Peak, and has been more recently active. Its last eruption was in 1798, and produced a stream of obsidian. The sides of Chahorra have an inclination of 28°.

The great outer crater is elliptical. Its longest diameter is eight miles. The inner walls consist of beds of lava-rock and conglomerate, appearing nearly horizontal in the cliff-sections, but showing a quaquaversal outward dip of from 12° to 15° where radial ravines have exposed them. The beds vary in thickness from 500 feet to a few inches, and are generally composed of a trachytic lava and its tuff, though some of the rocks are more augitic and ferruginous, passing through grey-stone into basalt. Greystone-lavas seem, on the whole, to predominate.

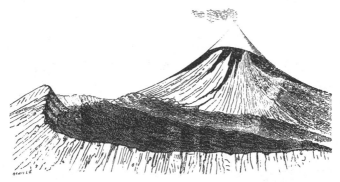

Peak of Teneriffe, seen from the edge of the surrounding crater-cliffs.

The trachytic rocks are often minutely laminated, and intersected occasionally with dykes of greenstone (hornblendic basalt) and of a black obsidian with white crystals of glassy felspar, identical with the recent streams of the Peak. Some of the tuff-beds are of a very white pumice-ash. The external flanks of the old volcano are dotted with minor parasitic ash-cones, each of which has given issue to a current of lava: these are mostly of fresh aspect. One wide opening, or baranco, penetrates the old crater-walls to the west, in the direction of Chahorra, whose lavas have nearly filled it up. Another opening, towards Orotava, on the north-east, has been similarly occupied by lavas that have flowed from the central volcano; and another large section of the old crater's lip, on the north-west, has been overflowed by these lavas, which reach thence down to the sea. The lavas generally of

Teneriffe are superficially broken up into a chaos of blocks. The crystals of glassy felspar are more numerous and larger in the recent than in the earlier lavas. The former generally stretch from the point of issue in long narrow ridges, looking like railway embankments. (Von Buch, Lyell, Piazzi Smyth.)

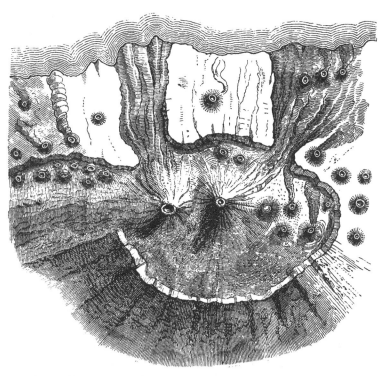

Plan of the Peak of Teneriffe and Chahorra, with the encircling cliff-range.
(From the photogram of a model in relief by Professor Piazzi Smyth.)

Palma.—This is an almost regularly conical volcanic mountain up to the level at which it is truncated by the magnificent central crater or 'Caldera,' nearly two leagues in diameter and 5000 feet deep, encircled by precipitous cliffs, with the exception of one point, where the basin is drained through a great cleft (the Baranco) into the sea. The internal cliffs are composed of alternating beds of basalt and conglomerate, which, though they appear nearly horizontal in the cliff-section, dip away from it on all sides towards the base of the mountain.   These beds, as usually is the case in the central masses of a volcanic mountain, are intersected by a network of dykes.   The lowest rock within the Caldera is trachyte, with its tuff and conglomerate.   It would seem therefore that the earliest products of the volcano were more felspathic than the later.

It is well known that Von Buch, in his volume on the Canaries, propounds the opinion that the basaltic beds composing this mountain were suddenly upheaved to the angle of inclination which they now exhibit, by a single impulse like that which blows up a bladder, they having been previously horizontal.   I will not here repeat the grounds on which I consider this hypothesis wholly untenable, as it is evidently in direct opposition to the normal laws of volcanic action represented in this volume.   Sir C. Lyell deals admirably with the question as relates to Palma, and amply refutes the Upheaval doctrine of Von Buch as applied to it. (Manual, 1855, p.498.) I am satisfied that Palma, on the contrary, offers a model type of an insular volcano, chiefly subaërial, truncated and hollowed out by a paroxysmal eruption, and drained through a principal radial fissure which torrents and debacles have since greatly enlarged, as in many other cases. (See p. 211.)

The island of *Great Canary* resembles Palma in many respects.   It is almost exactly circular in outline, and has, like Palma, a vast yawning central crater, showing trachyte at the bottom, overlaid first by tuff, and then by repeated beds of greystone and basalt, with their respective conglomerates— all sloping outwards with the usual quaquaversal dip of lava-streams that have flowed from the central or axial vent of a volcanic mountain.

Fuertaventura and Lancerote.—These two elliptical islands lie end to end, having their longer axes in the same N.N.W.-S.S.E. line,—Lancerote being the nearest of the whole group to the African coast, from which it is only distant twenty-eight miles.   They are both entirely volcanic, and studded by a vast number of cones, the highest of which attains 2740 feet.

M. Hartung, who visited and described the island of Lancerote in detail, divides its volcanic formations into four
classes :—1st, and earliest, syenitic greenstone, trachyte, and
basalt, without any scoriaceous matter, which has been presumably carried off by denudation; 2nd, 3rd, and 4th, basaltic
formations of different ages—the earliest showing heaps of
scoriæ of some magnitude, but no perfect cones or craters;
the next in age exhibiting very clearly cones, craters, and
currents of basaltic lava proceeding from them; the fourth
consisting of the products of the series of eruptions which, in
the years 1730–36, riddled the island with vents opened successively on a fissure running nearly its entire length, and
covered one-third of its surface at least with the lava-floods
to which they gave issue. The highest of these recent cones
is the Montagna di Fuego, 1750 feet. It still emits both
smoke and steam from its summit-crater; and the ground
there is so hot, that a stick thrust into a crevice 2 feet deep
is drawn up charred at the end. In 1824 another eruption
took place a little to the east of the Montagna di Fuego, but
of no very violent character. The general strike of the strings
of eruptive vents of all ages is the same, and corresponds with
the longer axis of the two islands. So much is this the case,
that, according to Von Buch, to an observer standing on the
summit of the Montagna di Fuego, the adjoining cone covers
those farther off. There are more than forty in all. The
lavas expelled from them mingled into a continuous horizontal
sea of basalt, serving as an example of the mode in which
many of the earlier traps were, no doubt, produced—extensive
surfaces being flooded by the abundant lava-streams emitted
contemporaneously, or nearly so, from numerous independent
but contiguous openings; beds of scoriæ and ash, separating
the successive flows, being the result of the accompanying
explosive ejections. The eruptions in this case were unquestionably subaërial—a fact worthy the notice of those geologists who are inclined to ascribe all flat extensive beds of
trap-rock exclusively to subaqueous volcanos.

It is evident that these outbursts of lava and steam occurred
from orifices successively opened upon a fissure traversing

the island in its length.   Such fissures were, indeed, seen to
open by the inhabitants at the commencement of the eruptions, accompanied by fearful earthquakes.

The lavas are, as has been said, basaltic, and contain much
olivine, the nodules of which are as large as a man's head near
the source, but lessen in size as the stream is traced farther,
until they are broken up into fine granular fragments.

The island of Fuertaventura is in all respects the counterpart and continuation of that of Lancerote, except that the
more recent class of eruptive cones and lavas is not of any
known date.   Many, however, are very fresh in aspect, and
certainly subaërial.

The older formations of the two islands are all covered
superficially by a calcareous deposit similar to that already
mentioned (p. 328) of Ventotiene and other Mediterranean
volcanic isles, having its origin, I believe, in the infiltration
among sandy volcanic ashes, or drift sand, of carbonate of
lime, proceeding from the decomposition, through a long
period, of land-shells in the upper layer of vegetable soil.   A
similar stratum is described by Mr. Darwin in the isles of
Ascension, St. Helena, &c.   The mollusks, no doubt, obtained
the lime from the volcanic matter superficially decomposed by
atmospheric agencies.

Cape de Verde Isles.—Next in order, proceeding southwards along the west coast of Africa, we find the group of
islands off Cape de Verde.   All are, it is believed, of volcanic
origin.   Few of them, however, have been visited or described
by geologists, except the Pic de Fuego and St. Jago.   The
former is at present in activity.   It appears, from the relation
of M. Duvalle (Bulletin de la Soc. Géol. iii. 1846), to rise in
a very regular cone to the height of 8800 feet above the sea,
from within a semicircular crater-cliff, just as Vesuvius is half
embraced by Somma.   Numerous cones of scoriæ stud its
flanks: the most recent were thrown up by eruptions in 1785
and 1799.

St. Jago has been in part described by Mr. Darwin.   He
speaks of a range of coast-hills about 600 feet high, with flat
summits, having steep escarpments towards the interior of the

island, which they almost encircle, but with intervals (that is
to say, valleys) radiating outwards between them. They have
all a gradual outward slope towards the sea, and are composed of basaltic beds of an early aspect, partly decomposed
and in the state of wacke, associated in places with calcareous
strata. These beds rest upon " a compact, fine-grained, ferruginous, felspathic, unstratified rock, generally in a state of
decomposition, appearing indeed like baked clay, or an altered
sedimentary deposit, yet containing all the elements of trachyte"—probably a trachytic tuff more or less altered. From
this description the hills would seem to be the remnants of a
vast crater-ring, belonging to a very early age. In the interior
of this circuit, towards the centre of the island, rise many higher
and more or less peaked volcanic mountains, from which have
flowed numerous lava-streams towards the sea, through the
valleys that separate the segments of the older mass already
described. Having threaded these defiles, they expand into
basaltic plains, often themselves cut up by wide valleys with
low cliff-formed sides, showing sections of basaltic lava resting
upon, and in places intimately mixed with, a calcareous marine
deposit. Some more recent eruptions appear to have broken
out from these coast-plains, as shown by scoriæ-cones—not,
however, of very fresh aspect. Mr. Darwin describes one
of these (Signal-Post Hill), where the generally horizontal
beds of lava and calcareous matter dip *downwards* under
the cone. It would seem that a certain amount of elevation in mass has affected this island since or during its latest
eruptions.

I know of no detailed report as to the other islands of this
group. Judging, however, by the excellent Admiralty Charts
of Captain Vidal, they are all volcanic, and contain numerous
cones with craters on their summits. The two most northern
islands, *St. Antonio* (7000 feet) and *San Vicente*, certainly
appear to possess each a vast central crater encircled by the
highest hill-ranges of either island.

A few degrees further south, in lat. 11° N., the small group
of islands called *Los*, close to the coast of Sierra Leone, are
reported to be volcanic, as well as a considerable mountain-

range, called *Loma*, in the interior of this part of the continent, in which both the Niger and the Senegal Rivers are supposed to have their source.

Again, near the mouth of the Calabar River, in the Bight of Benin, the group of islands, of which *Fernando Po* is the most northern, and *Annobon* the furthest south, are likewise of volcanic origin ; and on the mainland adjoining, one of the *Cameroon Mountains*, said to be 13,000 feet high, was seen in eruption, and emitted a stream of lava, in 1838.

Further to the westward, and more directly in the line of insular volcanos we have been tracing, we arrive at a region of the Atlantic close upon the Equator, and in from 20° to 22° of W. long., which, from appearances that have been frequently witnessed by persons in vessels passing across the spot, is certainly a site of frequent submarine volcanic eruptions. Earthquake-shocks (sensible to the crews of sailing-vessels), the rise of columns of smoke, fire, and ash, floating scoriæ, and the discoloration of the superficial water, are the phenomena observed on several occasions, and only to be explained on this supposition. (See p. 237.) This point is almost intermediate between the two nearest projections of the continents of Africa and S. America.

At a short distance to the south, the island of *Ascension* rises from the bosom of the Atlantic. As to this island, we have the advantage of Mr. Darwin's observations (Volcanic Islands). It is wholly volcanic. The component rocks are in great part trachytic, especially the central and lowest masses. These, however, are overlaid with black and rugged streams of basaltic lava, which can generally be traced to cones of red scoriæ, all opened towards the south-east, whence the trade-winds blow. The principal elevation in the island, Green Mountain (2800 feet), has a large elliptical crater on the north-east, with perpendicular sides 400 feet high, whence probably were shot out, in one of its latest paroxysms, the numerous loose fragmentary masses of pumice, trachyte, basalt, scoriæ, volcanic bombs, and some fragments of granite, which are profusely scattered over the surface of the island.

Much of the trachyte is a white, earthy, highly porous variety, resembling the white trachyte of Ponza, with sili-

ceous veins.   Much also is laminated, and passes into obsidian through pitchstone and pearlstone.   The zones of glassy
obsidian pitchstone and pearlstone alternate, and the latter
becomes crystalline,—the globular concretions or spherulites
multiplying till they compose the whole rock, in which felspar-
crystals then show themselves.   These, as well as the spherulites, have been dragged out, by the motion of the lava, into
laminæ, often contorted into innumerable flexures resembling
those of gneiss or some slates.   Ultimately the rock becomes
wholly lithoidal, but still retains traces of the laminated
structure.   I observed precisely similar varieties of trachyte
in Ponza.   (See pp. 140 & 329, *supra*, and Darwin's Volcanic
Islands.)

In 20° S. lat. and 28° W. long. a small volcanic island rises
from the Atlantic, called *Trinidad*.

St. Helena.—The margin of this island, twenty-eight miles
in circumference, says Mr. Darwin, is formed by " a rude
circle or horse-shoe of great black ramparts, composed of
basaltic beds dipping seawards, and worn into cliffs from a few
hundred to 2000 feet in perpendicular height." This ring (the
wreck, no doubt, of an enormous early-exploded volcano) is
open to the south, and breached in several other places, chiefly
on the east (*i.e.* the windward) side.   The interior has been
nearly filled up by the products of a newer central volcano,
which erupted felspathic lavas only, and which possesses at
present a vast summit-crater with an annular ridge, attaining
the height of 2750 feet above the sea, in some parts surmounted
by a parapet or wall perpendicular on both sides, the remnant
of a 'chimney-pot' ring, like that of Teneriffe.   Mr. Darwin
describes the older beds of this island as greatly decomposed,
and traversed by an infinite number of basaltic dykes, mostly
parallel and running in a N.N.W. direction.   They are
usually coated with a thin layer of pitchstone, and preserve a
remarkable uniformity of thickness through great distances,
both vertical and horizontal.   They frequently take a course
parallel to the strata of trachytic rock or conglomerate which
in other parts they cut through.   In some places the older
rocks have been considerably dislocated and tilted up—a fact
very rarely to be seen (as Mr. Darwin observes) in volcanic

districts. In the interior of the central crater, especially, some conical masses of clinkstone occur, which seem to have been injected into fissures of the tuff and scoriform lavas around, whose beds appear to have been more or less tilted up by the protrusion of these lava-masses, probably in a semi-solid condition—one not unusual with clinkstone lavas. (See p. 137, *supra*.) These are, apparently, the latest products of the central volcano.

The surface of many parts of the island, up to the height of 600 feet, is coated with thin beds of broken land-shells of a recent character, partly cemented together into a brown stalagmitic and sometimes oolitic limestone by the percolation of rain-water, like those of Lancerote described at p. 422. The upper beds contain the bones and eggs of birds. From the occurrence of gypsum and salt, as well as rounded pebbles, in the rocks forming the base of the coast-cliffs, Mr. Darwin considers that they were formed beneath the sea, and consequently that the island has sustained at some period a certain amount of elevation in mass.

In the interior of the West of Africa, near the Congo River, and again further south, below Walvich Bay, in the province of Damara, basaltic mountains and other traces of volcanic action are mentioned by some travellers; but it will be some time, probably, before any authentic information is acquired respecting the geology of these as yet unexplored regions.

A group of small volcanic islands, the largest of which is called *Tristan d'Acunha*, rises from the middle of the Atlantic, in S. lat. 37° 3'. It is about six miles in diameter, and has the form of a cone, truncated at a height of 3000 feet, with an upper dome rising 5000 feet higher, or 8300 feet from the sea. The mass of the outer cone is formed of augitic lava and tuff intersected by numerous dykes. The central dome or cone is of cellular lava and scoriæ, with ridges of lava radiating downwards from the edge of a summit-crater a mile in circumference. This is, no doubt, a recent volcano thrown up within the crater of an earlier one.

A little further to the south, an island called *Gough's* is also of volcanic origin.

In the *Southern Ocean,* eastward of the Cape, we find *Prince Edward's Island* and the *Crozet group* (lat. 46°–47°), both small conical hills with craters, whence streams of basalt have issued.

Still further eastward, and nearly in the same low latitude, *Kerguelen's Land* (examined by Sir James Ross) shows also several cones with craters, some rising 2600 feet, surrounded by beds of basalt and greenstone intersected by numerous dykes. Beds of coal were observed between some of these trap-rocks.

Northward of this, and somewhat more to the east, rise the islands of *Amsterdam* and *St. Paul,* both of which contain craters, and have been observed sending out vapour and fire. In the former, beds of pumice have been observed. The latter has a circular crater a mile in diameter, having one narrow opening to the sea, but everywhere else encircled by steep cliffs, while the outer slopes descend gradually, as usual, to the sea around, except where broken into cliffs by the waves. Sir C. Lyell gives a plan and sketch of it in his 'Manual' (1855, p. 512).

Madagascar.—There is some reason to believe in the existence of active volcanic vents in this great island. In the Mozambique Channel, which separates it from the east coast of Africa, the largest island of the Comoro group contains a volcano which has generally been seen in activity.

Bourbon.—Eastward of Madagascar, the two islands of Bourbon and the Mauritius are entirely volcanic. The first was ably described in the work of M. Moreau de Jonnés, published as early as 1778. It consists, as regards the western moiety, of the skeleton of a great early volcano, exhibiting the remains of one or more vast crateral cavities, nearly encircled by precipitous rocks composed of trachyte, clinkstone, and basalt, with their conglomerates, which are intersected by numerous dykes of the latter rock. All these masses have suffered largely from denudation. The principal summit of the group (the Gros Morne) rises nearly 10,000 feet above the sea.

At the eastern end of the island is the still active volcano, of which several notices have been given in the body of this work (see pp. 74, 133, & 197). It consists of a steep cone or dome 7000 feet high, chiefly composed of beds of a highly vitreous and viscous lava, which at the time of its greater eruptions has been seen to flow rapidly down the slopes of the mountain to the sea at its foot, but in the intervals is spurted out in continuous jets from vents on the top of the dome, which form several small and steep conical or pap-shaped hillocks, from 80 to 160 feet high, composed of overflowing waves or slobbered drops of highly viscid lava, coagulated together like the heap formed under a guttering tallow-candle. The great cone rises from the centre of a horseshoe-shaped ancient crater-ring, or 'cirque,' evidently blown out by some early paroxysmal eruption; and outside the rim of this circuit, on the land-side, are several separate cones and craters, showing that eruptions of recent dates have not been confined to the now active vent, whence one of great violence took place last year. Its lava-current reached the sea, and cut off all communication between the north-east and south-west sides of the island. The aëriform jets of this volcano are not usually violent; they throw up fine threads of viscid matter like spun-glass, often with pear-shaped drops attached to them like those drawn out from sealing-wax; and its lava and scoriæ are dark-coloured, highly glazed, ropy, and filamentous.

The volcano of Bourbon, nearly encircled by the cliffs of an early crater.
(From Bory de St. Vincent.)

The Mauritius.—This island is of an oval shape, and rises from the sea on all sides towards the ridge of an elliptic girdle of ramparts composed of basaltic beds, and surrounding a crater-like central area, which numerous streams of modern lava have almost filled up.   In this characteristic figure, indeed, it strongly resembles St. Jago, St. Helena, and several other volcanic islands already described.   The longer axis of this great crater-ring measures at least thirteen miles.   The rampart has many openings, through which the more recent lavas from the interior vents have poured down towards the sea.   There are also several fresh-looking cones and lava-streams that have been thrown up on the outer slopes, and some from below the sea, especially around the northern extremity of the island.   The earlier lavas are basaltic; the more recent contain a larger proportion of felspar; and some even fuse into a pale glass.

In sections afforded on the banks of some of the rivers, such lava-beds, of no great thickness, are seen piled one over the other in great number, with layers of scoriæ between them.   The mountain called the *Piton* was probably the chief vent of these central and latest eruptions.

VOLCANOS OF THE WESTERN HEMISPHERE.

Western Atlantic.—Returning to the Atlantic Ocean, we cannot but be struck by the almost total absence of any traces of volcanic action on its Western shores, through their whole extent from Davis's Straits to those of Magellan, with the exception of a small portion of this vast space, viz. between the 10th and 18th degrees of north latitude, where the American continent is reduced to a comparatively narrow and low isthmus by the sweeping recess of the Caribbean Sea, the entrance to which is studded by a chain of volcanic islands.   And not only does this non-volcanic character apply to the Eastern shores of both the Americas, from north to south, but also (with an insignificant exception which will be shortly noticed) to their entire breadth likewise, from those shores up to their axial ranges, the Rocky Mountains and the Andes, which border the Pacific.

A general fact upon so vast a scale, applying, as it does, to an entire hemisphere, cannot be looked upon as a mere accident. It must be the result of some general law ; and seems to afford a strong confirmation of the suggestion made in the body of this work (p. 273), that volcanic vents act as safety-valves to a certain geographical area around or on one side of them, permitting the outward escape of the same subterranean heat or heated matter, with little or no superficial disturbance; while in neighbouring areas, where no such vents have been formed, the subterranean expansive force has elevated the overlying strata in wide continental masses above the mean level of the earth's surface.

Volcanos of the Caribbean Sea.—The islands of the West Indian Archipelago are generally classed as the Greater and the Lesser Antilles. The former consist of the four large islands, Cuba, Jamaica, St. Domingo, and Porto Rico, which form two parallel elevated ranges, stretching from the north-eastern extremity of Yucatan, at the mouth of the Gulf of Mexico, eastwards towards the Atlantic, where the chain of smaller islands, called the Lesser Antilles, meets them almost at a right angle, running thence southwards to the coast of Venezuela in the South American continent. The first, or east to west islands, are composed chiefly of plutonic crystalline rocks and sedimentary strata of the secondary and tertiary periods. In Jamaica and Porto Rico alone have true volcanic (trap) rocks been noticed. The latter, or north to south islands, are for the most part of comparatively recent volcanic origin ; and several of them contain habitually active vents. They consist of the following, commencing from the south :—

1. *Trinidad.*—Although the main portion of this island is granitic—a prolongation of the elevated east to west continental coast-range of Caraccas (so liable to earthquake movements)—Professor Jukes observed in it black lava-rocks associated with sandstone strata containing recent shells. Its well-known pitch-lake and mud-volcanos also indicate that it partially overlies a volcanic fissure.

2. *Grenada.*—The mountain called Le Morne Rouge is an

extinct crater, composed of scoriæ and vitrified matter. Some of the heights are crowned with columnar basalt; and boiling springs attest the very recent or continued activity of the volcanic forces beneath.

3. *St. Vincent*, the next island towards the north, contains an active volcano, called the Morne Garou, which rises 4940 feet above the sea. It has long been in the condition of a solfatara, but occasionally breaks out into greater activity, as in 1718; and again in 1812, when its eruption followed by twenty-two days the great earthquake which overthrew the city of Caraccas on the neighbouring continent, and the shocks of which were felt severely in several of the adjoining islands. During that eruption the volcano of St. Vincent threw out, in a perpendicular black jet, prodigious volumes of grey ashes (pumiceous and augitic), with much organic matter (derived, probably, from a crater-lake), which nearly overspread the island, injuring its soil to an extent from which it has not yet recovered. A vast stream of lava issued from the summit of the mountain, and in the course of four hours reached the sea.

4. *St. Lucia*, a cone about 1400 feet in height, contains a very active solfatara. There are several intermittent springs of boiling water in the crater, which fill some small basins, resembling those of the Icelandic Geysers. This volcano is said to have been in eruption in 1766.

5. *Martinique* is not exclusively volcanic, since coralline strata are found resting on its felspathic lava-rocks, some of which rise into broken eminences of considerable height. One mountain, the Montagne Pelée, seems to be a cone of pumice. Some basaltic plateaux occur of seemingly earlier age. There are here also several hot-springs.

6. *Dominica* is wholly volcanic. It contains several solfataras; but no eruptions from them are recorded. Trachytic rocks occur among the mountains, which rise to the height of 5700 feet.

7. *Guadaloupe* is a double island; one part consisting of a stratified limestone of very recent origin, containing shells identical with those now living in the surrounding sea, overlaid by a clayey conglomerate with rolled pebbles of lava. The

other and larger island is wholly volcanic, and comprises at least fourteen extinct craters, besides one existing always as a solfatara, and occasionally in full eruptive activity : its summit is 5000 feet above the sea. In 1797 a violent eruption took place from this high point, which ejected vast quantities of pumice, ashes, and sulphureous vapours. Again, in 1836 there occurred an eruption of the same character from another crater on the eastern side of the mountain, followed, after a few months, by the emission of a deluge of mud from its north-western side—probably from the bursting of a crater-lake in that situation by an earthquake, several of which were felt severely at the time through the whole island. It is said that no lava was emitted on these recent occasions ; but as the lower part of the mountain is composed of basalt, and the upper of trachyte, lava must have been frequently pro-duced by its earlier discharges. The trachyte contains grains of quartz, with several varieties of felspar—labradorite, rhy-akolite, and sanidine (Dufrénoy, Comptes Rendus, t. iv. 1837).

8. *Montserrat* is a volcanic mountain, having a crater in the state of a solfatara. Sulphuretted hydrogen also issues from fissures in several points. Its lavas consist of highly porphyritic trachyte, with large crystals of felspar and horn-blende, often much decomposed by sulphureous exhalations.

9. *Nevis* contains crystalline trachytes, and much clay de-rived from their decomposition. It has a solfatara, and several thermal springs, the water of which holds silica in solution, which it deposits in crusts of sinter and fiorite on cooling.

10. *St. Christopher's* has a central mountain of consider-able height, with a very perfect crater on its summit, from which an eruption proceeded in 1692. Its lavas are trachytic.

11. *St. Eustachia* contains the most perfect crater possible, called ' The Punch-bowl.' Its banks are composed of pumice, and its lavas are consequently no doubt highly felspathic tra-chyte.

Several small islands lying at short distances to the east-ward of the above volcanic chain, viz. Antigua, St. Bartholo-mew, St. Martin, St. Thomas, Margarita, Curaçoa, &c., are composed of recent calcareous strata, with existing shells, or

2 F

coralline limestone, resting upon a volcanic conglomerate of trachytic tuff, containing much silicified wood, with agates and jaspers.

Coasts of America.—The only other known instance of volcanic development on the Eastern coast of America is found in the small island of *Fernando Noronha,* opposite Cape St. Roque, just at the narrowest part of the Atlantic. There appear to be no signs of recent activity. But Mr. Darwin (S. America, p. 145) describes several great conical protuberant pinnacles of columnar clinkstone, associated with layers of white tuff intersected by dykes and beds of basalt, trachyte, and slaty clinkstone. He also speaks of a patch of trappean rocks near the mouth of the Plate River; but they belong apparently to a very early period, and are probably coeval with the porphyries of Patagonia, of the Oolitic age.

In the extreme Southern Atlantic, the *South Shetland Isles* are said to be volcanic. One of them, the Isle of Deception, consists solely of a vast crater-ring of perpendicular cliffs, inclosing a basin eight miles in diameter, into which the sea penetrates by a breach to the south. The cliffs are described as composed of alternate layers of *ice* and lava, having the usual gentle slope away from the central caldera (Journ. of Geogr. Soc. i. p. 64).

The terminal extremity of South America, in Tierra del Fuego, is chiefly of clay-slate of cretaceous age, but intersected by dykes of greenstone. Much basalt and porphyritic lava show themselves also on many points of the south-western coast, together with scoriæ-conglomerates. Hence it is certain that the great western volcanic train of this continent reaches at least to its extreme southern limits.

Mr. Darwin describes the shores of the Pacific, from the Straits of Magellan northwards, as composed of a base of metamorphic schists and clay-slate resting on and penetrated by plutonic rocks, chiefly a variety of granite called Andesite, its principal element being white albite-felspar. These rocks are overlaid by an immense formation several thousand feet in thickness of porphyries and porphyritic conglomerates, often scarcely distinguishable from recent trachytes, and which appear to have been poured forth from a vast train of submarine volcanic vents or fissures along the whole length of the Cordilleras, in or about the Oolitic age, since they are covered by great stratified deposits of sandstone, limestone, and gypsum containing shells of that or the subsequent Cretaceous period. These strata are themselves mixed up with much volcanic ash. "If," says Mr. Darwin, "we picture to ourselves the bottom of the sea with numerous craters in more or less activity, the greater number in the state of solfataras, discharging calcareous, siliceous, ferruginous matters, and gypsum or sulphuric acid to an amount surpassing the existing sulphureous volcanos of Java, we shall probably understand the conditions under which this singular pile of varying strata was accumulated." (South America, p. 239.) Their great thickness (6000 or 7000 feet at least) indicates, he thinks, that the bed of the ocean was slowly subsiding at the time of their deposition. All these formations were afterwards upheaved and dislocated by a general elevatory force, acting probably at different periods, or by slow degrees, and with occasional intervals of subsidence, and accompanied by further volcanic outbursts, producing deluges of lava and tuffs that overspread the floor of the sea towards the east, and now underlie the extensive tertiary plains of the Pampas of Patagonia. These tertiary tuffs were traced by Mr. Darwin from the shores of the Atlantic up the valley of Santa Cruz to a height of 3000 feet on the eastern slope of the Cordilleras:

This immense north and south mountain-range, or rather series of parallel ranges, so composed, is pierced very generally through its entire length from Tierra del Fuego to Mexico by volcanic orifices of still more recent date, many of them now in activity, to the description of which I shall shortly proceed. But what has been said suffices to show that volcanic eruptions have taken place along this great north and south fracture of the earth's crust almost continuously from the earliest times, accompanied by oscillatory up and down movements of the surface on an equally extensive scale, which have left the old sea-bottom now several thousand feet above the actual level of the ocean. The loftiest peaks of the Cordilleras consist of active or dormant volcanos. The flatter masses next in height are formed of the gypseous and porphyritic strata, thrown into vertical or highly inclined positions. The intrusion of the andesitic rocks from beneath has, in Mr. Darwin's opinion, been the cause of the upheaval of the latter strata, and he considers it probable that they form a great axial ridge or longitudinal dome beneath the entire range (S. America, p. 241). These, then, are rather plutonic than volcanic. The modern lavas would seem to have originated in or beneath them, wherever fissures were broken sufficiently deep through the overlying rocks to permit the eruption of portions of the heated interior matter. The earthquakes of almost daily occurrence in the present day, and the rise of the coast-lines along the shores of the Pacific show that this elevatory action is still going on ; and so likewise are the attendant volcanic phenomena from the high peaks of the Cordilleras.

Recurring to these last, as our proper subject, I may observe that it is not clearly made out to what extent volcanic action is now taking place, or has been so during recent times, in the extreme southern angle of the American continent. Captain Hall saw what he considered a volcano in activity in lat. 55° 3′, to the north of Cape Horn ; and another is noted in the map of La Cruz, in lat. 51° 4′ S. Vast beds of basalt have been observed, ranging about 1000 feet above the sea, over large areas from the 45th to the 46th parallel, opposite to the peninsula of Tres Montes, which, according to Mr. Darwin, is itself granitic. And from this point northwards, the ridge of the Cordilleras is certainly studded with volcanic peaks mostly in habitual activity through a range of 16 degrees of latitude, up to the parallel of Coquimbo, lat. 30°. No less than twenty-four distinct volcanos, of which thirteen have been seen in eruption, are reckoned in this train. The most important and active are those of *Yantales*, 43° 29′, 8000 feet high ; *Corcovado*, 7500 feet; *Osorno*, 41° 9′ ; *Michinmadom*, 8000 feet; *Antuco*, 37° 7′, 16,000 feet,—a trachytic cone, surrounded by an earlier basaltic crater-ring, and having a crater emitting sulphureous vapours, though the lavas usually break out from its foot. The same author, Pöppig, to whom we

are indebted for this account, describes two other volcanos in
the same latitude, in a parallel chain of the Andes far to the
eastward of Antuco. *Peteroa*, lat. 35° 15', is now moderately
active; and a paroxysmal eruption took place from it in 1762,
a new crater being formed, and a great rent in the side of the
mountain. The peak of *Tupungato* is upwards of 22,000 feet
high. *Rancagua*, lat. 34° 15', is said to be in constant eruption.
Maypu, 33° 53', 17,620 feet high—a truncated cone with a
summit-crater emitting vapour and flames (red-hot scoriæ
probably; their light being reflected from clouds of vapour)—
is described as rising out of rocks of Jura limestone associated
with dolomite, vast beds of gypsum, and salt-springs, which
attain a height of 9000 feet.

Aconcagua, east of Valparaiso, lat. 32° 39', said to be above
23,000 feet in height, and therefore one of the most lofty
mountains of the New Continent, is still active. The city of
Mendoza, capital of the province of that name belonging to
the Argentine Confederation, and seated on the eastern slope
of the Cordillera in this latitude, was destroyed in March
1861 by an earthquake, and 10,000 persons killed. Probably
Aconcagua broke into eruption at the moment, since travellers on the neighbouring pass of Uspallata met with showers
of ashes. The earthquake was local only, the western side of
the chain (Chili) being undisturbed.

Three other volcanos follow, *Ligui*, *Chuagui*, and *Limara*,
nearly in the latitude of Coquimbo; northward of which,
up to lat. 21° 50', through an interval of 560 miles, it is
said by Humboldt that no active volcano exists. One, however, is reported by Philippi in lat. 22° 16', near Copiapo;
and it is probable that many dormant or extinct volcanic
mountains may be found on fuller examination of this portion
of the Andes.

A submarine eruption broke forth in 1835 close to the shore
of the island of Juan Fernandez, at the moment that the opposite coast of Chili was shaken by a violent earthquake and
its shores deluged by the influx of an extraordinary sea-wave.
The depth of the sea at the point of eruption was sixty-nine
fathoms; notwithstanding which, the column of fiery ejecta

thrown up above the sea-level was so vivid as to light up the adjoining island through the night (Phil. Trans. 1826).

Earthquakes of a formidable character are frequent along this part of the coast. Copiapo was overthrown by one in 1819, and others of equal violence occurred in 1773 and 1796. A very considerable elevation of the land has also taken place along the whole extent within recent times, as shown by terraced beaches of shingle and shell at various heights. To this I may add, that one of the latest and best observers, Mr. D. Forbes, in his paper on the Geology of Bolivia and Southern Peru (Quart. Journ. Geol. Soc. vol. xvii.), describes the volcanic range of the Cordillera as almost continuous through this interval. He mentions the following volcanos, proceeding from south to north, as still in occasional activity : *Lullayacu*, 25° 15′; *Joconado*, 23° 10′; *Licancau*, 22° 50′; *Atacama*, 22° 30′; *Calama* ; *Isluga*, 19° 20′; *Tucalagua, Tutapaca, Co-quina, Gualitieri*, and *Sahama*, 18° 7′ (23,914 feet),—the last a truncated cone of most regular form, and nearly 1000 feet higher than Chimboraço, long supposed the loftiest of the Andes.

At this point a large number of volcanic peaks are grouped together. And here the ridge of the Cordillera, which hitherto ranged nearly due north and south, takes a bend to the west; its breadth too increases, and a second parallel ridge makes its appearance on the east; the intervening depression—a wide alpine valley lying 13,000 feet above the sea-level— being partly occupied by the great Lake Titicaca. The eastern range is plutonic; its axis granite, supporting highly inclined beds of Silurian, Devonian, and Triassic rocks, and penetrated by metallic veins, as at the celebrated mines of Potosi. The highest points of this range, Sorata and Illimani, rise more than 24,000 feet above the sea. Some recent volcanic rocks cut through the Carboniferous and Devonian series on the extreme western limit of this ridge, near the Lake Titicaca; they are described by Forbes as "true greystone and trachytic lavas, characterized by a peculiar ribboned structure."

The western range of the Cordilleras is almost wholly volcanic. The lavas are chiefly trachyte—many of a character closely resembling the domite of Auvergne—composed of quartz, black or brown hexagonal mica, and glassy felspar, and accompanied by a whitish trachytic tuff with abundant imbedded fragments of pumice. The tuffs are generally compact, so as to form an excellent building-material, and are often with difficulty to be distinguished from true trachyte. The eruptions are described as having broken through and overflowed the oolitic strata near the coast, which contain interbedded and intrusive porphyries and diorites (greenstone). Among the most recent lavas there is much greystone (trachydolerite), of a darker colour than the trachyte, with numerous crystals of black or dark-green augite. Basalt also occurs, of a very fine-grained compact character. But the mass of the volcanic rocks consists of a " crystalline felspathic lava with a striated or ribboned structure, similar to the striæ in particoloured glass." (Forbes, *loc. cit.* p. 27.)

These lavas generally appear to have burst forth from lateral fissures on the flank of the lofty snow-covered volcanic peaks which crown the range. In the southern part of Bolivia such "lateral eruptions have covered the ground with trachytic lava for more than 300 miles continuously." Mr. Forbes thinks that some single fissures have produced outflows of lava through a continuous length of fifty miles. (See p. 135, *supra.*) It must, however, be remarked of these South American volcanos, that although the peaks are frequently seen to emit vapour and ashes, and contain many solfataras, we rarely hear of their giving issue to any great currents of lava in recent times, as is the case in other sites of ordinary volcanic activity.

The volcano of *Miste*, near Arequipa in Peru (18,877 feet), exhibits several small craters, frequently in gentle eruption. Another adjoining peak, called *Chacani*, has a large crater; *Vejo*, the same, with lava-streams and much pumice; *Omato*, 16° 50', had a violent eruption in 1667; and others, whose names it is needless to give, are mentioned by different writers.

From this point, in lat. 16° S., the main chain of the Andes appears to be non-volcanic, or, at least, to show no active volcano for the long interval of 960 miles; the first that occurs, proceeding northwards, being *Sangay*, south of Quito, in lat. 2° S. From thence to the latitude 2° N. a crowded group is found of eighteen or twenty volcanos of great height, half of which at least are supposed to be in activity. Sangay, 16,040 feet high, is believed to be in permanent eruption, like Stromboli. M. Sebastian Wise, who reached its summit, counted 267 explosive jets of ashes and scoriæ in an hour. The black ejected lapillo forms on the slopes of the mountain, and to a distance of twelve miles round, beds 300 or 400 feet thick. Humboldt dwells frequently on the rolling thunders of Sangay, which are heard at great distances all round. The fact is common to other lofty volcanic mountains during their eruptive phases, and no doubt proceeds from the explosions of steam taking place within the depths of their craters,—the sound being often propagated not through the air only, but the earth likewise. The detonations of ordinary artillery may, we know, be heard at distances of fifty or more miles; it is not, therefore, to be wondered at, if the infinitely more violent explosions of some volcanic vents should occasionally be audible five or even ten times as far. It is to be suspected that the mysterious 'Bramidos,' or subterranean thunders mentioned by Humboldt as often heard in the Andes, always have this simple and natural origin.

The city of Riobamba, at the foot of *Tunguragua*, in lat. 1° 41' S. (16,424 feet), was destroyed by a violent earthquake, February 4th, 1797. This volcano, together with the others adjoining, of the same group, *Carguairazo, Chimboraço, Cotopaxi, Antisana, Pichincha, Imbaburu*, and nine more, all within an elliptical area measuring about 120 miles in its longest diameter, may be considered as the several vents of one volcanic mass, rather than distinct volcanos. Some of these vomit torrents of mud (ash and water) at their eruptive crises, as did Carguairazo in June 1698, and Imbaburu in 1691. In this mud are found multitudes of small fish (*Pimelodes Cyclopum*). The probable origin of the water, and of the fish and infusoria which are found in it,

is shown in the crater-lake of the volcano that rises above
the town of Pasto, in lat.1° 13′ N.   Chimboraço (21,420 feet)
is a regular snow-covered dome, the highest visible rock being
prismatic trachyte : it has not been seen in eruption.   Cotopaxi (17,662 feet), on the contrary, has been in frequent agitation since 1742, its eruptions being often accompanied by
torrents of water—in this case proceeding rather from the
sudden melting of its snow-covering, the whole of which,
according to Humboldt, has been known to disappear in this
manner in a single night.   The debacle this must have occasioned will account for any quantity of alluvial conglomerate
about its base.   The cone of Cotopaxi is described as of re-

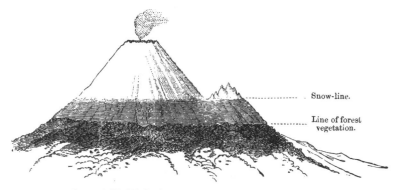

Snow-line.

Line of forest
vegetation.

Cotopaxi (17,662 feet), seen from a distance of ninety miles.
(From Humboldt's ' Vues.')

markable regularity, notwithstanding the tremendous igneous
and aqueous phenomena to which it is subject. (See the woodcut annexed, and p. 182.)   It has emitted vast currents of
perfectly glassy pumice, precisely like those of Lipari : they
are quarried for building-stone, and blocks extracted easily
twenty feet or more in length, and five or six in thickness.
(Humboldt, ' Kosmos.')   The summit of Carguairazo is said
to have disappeared (probably blown off by paroxysmal explosions) during the eruption of 1698.   *Sinchulagua* was in eruption in 1660 : its height is 15,420 feet.   Antisana (19,160 feet)
springs, as a conical peak nearly 5000 feet high, from an oval

plain, probably a vast filled-up crater, across whose surface several currents of black obsidian and pitchstone have flowed from the higher volcanic vent. At a still lower level, on the eastern declivity of this mountain, similar black lava-streams have issued from two small craters, now lakes. These lavas have split on cooling into loose massive blocks—a not unfrequent circumstance (see p. 71), but which Humboldt seems to have had much difficulty in comprehending. ('Kosmos,' iv. Sab. Transl. p. 313.)   The slopes of Antisana are covered with pumice and fragmentary pitchstone. It was eruptive in 1590, and again in 1728.

Pichincha (17,644 feet) has been frequently in eruption, at least from 1539, and was in full activity in 1831. It has on its summit two large funnel-shaped craters; one of them contains a central minor cone, from the surface of which numerous vents are seen to emit steam, sulphuretted hydrogen, and sulphureous acid gases, with vast columns of black ashes and pumice. The other great crater is at present inactive. *Cumbal, Chiles, Pasto, Sotara,* are the names of other volcanic peaks rising above the limit of perpetual snow. *Puracé* (16,000 feet), near Popayan, is a truncated cone, composed of obsidian on the summit: the 'Vinegar' River, strongly impregnated with sulphuric acid, has its source in this mountain. *Tolima*, also a truncated cone, in lat. 4° 35', and west of Santa Fé di Bogota (18,000 feet), is the loftiest peak of the Andes north of the equator. It now emits only steam and gases, but is known to have been in violent eruption in 1595, and again recently in 1826, at which time the whole surrounding province of New Granada was powerfully agitated by earthquakes; loud detonations were heard, and rents formed in the surface-soil, which emitted carbonic and other acid gases, as well as mud smelling strongly of sulphuretted hydrogen.

In connexion with this volcanic band of the Cordilleras, I may here notice, although distant about 500 miles, yet immediately opposite to the coast of the province of Quito, the insular group of the Galapagos Isles, in which volcanic energy has been very actively developed. Scarcely anywhere have so

many cones and craters (generally extinct) been observed
within so small an area (about 120 miles in diameter).   Mr.
Darwin estimates their total number at 2000 (!).   When he
visited the islands with the expedition under Captain Fitzroy,
two craters were simultaneously in fiery eruption.   In all the
islands streams of lava may be seen separating into branches,
and often reaching the sea.   The lava is generally augitic,
with olivine and large crystals of albite.   True pumice is entirely wanting.   The cones are usually formed of a hardened
brown tuff, covered, like that of Naples, with looser arenaceous strata of the same.   The layers have always the usual
quaquaversal dip away from the craters, at angles of 20 to
30 degrees.   The craters are invariably open towards the east ;
the materials having either accumulated in greatest abundance, or resisted the erosion of the sea-waves best, on the
leeward side of the vent.   The highest cone does not exceed
4536 feet.   The group is generally remarkable for its vast
number of separate independent points of eruption, and the
absence of any great predominant volcanic mountain.

Tolima is the most northerly of all the active volcanos of the
Southern continent.   It rises on the central of the three mountain-ranges into which the Andes are here divided, called the
Sierra di Quindiu.   The western branch has in this latitude
no eruptive vent.   One rests upon the eastern declivity of the
knot or 'ganglion' in which this trifurcation has its origin,
near the source of the Rio Fragua.   It is more distant from
the Pacific (152 miles), as Humboldt remarks, than any other
active volcano of the Southern continent : it is still smoking.
North of this point, between the ninth and tenth parallels,
each of the two lateral ranges takes a somewhat sudden bend.
The eastern one is continued due east, in the elevated coast-
range of Caraccas, Cumana, and Trinidad (from 8000 to 9000
feet), where it meets almost at a right angle the N.-S. volcanic train of the Carib Islands already noticed.   The other,
which is much broken and lowered in elevation, turning to the
west in the Isthmus of Darien, also shortly after bends again
nearly due north ; and here, at the bottom of the Golfo Dolce,
near Baruca, occurs the most southern known volcano of

North America. In the intervals, however, traces of early volcanic action show themselves along both ranges. The Western Cordillera has many beds of intrusive augitic porphyry and diallage-rock interstratified with or penetrating its sedimentary rocks. Along the course of the Eastern, or coast-range of Venezuela, mud-volcanos, petroleum-springs, and mineral pitch frequently occur. Near Valencia there is a solfatara, and several sources of water heated above the boiling-point.

VOLCANOS OF CENTRAL AMERICA.

From Baruca, a train of more than thirty volcanic vents, chiefly bordering the Pacific, ranges N.W. up to the parallel of 16° N. in the province of Guatemala. There are few other parts of the globe in which so many active volcanos lie in so small a compass, together with numerous mountains of volcanic formation, which, though now quiescent, have evidently been eruptive within very recent times. Next to Baruca, two volcanos rise upon the shore of the Atlantic, near the Gulf of Chiriqui. Another, called *Irasu*, near the town of Cartago, reaches the height of 11,900 feet: it is easily accessible to the summit, from whence both oceans are visible. The active cone of cinders and lapillo, about 1000 feet high, rises in the centre of an encircling crater-ridge. There is another crater, 8000 feet in circumference, on the north-east of this. This volcano had active paroxysms in 1723, 1726, 1821, and 1847, attended by earthquakes which did great damage to many of the towns between Nicaragua and Panama.

El Reventado (9486 feet) has a deep crater, breached to the south, formerly filled with water. *Barba*, north of San José, the capital of Costa Rica, has a crater enclosing several small lakes. Hitherto the main chain of volcanic heights runs S.E.-N.W.; but at this point it is crossed by a transverse east and west train of vents. On this cross-fissure rise four or five active peaks: one called *La Vieja*, which is said by Mr. Squier to break out every spring at the beginning of the rainy season with a discharge of ash; another, the volcano of *Votos*, rich in sulphur; and one to the north of Cartago.

Rincon, Miravaya, and *Orosi* are more or less active volcanos bordering the Lake of Nicaragua to the south-west.

"Perhaps," says Mr. Squier, "no similar extent of the earth's surface shows so many and such marked traces of volcanic action as the part of Nicaragua intervening between its lakes and the Pacific Ocean. Studding the lake itself are clusters of innumerable islands, all of volcanic origin, rising in the form of cones from 20 to 200 feet high. And in the hills around, besides some hundred yawning craters, there are numerous crater-lakes of volcanic origin, shut in by burnt, blistered, and precipitous walls of rock, without outlet, often of great depth, their waters being salt and bitter." One of these lakes, *Slopango,* is described by the same traveller as no less than twelve miles in length by five in width. It receives no tributaries, but has one small outlet. The surface of the water is 1200 feet below the level of the surrounding country. These lakes generally are found at the base of a volcanic hill or mountain. Such is the Lake of *Masaya,* at the foot of the volcano of that name, which has been in permanent eruption since 1853. Previous to that year, it had remained dormant from the year 1670, before which date it had been, as now, continually active, and is described as such by Oviedo, who visited it in 1529. (See p. 33, *supra.*) The eruption of 1670 poured forth a flood of lava to the north which reached a distance of twenty miles, and still looks, says Mr. Squier, like an ocean of ink suddenly congealed during a storm.

Nindiri is a twin volcano attached to Masaya, which in 1775 emitted a vast stream of lava into the Lake of Leon at its foot. *Mandeira* and *Omotepec* are other twin volcanos, rising as an island from the Lake of Nicaragua, as also does the extinct cone and crater of *Lapatera.* On the western shore, near the town of Granada, is that of *Mombacho. Momotombo,* still smoking and continually detonating, shows on its flanks black floods of lava. (Squier.)

From thence to the Gulf of Fonseca there extends a chain of heights ranging S.E.–N.W., and consisting of six volcanos, called *Los Morobios. El Nuevo* was in eruption in 1850, when a stream of lava issued from the plain at the foot of the mountain. *Telica,* above the town of Leon (3517 feet), has a crater on its summit 300 feet deep, in the condition of a solfatara. *El Viejo* (6000 feet) was very active in the sixteenth century, and is still seen to eject red-hot scoriæ. *Coseguina,* forming the southern promontory of the Gulf of Fonseca, was active in 1812, but is specially celebrated for a terrific eruption, preceded by an earthquake, which began on the 20th of January, 1835. Its detonations were heard in Jamaica, and also near Bogota, 560 miles distant. The ashes ejected into the air produced darkness for two days over all the surrounding country, which was covered by their fall to the depth of many feet: they fell thickly on the sea to the west through a space measuring forty miles north and south, and through at least twenty degrees of longitude east and west. The amount of fragmentary matters thrown out by this eruption was prodigious. (See p. 204–5.) We have no account of the present figure of the mountain; but it is probable that it was on this occasion truncated, and a crater of great size hollowed out of its entrails. The neighbouring volcano of *San Vicente* was likewise eruptive in the same year (1835).

The northern horn of the Gulf of Fonseca, directly opposite to Coseguina, is formed by another volcanic mountain, *Concagua* or *Amalapa.* And here the volcanic train, which has ranged generally N.W.–S.E., is again crossed by one which has nearly an east and west direction, corresponding to the similar bend in the coast of the Pacific, and also to the great eastward expansion of the continent that forms the provinces of Honduras and the Mosquito Coast, the line of which is continued in the elevated axial ranges of Jamaica and San Domingo. The two volcanos near the respective towns of S. Miguele and S. Salvador are said to be very active. *Apaneca,* not far from Sansonate, is perhaps extinct. *Isalco,* yet nearer to that town, is said by Mr. Squier to have been "thrown up from a plain near the extinct volcano of *Santa Anna* in 1770. Earthquakes were felt through the previous months; and on the 23rd February the earth opened and sent out lava, accompanied with fire, smoke, and vast quantities of ashes, cinders, and stones, the ejection of which continued for months and years after the cessation of the lava-flow. These products formed a *continually increasing* cone round the vent or crater. It has indeed remained ever since in a state of constant eruption, and is called thence the Faro (lighthouse) of San Salvador. Its explosions, like those of Stromboli, occur regularly at intervals of ten to twenty minutes, and throw up a dense smoke and cloud of ashes and stones, which, falling, add to the height and bulk of the cone. It is now about 2500 feet high." (Squier, Central America, p. 296.)

An adjoining volcano, *Atitlan*, was in violent eruption in 1828, and again in 1833, emitting vast quantities of scoriæ and ash, which covered the coast for many leagues : the whole country for thirty miles round was darkened for two days. Another neighbouring volcano near Amatitlan, called *Pacaya* by Humboldt, is in permanent activity. Its most violent recorded eruption occurred in 1776, when its ejections overwhelmed villages *nine miles* distant. The mass of lava which then flowed down its sides is in many places more than 100 feet thick, and still looks as fresh as if poured out yesterday. The Lake of Atitlan lies at the base of these two volcanos : it is thirty miles long and ten in breadth, and its depth exceeds 1800 feet. It is shut in by dark precipitous volcanic rocks, and has no visible outlet. Several hot-springs rise on its shores. (Squier, p. 490 : see also p. 225, *supra.*)

At but little distance, and near the old town of Guatemala, are two other volcanic mountains. The *Volcan de Agua*, a trachytic cone (14,903 feet) higher than the Peak of Teneriffe, rises into the region of perpetual snow : hence its name,— the eruptions that occur near its summit melting the snow, and sending down torrents of water. Such an inundation destroyed the first city of Guatemala in 1541. The *Volcan de Fuego* is twenty miles W.N.W. of this, near Acatenango. It is still active, though less so than formerly. Nine great eruptions from it are recorded between 1581 and 1799. The last was in 1852, when it sent forth a stream of lava towards the Pacific. It is 14,665 feet in height. The constant earthquakes to which its neighbourhood was subject caused the removal of the city of Guatemala a second time, in the middle of the last century, to a fresh site.

Two or three more volcanic mountains are spoken of still further north. But the last of this thickly-sown group of Central-American volcanos is *Soconusco*, twenty-eight miles south of Ciudad Real, in lat. 16° 2′.

Proceeding northwards, an interval occurs without any active vent; though there is reason to believe that the central range of this part of the continent is of volcanic origin, especially one peak, *Cempultepec* (17,000 feet high). It is only in the parallel of Mexico that we again meet with recent eruptions. Here a chain of six or more volcanos crosses the entire continent from east to west—consequently in a transverse direction to its general axial range,—a fact we have already noticed, though on a smaller scale, in other instances to the north of Panama. It must be referred to the production of transverse fissures crossing the main lines of dislocation along which the primary crystalline ranges and high platforms of the northern continent were elevated.

The most eastern of these volcanos, that of *Tuxtla*, forms a promontory on the coast of the Atlantic fifty miles to the south-east of Vera Cruz: it was in paroxysmal eruption in 1793. Further to the west rises the magnificent cone of *Orizaba* (17,900 feet) (see fig. 38, p. 182), adjoining which, to the north, M. de Saussure found numerous basaltic lava-streams flooding the high Mexican plateau.

Orizaba was ascended in 1851 by Alexandre Doignon, a young Frenchman, and again very recently by Baron Müller. It has a crater on the summit, estimated as 6000 yards in circumference, and broken down on the south and east. The interior is deep, crusted with sulphur in many places, the product of numerous fumaroles. An eruption that took place from it in 1569 is said to have lasted for twenty years: since that epoch it has been in its present state of comparative quiescence. The base of Orizaba is surrounded by smaller cones, mostly truncated: some have thrown out lava-streams; some, mud and ashes. Eastwards is the volcanic mountain of *Acatepec*. On the west occur several sulphureous springs and fumaroles, as well as a group of hills, called ' Los Derrum-batos,' one of which has a crater which is frequently in activity.

To the north also stretches the *Cofre de Perote*, an isolated rocky ridge (13,550 feet high), composed of dark dioritic trachyte, or greystone, and rising from 6000 to 7000 feet above the plateau. It would seem, from recent reports, to be the remaining side of a great crater, the other having been broken down or blown off. The central Mexican platform itself, having an average elevation of 6000 or 7000 feet above the sea, consists in great part of trachytic porphyry full of glassy felspar, the product of ancient, perhaps submarine, eruptions. The deep gorges (barancos) that intersect it exhibit in their precipitous cliffs numerous thick horizontal beds of this rock and its conglomerates. On many points of this high plain, however, wide fields of scoriaceous lava, or of basalt, or of pearlstone and obsidian, are met with, of very recent aspect, and uncultivable, whence they are styled *Malpais*. These lavas probably issued from some lateral vents of the great volcanic mountains, or from fissures in the plain at their foot. (See 'Kosmos,' iv. p. 305.) Further westward, and immediately to the south-east of the Lake of Mexico, is *Popocatepetl*, an equally vast snow-covered volcanic cone (17,720 feet high), still in constant activity, throwing up vapour and ash, but not emitting lava-streams at present. Its principal component rock is an augitic trachyte, or greystone. Some of the beds, however, are of pitchstone. Another volcano, called *Istaccihuetl*, closely adjoins Popocatepetl on the north; and other mountains surrounding the lake-basin are, no doubt, volcanic. To the south-west of the city rises the great snowy volcano, *El Nevado di Toluca* (15,168 feet), not now in activity. Its lavas, as well as those of Orizaba, contain hornblende.

Very much further westward, in lat. 19° 11', W. long. 101° 20', we come to the celebrated volcano of *Jorullo*, first described by Humboldt, and proclaimed by him as a typical example of the upheaval by volcanic force of pre-existing strata. I have already (p. 81–83) noticed this unhappy misapprehension of the great traveller, to which was owing, probably, the

2 G

birth of the mischief-working Elevation-crater theory of Von Buch. There can be no doubt now, since the visit paid to the spot by M. de Saussure, that (as I had the boldness to proclaim in 1825, in the first edition of this work—drawing my inference solely from the facts and appearances recorded by Humboldt himself) there occurred in 1769 *no upheaval whatever* of superficial strata, "in shape of a bladder" or otherwise, but simply the ordinary phenomena of a normal subaërial volcanic eruption,—five vents having been opened in a low plain, or rather valley, on a N. and S. fissure-line, over each of which a common cinder-cone was thrown up by the continuous ejection of scoriæ, while from all these probably, but chiefly from the largest and central cone of Jorullo proper, copious streams of an imperfectly liquid basaltic lava were poured forth, which, not flowing to any great distance, accumulated one upon the other into a high convex platform—the *Malpais*, or '*plaine bombée*' of Humboldt. The latest emission of lava took place from the breached crater of Jorullo itself, and, owing to its extremely imperfect fluidity, formed a massive promontory or buttress, still seen projecting from the side of the cone (see fig. 19, p. 83), its base mingling with the lava of the Malpais. Vast ejections of black ash were among the last-recorded phenomena of the eruption; and the rough protuberances over the fumaroles of the lava-current, covered by a stratum a foot or two in thickness of this fine fragmentary matter, which in consolidating had assumed a globular concretionary structure, formed the haycock-like '*hornitos*' (described by M. de Humboldt as so perplexing to him). M. de Saussure concludes his paper by remarking,—" The volcano of Jorullo has certainly *not* been formed *by upheaval*; and its phenomena, so far from pleading in favour of the upheaving action of the volcanic force, show, on the contrary, that the most powerful outbursts can take place without the slightest disarrangement of the superficial beds." (Bulletin de la Société Vaudoise des Sciences Naturelles, 1859, No. 45, vol. vi.)

The lava of Jorullo is greystone, approaching to basalt, and containing grains of olivine. It is the more singular that M. de Humboldt should have so misconceived the character of this eruption, since he himself describes a portion of the old surface, upon which previously stood, and still stand, "some old farm-buildings, and aged trunks of Cactus and Guava trees," as having remained undisturbed in the middle of the Malpais, at the foot of the cone of Jorullo. This was, of course, a knoll, which (as very often happens) had been surrounded, not covered, by the lava-flood. And the undisturbed condition of the farm-buildings at least should have led M. de Humboldt to doubt that there had been a sudden blister-like upheaval of the whole surrounding plain, four square miles in extent, to the height of 500 feet, and of the cone of Jorullo itself to that of 1680 feet! The extraordinary determination to find something new, strange, and unexampled in the phenomena of Jorullo is remarkably evidenced in a passage of a letter from Von Buch to Humboldt, quoted by the latter (Kosmos, iv. Sabine's transl. p. 303):—"Your hornitos are not cones heaped up by the fall of erupted substances; they have been *upheaved directly* from the interior of the earth"! And he goes on to compare the origin of Jorullo with that of the Monte Nuovo. The parallel is correct; but the origin attributed to both, by the Upheavalists, is equally erroneous. (See p. 325.)   The summit of Jorullo is only 4265 feet above the sea.

Westward of Jorullo, and almost on the shore of the Pacific, is the volcanic mountain of *Colima* (13,000 feet), in 104° W. long. Colima was ascended by Pieschel in 1852. It has a double summit, one having a crater whence vapours and sulphuretted hydrogen gases were issuing. A great eruption of ashes is recorded from it in 1770, and in March 1795 a column of glowing scoriæ.

It is remarkable, that if the line of transverse fissure on which these great Mexican volcanos occur is prolonged across that ocean, it will strike, at a distance of 440 miles to the westward, the group of the *Revillagedo Isles*, in 110° W. long., which are believed to be volcanic, much floating pumice having been seen around them.

Pieschel describes a volcanic branch-fissure, studded with extinct craters and lava-streams, as running from Colima

parallel with the shores of the Pacific, north-west from Guadalaxara. This north and south train perhaps terminates in
the neighbourhood of Durango, in lat. 24°, where a group of
basaltic rocks, covered with scoriæ, rises from the middle of a
level plain; and on the summit of one of the neighbouring
hills a crater has been observed. It is therefore probable that
on further research other traces of volcanic action would be
discovered, especially as several thermal springs are known in
the province of Guadalaxara.

From this parallel of latitude, the *Sierra Madre* (the south-
western extremity of the great axial granitic range of the
Rocky Mountains) begins to swell upwards and commence its
long course to the north, ranging at a distance of from 450 to
800 miles from the shores of the Pacific. It has no active
volcanic vents, but on both its eastern and western declivities
are occasional traces of volcanic action of recent date, consisting of cinder-cones, craters, and lava-streams. The exploring expedition of General Fremont discovered two large
tracts of volcanic formation,—one on the route from Bent's
Fort, on the Arkansas River, to Santa Fé in New Mexico,
where are three extinct volcanos—the *Raton Mountains*,
Fisher's Peak, and the hill of *El Cerrito*. The lavas of the
first cover the whole country between the Arkansas and the
Canadian River. Peperino and basaltic scoriæ abound in
the prairies eastward of the Rocky Mountains; and there is
reason to believe the great *Spanish Peaks* (in lat. 37° 32′ N.)
to be volcanic mountains. This eastern volcanic district
covers an area at least eighty geographical miles in diameter.
On the western slope of the Rocky Mountains the evidences
of eruptive activity occupy a still wider space. The area
begins on the south, about lat. 33° 48′, and embraces two
north and south ranges—that crest of the Rocky Mountains
which is cut through by the Pass of Luñi, and a more western
division, called Sierra di San Francisco. The first has one
conical mountain, *Mount Taylor*, 12,256 feet high, whence
radiate vast lava-streams, stretching to great distances from
the base of the volcano in barren fields of lava covered with
scoriæ and pumice. About seventy miles to the west of Luñi

rises the other lofty volcanic ridge of *San Francisco*, the highest peak of which is estimated at 16,000 feet. It has a great summit-crater, surrounded with vast currents of greenstone, basalt, trachyte, obsidian, and their conglomerates. Several other mountains of the same character continue the range northwards; and the volcanic territory extends on the west of the Great Colorado River in lat. 34° 25' N., where many extinct cones and open craters can be recognized near the Soda Lake.

Further to the west, and also 500 miles to the north, about the forty-third parallel of latitude, and between the meridians of 110° and 112° W., north of the Great Salt Lake, occurs a vast volcanic group, consisting of *Fremont's Peak* (13,568 feet), the ' *Three Mountains*,' and the ' *Trois Tetons*,' of nearly equal height. They are conical peaks, surrounded with wide-spread banks and fields of black lava, with scoriaceous surfaces, and therefore belonging to a recent eruptive era, though now seemingly extinct.

All the volcanic mountains of this main axial range of North America are remarkable for their extreme distance (from 500 to 600 miles) from the nearest sea. But besides these, several lines of volcanic openings run parallel with the Rocky Mountains, nearer to the shores of the Pacific. The peninsula of Old California itself is believed to be in great part of volcanic formation. Three high peaks, called respectively *Loretto, Gigantas*, and *La Vergine*, are particularly mentioned as of this character. Mr. Farnham describes the whole peninsula as " a pile of volcanic ashes, scoriæ, and lava, almost destitute of vegetation." So also is the continuation northwards of this chain, called the Coast Range, or *Sierra Nevada*, up to the latitude of San Francisco, where is the *Monte del Diablo*, an extinct volcano, 3672 feet high, and further north, in the valley of the Sacramento, a wide trachytic crater, called the *Sacramento Butt*. Yet more to the north, about the source of that river, the *Shasty Mountains*, whose summits are covered with perpetual snow, are composed of basaltic lavas; and yet further, in Oregon, the *Cascade Mountains* contain several peaks, equally snow-covered, and rising to heights of 15,000 and 16,000 feet, which are known to be of volcanic nature, some being still active. Humboldt gives the following list of these, commencing, in lat. 42° 30', with *Mount Pitt*, 9569 feet high; *Mount Jefferson* or Vancouver, lat. 44° 35' (15,700 feet), a conical mountain; *Mount Hood*, 45° 10', certainly an extinct volcano, covered with cellular lava. This and the adjoining, *Mount St. Helen's*, are nearly 16,000 feet high. *Saddle Hill*, S.S.E. of Astoria, has a wide crater. Mount St. Helen's continually sends forth smoke from its crater: it is of a very regular conical figure, but covered with perpetual snow. In 1842 an eruption took place, which spread ashes and pumice to a great distance over the surrounding country. *Mount Adams*, 46° 18', almost due east of St. Helen's, and 112 miles from the coast, not now in activity. *Mount Reignier*, 46° 48', a still burning volcano, 12,230 feet high, in violent eruption in 1841 and 1843. *Mount Olympus*, 47° 50', south of the Strait of San Juan de Fuca; and to the east *Mount Baker*, 48° 48' (10,700 feet), a great and still active volcano, of very regular conical form, which forms an imposing feature in the scenery of the straits Quitting the Coast Range for the interior, we find the *Pyramid Mountain, Mount Brown* (16,000 feet), and *Mount Hooker* (16,730), 52° 25',—lofty volcanic mountains of trachyte, in British Columbia, rising upon the main central ridge of the Rocky Mountains, near the sources of the Columbia River, and at distances of more than 300 miles from the coast of the Pacific. There is reason to believe in a great development of volcanic formations about this point. Between this range and Fraser's River, the Black Mountain is the centre of another volcanic district. Indeed, "the great Columbian desert-plain," says Mr. Hector (Journ. Geol. Soc. 1861), "is occupied over almost its whole vast surface by a series of horizontal basaltic lava-flows, without any mountains or peaks near, to which their origin can be referred. The plain is cut up by chasms 500 to 600 feet deep, the sides of which expose stratum after stratum of thin lavas intercalated with softer tuff-beds. The lava-beds have often a columnar structure. There are depressions in the plain occupied by lakes, which, possibly, mark the position of ancient craters. At some points limestone with tertiary fossils un-derlies these basalts. The Columbia River flows for a great distance through an enormous chasm in these strata of lava and tuff, giving rise to the most wonderful scenery."

The volcanic Coast Range is believed to be prolonged north-wards, and the adjoining islands are probably of the same character. It is to be hoped that the progress of settlement in this new territory will before long afford us some more certain knowledge of its geological structure. The small island called Lazarus, near Sitka, lat. 57° 3' N., has a volcano called *Mount Edgecumbe*, 3040 feet high, which was seen in violent eruption in 1796, when it discharged abundance of pumice. In 1806 it was found to have a lake within its crater, then tranquil. Hot-springs break out from granite in the vicinity. *Mount Fairweather*, lat. 58° 45', 14,610 feet high, is covered with pumice, and is said to be in frequent activity. In Cook's Inlet, 60° 08', there is an active volcano, according to Wrangel, 12,064 feet high; and *Mount Elias*, 60° 17' (18,000 feet), has been certainly seen in frequent eruption. Another is reported to be seen occasionally in activity very far inland, about lat. 62°.

On the whole, enough is known of this scarcely explored region to assure us that volcanic agency has been and still is largely developed along the whole western coast of North America from the Rocky Mountains to the sea, the trains of habitually eruptive vents generally having a N.W.–S.E. direc-tion, parallel to the main western axis or spine of the Con-tinent.

About the parallel of 60°, where Mount Elias marks the boundary between British and Russian America, the Coast Range, which had hitherto been trending more and more west-wards from its general N.N.E. course, turns suddenly round to the S.E., forming the lengthened peninsula of Alashka, which is continued in the curving chain of the Aleutian Islands, across the whole breadth of the North Pacific Ocean—there, however, narrowed to less than 15 degrees of longitude. We possess no detailed accounts of the structure of this remarkable chain (960 miles long) of peaks of so many submarine mountains; but it is believed that the greater number, if not all, are of volcanic origin. Upwards of thirty-four distinct eminences, it is said, have been seen *in activity* within recent historic times. Beginning on the eastern or American side, two volcanos are reported in the peninsula itself of *Alashka,*—one of them, *Ilaman,* a peak 11,600 feet high. The island of *St. Paul,* in Behring's Sea, is entirely volcanic, with much recent lava and pumice. The adjoining island of *St. George* is, however, said to be granitic. On the island of Unalashka a volcano occurs, called *Matuschkin* (5474 feet high); its trachytic lava-rock contains much hornblende and black pitchstone-porphyry. Near to the north point of *Unimak,* in May 1796, a submarine eruption, described by Kotzebue, produced an insular volcano which continued in eruption for eight years. It had grown in 1819 to be about sixteen miles in circuit and 2200 feet high. Unimak is the loftiest of all the volcanos of this group, being 8076 feet high. *Atcha,* one of the Andrejanowski Islands, has three smoking vents; and in the island of *Tanaga* is a very large volcano. *Attou,* further east, continues the chain, though at a considerable interval. Behring's Isle forms the last link of the chain of the Aleutian Isles towards the Asiatic side. Its direction, if prolonged, would cut the main axis of the Kamtschatkan peninsula nearly at right angles, which itself, through its length, is believed to consist of a range of volcanic heights.

We have here therefore another and most striking example of a transverse volcanic fissure meeting at right angles, or nearly so, not one only, but two others of the same character. The line of Kamtschatkan volcanos extends northwards of the point at which that of the Aleutian chain appears to cut it, which would be about Cape Kamtschatka, in lat. 56°. At this precise point, however, there is to be seen a remarkable group of volcanos :—*Krestowik* (in lat. 56° 4'). *Klutchewsk* (16,500 feet high), which was violently eruptive from 1726 to 1731, and again in 1767 and 1795. In 1825, Erman was an eye-witness on the 11th of September to the eruption of glowing stones, ashes, and vapour from its summit, while, far below, a considerable stream of lava poured forth from a fissure on its western declivity : its lava is trachytic, and rich in obsidian. *Uschinskaja Sopka* (*i. e.* conical peak), lat. 56°, is nearly connected with Klutchewsk. *Tolbatschi*, S. lat. 55° 51′ (8313 feet), discharges at intervals smoke and ash from frequently shifting vents. *Schiwelatsch*, to the north of this group, 56° 40′ (10,544 feet), has two summits. Its crater was smoking when visited by Erman in 1829. Great eruptions took place from it in 1739 and between 1790 and 1810 : on the last occasion large rocky fragments were ejected. In February 1854 an eruption began, and continued for some months, which produced torrents of lava, and destroyed the northern summit of the mountain (C. von Dittmar). Three or four other volcanos are spoken of at no great distance from this, all more or less active. On the north-west, in 57° 20′, in the central range of the peninsula, on the Baidar Plains, is to be seen a very ancient crater, about a league in diameter, at a distance from any conical mountain, from a fissure in the floor of which lava and vesicular scoriæ of a brick-red colour had been erupted in abundance. Still further north is a circular crateriform lake, about fifteen miles in diameter, surrounded by the Palan Mountains. Southward of Klutchewsk is a group of high volcanic peaks surrounding another large elliptical lake-basin, called Kranosk. Both of these are probably wide crater-lakes of the character of those which were described in Mexico, the Lakes Bolsena and Bracciano in Italy, &c.

Further south, in 53° 32′, we come to *Jupanowa* (9055 feet), having a truncated summit sending forth continual smoke. It is probable that many of these ' smoking' mountains may be in the condition of solfataras. *Koriatskaja*, 53° 19′ (height 11,210 feet), rich in obsidian, which the natives used till lately to point their arrows. *Awatska*, 53° 17′ (8910 feet high), was in violent eruption in 1837. Messrs. Postel and Lenz, who visited it in the following year, describe a vast current of trachytic lava as having descended from the rim of the great crater, and now projecting like a massive buttress from the flank of the mountain. At the base it spreads out in a high platform, from the surface of which, covered by a thick bed of ashes, rise numerous conical knolls averaging 10 feet high, evolving at the time of heir visit

streams of hot vapours with an odour of sulphuretted hydrogen; thus presenting a perfect parallel to the *hornitos* of the
Malpais of Jorullo as seen by Humboldt at the time of his
visit to that volcano, twenty years after its eruption. Some
six or seven other lofty volcanic cones follow in a string
along the eastern coast-line down to the last, *Opalinskaja
Sopka*, just above Cape Lopatka, the southern point of the
peninsula, which was exceedingly active at the close of the last
century. It overhangs on the north another large oval lake,
of about the size of the two already mentioned. There are
likewise three or four peaks, some always smoking, on the
western slope of the peninsula. In all, no less than fourteen
active volcanos have been reckoned in the entire length of
the peninsula of 420 geographical miles; and it would seem
probable that at least an equal number of mountains of volcanic origin might be counted, now extinct, or, perhaps, only
dormant.

Kurile Isles.—Immediately to the south of Cape Lopatka,
the first island of the Kurile Group, *Paramouchir*, contains
an active volcano. Another, *Alaid*, somewhat to the east,
but in the prolongation of the line of more easterly volcanos of
Kamtschatka, above 12,000 feet high, was in violent eruption
in 1770 and 1793. The other islands of the Kurile chain,
extending 720 miles in length (evidently the summits of a
train of volcanic mountains rising from the depths of the
ocean), are said to show eight or nine vents, more or less
in activity. Next follow, in continuation of the same remarkable line of habitually eruptive volcanos, those of Jesso, and
the other islands of the Japanese group.

The first (*Jesso*) is as yet but little known to Europeans.
At its north-eastern angle an insular volcanic peak (*L'Angle*)
rises from the sea to the height of 5350 feet. And the main
island is believed to be intersected by a range of volcanos
from the Northern Cape to Volcano Bay in the south-east,
which is bordered on either side by two lofty and active
cones. Indeed, Siebold counts no less than seventeen conical
mountains in the entire length of the island, all supposed to
be of volcanic character. One is called by the Japanese

Usuga-talee, or Mortar Mountain, from its deep crater. This and *Kajo-nori* are still burning.

In the other great islands of *Japan* six volcanos are recorded as eruptive by the local historians, two in *Niphon,* and five in *Kiusiu.* Besides these, however, a range of conical mountains threads the entire length of the three islands, many of which are characterized by clearly-marked craters, and are unquestionably volcanic. In Niphon nine such are enumerated, productive of trachytic lavas; two of them are estimated at above 12,000 feet in height.

The known active volcanos are, beginning at the north, *Jakejama,* lat. 41° 20′, at the north-eastern extremity of Niphon; another of the same name (probably it is Japanese for 'burning mountain') in lat. 36° 33′. *Assamajama,* north-west of Jeddo, in the interior of the island, lat. 36° 22′, which was in violent and destructive eruption in 1783, and has ever since been permanently active. *Fusiyama,* lat. 35° 18′ (12,443 feet high), a beautifully regular cone, truncated only near the top, where is an oval crater measuring about 1600 yards by 600, and 350 in depth. The mountain is an object to the Japanese of great veneration and frequent visits. It is reported by their chronicles to have been first formed by a fierce eruption in the year 286 B.C. Its historically-known eruptions took place in 799, 863, 937, 1032, and 1707 A.D., since which last date the volcano has been in repose. No doubt the summit-crater dates from the last outbreak.

In the island of *Kiusiu* are counted as active volcanos, *Wunsen,* 32° 4′ (4110 feet), tremendously eruptive in 1793, when 53,000 persons are said to have been destroyed. The summit of the mountain '*fell in*'—more probably was blown off. *Asoyama,* 32° 45′ (both to the E.S.E. of Nagasaki); *Kirisima,* 31° 45′; *Mitake,* an island in the Bay of Kagosima; and *Ounga,* on the west coast, S. of Nagasaki.

In addition to these, European navigators have observed several islands adjoining the larger, with smoking peaks and craters: viz. *Iwosima* (or Sulphur Island), south of Kiusiu, in Van Diemen's Strait, lat. 30° 43′, 2366 feet high; *Ohosima,* lat. 34° 42′, long. 139° 26′ W., from which Broughton, in

1797, saw smoke rising from a crater which had been recently in eruption. A range of volcanic islands, according to Postel, runs southwards from Ohosima to Fatsi Sjo (33° 6′), and thence to Bonin Islands (26° 30′ N.), more than 12 degrees to the eastward. The chains of the Marianne and Caroline Isles, still further south, which are likewise volcanic, appear to be the prolongation of this line.

On the continental side of the Japan Sea, in the peninsula of Corea, no volcanos are known; they seem to be confined to the adjoining islands. That of *Quelpaertz* shows several conical peaks indicative of volcanic origin; and an island-volcano on the coast of Corea, *Tsinmara*, is said to have risen from the sea in the year 1007.

The volcanic train of the Kuriles and Japanese Islands is continued southwards through a string of small islands to the *Loochoo* Group, and thence to the great island of Formosa, where, in lat. 24°, near the east coast, Lieutenant Boyle observed in 1853 an eruption from the sea. Among the smaller islands of the chain, *Suwasesima*, in 29° 39′, was seen in active eruption by Captain Belcher: its height is 2803 feet. Captain Basil Hall describes it as having a crater in the state of a solfatara. Its steep sides are formed of beds intersected by dykes.

In *Formosa* itself a high chain of mountains ranges centrally through the island, which are probably volcanic, since abundance of salt and sulphur are met with, and hot-springs; flames are also said to rise out of some of the lakes and from the ground. Tradition indeed speaks of some of these mountains as having been eruptive. Two volcanos are described by M. Stanislaus Julien in the 'Comptes Rendus' for 1840.

Some of the lesser islands which connect Formosa with the *Philippines* have been seen in eruption. In this latter group no less than nineteen lofty insulated conical mountains, all called in the country 'volcanes,' are enumerated by Von Buch.

In Luzon or Manilla, the most northern island, the largest volcano is *Mayon*. It is described as perfectly conical, 3200 feet high, and as constantly sending out a column of vapour from its summit, and sometimes flames (projected scoriæ) : its detonations are heard at great distances. The surrounding country is covered with its ejections. An eruption in 1767 lasted ten days, and poured forth a destructive river of lava for two months, followed by enormous deluges of water, which devastated the country and destroyed several towns and villages. Mayon was also in eruption in 1800, and again in 1814.

Another volcano in Luzon is that of *Taal*—an island within a lake, containing within its crater (two miles in diameter) another lake, from which rises another conical volcano. The cliffs encircling the crater are described by Lopez as from 900 to 1200 feet in perpendicular height, and at the time of his visit millions of jets of inflamed (hydrogen) gas issued from their crevices. The waters of the lake are impregnated with sulphuric acid. In 1716 a terrific eruption took place from Taal, and in 1754 one still more violent. The detonations of the volcano were heard at a distance of 300 leagues. The explosions were terrific, and lasted ten days, producing darkness from the clouds of ejected ashes, which covered the roofs of houses at Manilla, twenty leagues distant. In the immediate neighbourhood of the mountain rocks of large size were thrown up, and *streams of sulphur and bitumen* overran the district of Bongbong. The alligators and fish in the adjoining river were destroyed in vast numbers.

No less than eleven volcanos are said to be ranged in a line along the narrow peninsula of Camarines, the south-east extremity of this island (Luzon). One of them, 3200 feet high, was in violent eruption in 1800 and in 1814. In the little island of *Mindoro*, opposite to Manilla, is a volcano in never-ceasing activity.

In the island of *Mindanao*, the most southern of the larger Philippines, at Bukayan, is a volcano which was in eruption in 1640, and flung out large masses of rock to a distance of two leagues. The ashes fell throughout the Moluccas and Borneo. Dense darkness covered the nearer islands. The mountain whence the explosions proceeded *disappeared,* and a lake was formed in its place (a crater-lake), the waters of which were long white with ashes. (Bowring's ' Philippine Isles,' 1861.)

This large island, Mindanao, it will be observed, has a forked shape, throwing out two great promontories, one to the south, the other to the south-east; and it would seem that the great volcanic chain or fissure is here divaricated, and branches off in these two several directions : the south-eastern threading the *Sooloo* chain into Borneo; the southern ranging through that of *Sangir* and *Sivao* (both volcanos recently in eruption), across Banka Strait, to the north-east extremity of Celebes.

The long-extended group of the *Sooloo* Islets, probably fully a hundred in number, which connects Mindanao with Borneo, is partly volcanic, partly composed of coral-reefs.

In *Borneo* itself we have no certain knowledge of any active volcano; we are, indeed, only acquainted with some narrow strips of its coast; nor are its loftiest mountains supposed to be volcanic. Rajah Brooke, however, mentions a mountain in Sarawak as bearing the name of *Goonung Api* ('fire-mountain' in Malay), and as surrounded with scoriæ—without doubt, therefore, of volcanic origin.

The other or eastern branch of the volcanic train would seem to range through *Celebes*, in which Dr. Junghuhn counts eleven active volcanos, in a line pointing due south towards *Flores*, where it would meet and cross another train of the same character, which, diverging from the north of Celebes, passes with a sweeping curve through *Tidore* and *Ternate* (both volcanos) into the Moluccas, Ceram, and Amboyna (sending off also another branch to the coast of New Guinea, where Dampier saw a volcano), until it falls into the due east and west train of Timor, Flores, Sumbawa, Java, and the intermediate islands of this remarkable linear range.

Six volcanos are said to range close together along the narrow peninsula of *Mucado*, in the north-east of Celebes. The island of *Ternate*, to the west of the larger island of Gilolo, one of the Moluccas, consists of a volcanic cone (5755 feet high), whence violent eruptions broke out from 1608 to 1673, and again from 1838 to 1849, after an interval of a century and a half of quiescence. Another volcano opened on the west side of *Gilolo* in 1673, and threw up immense quantities of pumice. *Amboyna* is wholly volcanic; it had a fearful eruption in 1694, and another in 1820; it now emits sulphureous vapours and eruptions of hot mud. The island of *Sorea*, the most southern of the Moluccas, was entirely desolated in 1693 by an eruption, having been previously very fertile, populous, and well cultivated. The bulk of the mountain was destroyed, and a lake of incandescent lava took its place. In that of *Machian*, in 1646, a lofty volcanic mountain was rent in two from top to bottom by a violent eruption. The little island of *Banda*, or *Goonung Api*, south of Ceram, 1800 feet high, was always burning from 1586 to 1824. The great peak of *Timor* once served, like Stromboli, as a lighthouse to mariners, being, from its vast height, visible at a distance of 300 miles. In 1638 a prodigious eruption blew off the greater part of the cone and replaced it by a large lake. It will be remarked that this is an event of frequent occurrence among the powerful volcanos of this quarter of the globe. On the small island of *Pulu Batu*, north of Flores, a volcano in 1850 was seen to pour down a glowing stream of lava upon the sea-beach.

Indeed volcanic activity is so rife in this Archipelago, that Dr. Junghuhn reckons, from actual enumeration within the wreaths of islands surrounding the great continental mass of Borneo, no less than 109 lofty fire-emitting mountains and ten mud-volcanos, all in activity at present.

The large island of *Flores* contains at least three active volcanos. *Sumbawa* is noted for an eruption of paroxysmal violence which took place from the mountain of *Tomboro* in 1815, of which we have an authentic account from Sir Stamford Raffles. It commenced on the 5th of April with detonations which were heard at Sumatra, nearly 1000 miles off in a straight line, and were mistaken there for the firing of artillery. Three distinct columns of flame (red-hot scoriæ) were seen to rise from the mountain to a vast height, and its whole surface soon appeared as if covered with fiery lava-streams, which spread to great distances on all sides; stones fell for miles around—many as large as a man's head; and the black fragmentary matter carried into the air caused total

darkness.   A whirlwind is said to have accompanied the commencement of the eruption, by which the roofs of houses,
trees, and even men and horses were carried into the air.
The shore about the town of Tomboro sank permanently to a
depth of eighteen feet.   The explosions lasted thirty-four days.
The abundance of ashes ejected was such as to cause complete darkness at mid-day in Java, 300 miles distant, and they
covered the ground and roofs of houses *there* to the depth of
several inches.   In Sumbawa itself the part of the island
adjoining the mountain was entirely desolated, all the houses
destroyed, with 12,000 inhabitants, and the trees and herbage
overwhelmed to a great depth by pumice and ash.   At Bima,
forty miles distant from the volcano, the weight of these fallen
matters was so great as to break in the roofs of houses.   The
floating pumice upon the sea around formed a layer two feet
thick, through which vessels with difficulty forced their way.
(Raffles's ' Java,' vol. i. p. 28.)   It is evident that the void left
in the mountain by the abstraction of so much matter will
account for a crater of the largest known area. (See pp. 171
& 204.)   Sumbawa is connected with Java through the island
of *Lombok*, in which rises one volcanic mountain, 7500 feet
high, and another, *Bali*, which was in eruption in 1803.

The single island of *Java* contains a larger number of active
volcanos than any other district of equal extent on the face of
the globe.   We owe our knowledge of them in great part to
Dr. Horsfield, who accompanied Sir Stamford Raffles during
his residence in the island, and more recently to Dr. Junghuhn, who spent twelve years there in the examination of its
natural phenomena.   The latter author measured and describes no less than forty-five volcanic cones, which range
through the entire length of the island, composing its axial
heights.   About half of these, twenty-eight in number, are
still in activity, that is, in continual or occasional eruption,
They are generally far inferior in elevation to the great conical
volcanos of Central and South America, but are yet of imposing height and magnitude.   The highest, *Gunung Semera*,
was ascended by Junghuhn in 1844, and rises 12,235 feet
above the sea.   Four others are between 10,727 and 11,116

feet, and seven between 9000 and 10,000 feet high. The largest known crater on any of these mountains is that of *Gunung Tengger* (8700 feet high). It is 1865 feet deep from the ridge of the surrounding escarpment, and three and a half miles in diameter. In its centre rise four small cones, each with its crater, and all burning except one, called *Bromo*, which has a lake of warm and sulphureous water in its interior. It was visited and described by Professor Jukes. Another mountain, *Gunung Raon*, 10,180 feet high, has a still deeper crater, the bottom being 2400 feet below its encircling ridge, and the sides precipitous: its diameter is not quite two miles. The view into this crater is described as awfully grand. According to Junghuhn, the active cones and craters of the island are surrounded by old crater-walls— sometimes by a double enceinte, the outer one being five or six miles in diameter.

The most destructive eruption on record of the Javanese volcanos took place in 1772, from the mountain (or Gunung) called *Papandayang* (7034 feet high). At the same moment two other volcanos of the same island, situated respectively 184 and 352 geographical miles in a direct line from Papandayang, also broke out, although several intervening volcanic cones of the chain remained undisturbed. This fact, and some other parallel ones that might be quoted, indicates the complex character of the communication or connexion which must exist between the eruptive fissures through which the volcanic matter finds its way. Humboldt, in reference to this, very justly recalls the fact, that on the morning of the 4th February, 1797, when a dreadful earthquake destroyed the town of Riobamba, at the foot of Cotopaxi and Tunguragua, neither of those volcanos seemed to be influenced by the commotion, although that of Pasto, at the distance of 120 miles, showed its sympathy by the sudden cessation of its habitual column of vapour. Probably the superficial crust is penetrated by a network of fissures, some reaching to far greater depths than others—some firmly sealed up by consolidated lava—others empty, or in which the filling is still fluid, and therefore more yielding. (See pp. 272-5.)

2 H

The great eruption of Papandayang has been often quoted as affording an instance of the engulfment of the summit of a volcanic mountain, and its giving place to a lake or low plain. Dr. Junghuhn refutes this misrepresentation, and describes the occurrence to have been, on the contrary, of the character which in the body of this work I have represented as the ordinary and normal result of a great paroxysmal eruption from a volcanic cone, namely the *blowing off* of the entire summit by long-continued explosions, and the scattering of its materials in fragments over the surrounding country— the finer and lighter or more triturated particles being borne away by the winds to enormous distances (see p. 195). The forty villages which, with their inhabitants, were generally reported as having been " *swallowed up by the opening ground,*" were, on the contrary, overwhelmed by these fragmentary ejecta. And the area of fifteen miles by six, which was supposed to have "sunk deep into the earth," is the hollow left by the explosions of this paroxysmal eruption,—a true crater, in short, but in magnitude comparable to the very largest of those whose vast dimensions have caused many geologists hitherto to question their explosive origin.

The volcanos of Java throw up prodigious quantities of fine ash, often in the state of mud, from mixture with the contents of crater-lakes, and of these substances the external slopes of the mountains are consequently for the most part composed. They are hence very subject to erosion by tropical torrents of rain acting on their loose or soft materials, and are thus scored deeply on all sides with radiating ravines of a regularity compared by Junghuhn to the depressions between the sticks of a half-opened umbrella. When the summit of such a cone has been blown off by an explosive paroxysm (and many of the cones are so truncated), the upper parts of these ravines form, of course, so many notches cutting into the sharp ridge-rim of the crater all round its circuit. The eruption of lava is spoken of as unfrequent in comparison with that of fine ash and mud, which hardens into a kind of trass or Moya. Dr. Horsfield, however, describes, upon the volcano of *Gunung Guntur*, five great streams of lava which flowed from its summit and reached to the foot of the mountain at different known periods—the last in 1800. Three other volcanos are also mentioned by Jung-huhn as having poured forth black basaltic lava-streams. Pro-bably they are not so rare as is supposed, but concealed from view by the abundance of ash and trass. The lavas of Java are greystone chiefly, and often contain much leucite, bearing a great resemblance to those of Vesuvius. There are dykes of pitchstone. Trachyte seems rare, but composes one volcano called *Tilu*: it is a mixture of glassy felspar and hornblende. Pumice and obsidian, as we might expect, are to be seen only in this locality. Beds of basalt or greystone are met with wherever the torrents have eaten deeply into the plains at the foot of the volcanos, and are sometimes interbedded with ter-tiary sedimentary rocks. These, indeed, compose the greater part of the superficial area of the island, especially the south-ern side.

The eruption from Guntur in 1800 threw out, in addi-tion to the lava already mentioned, a vast torrent of white, acid, sulphureous mud—the contents, no doubt, of a long-seething solfatara—which devastated the wide surface of a previously fertile valley. Sulphureous vapours are abundant in many parts of the island. Mount Idienne, towards the east, has a crater-lake of acid waters surrounded by banks of pure sulphur. The river that issues from it contains no fish. Hot saline springs with petroleum are frequent: some throw up volumes of black mud, in great globular bubbles, 20 or 30 feet high (mud-volcanos). One valley (probably a crater) is called the Poison-valley (Guevo Upas), from the fact that every living thing that crosses it is stifled by the carbonic acid gas that fills its hollow. Bleached bones of men and animals are strewn over its surface.

Sumatra.—This great island has been less visited by natu-ralists than Java, with which it is continuous, with the in-terval of a narrow strait. It is known, however, to be almost equally volcanic. Marsden describes four active volcanos in its principal range. One called *Priamang*, twenty miles inland from Bencoolen, emitted continual vapour from a vent one-fourth-way down from the summit. It was observed to send forth flames (probably red-hot stones) at the same moment at which an earthquake was felt. The inhabitants of the island connect the occurrence of the more destructive earthquakes with the quiescence of their volcanos, which they consequently rejoice to see in activity. *Gunung Dempo*, the highest mountain in the island (12,000 feet), almost constantly emits vapour. Hot-springs are frequent, and much basaltic rock has been observed. It would seem, however, that granite is found in some parts of the island, as well as limestone of a coralline origin, especially along the northern coast.

The *Nicobar* and *Andaman Isles* appear to be the prolongation of the volcanic range of Sumatra northwards. One of the latter, *Barren Island*, has been already mentioned (p. 199)

Barren Island, in the Bay of Bengal.

as a typical example of an insular volcano, consisting of an active cone, encircled by the cliffs of a wide ancient crater, into which the sea enters by a breach. The explosions of this volcano occur regularly, at intervals of about ten minutes. The island of *Narcondam* (13° 24′ N.), north of this, has also shown volcanic activity. It is a cone 700 feet high, with streams of lava visible on its flanks.

Another island, *Chedooba* (18° 40′), and a neighbouring one, *Rhamree* (19°), near the coast of Arracan, are likewise volcanic. The latter is said to have burst into violent eruption in March 1839, at the moment that the neighbouring peninsula was disturbed by a tremendous earthquake, ranging north and south, which was felt in the Andaman Isles on one side, and in China on the other. Much change of level appears also to be going on upon this coast. Certain islands, said to have existed in 1554, are now submerged ; and in 1843 a new island was formed off the coast of Arracan. Remarkable hot-springs, as well as inflammable gases, rise from the ground in the neighbourhood of Chittagong.

Here seems to terminate (unless the future exploration of the Burmese territory should bring to light its further continuation towards the Himalayas or Hindostan) the most remarkable train of volcanic vents visible upon the surface of the globe, and which we have now traced through 60 degrees of latitude,—from the north of the peninsula of Kamtschatka, beyond the point where it meets the transverse chain of the Aleutians, threading the Kurile, Japanese, and Loochoo insular ranges, almost touching the coast of China in Formosa, then stretching due south through the Philippines, whence several loop-lines appear to branch off, through Borneo, Celebes, the Moluccas, and New Guinea, in sweeping and almost concentric curves. These again unite on the south in the great east and west chain of almost continuous volcanic heights, from Timor Laut, through Flores and Java, bending once more northwards in Sumatra and the Andamans. The interior of this grand curvature is occupied by the as yet unexplored great peninsula of Cochin China and island of Borneo, whose rounded coasts repeat it with parallel concentric outlines—a concordance which can hardly be a mere accident. It is, indeed, difficult to escape the conviction that the plutonic elevation of the axial mass of this great southern embranchment from the high central Asiatic platform had the collateral effect of rending the crust of the earth, at some distance around it, from east to west by south, with a loop-like series of concentric fissures, which allowed the extravasation of the subterranean intumescent lava in those curved ranges of volcanic apertures we have been lately engaged in tracing.

VOLCANIC ISLANDS OF THE PACIFIC.

From the island of Gilolo, in the centre of the group of the Moluccas, it has been already noticed that another train of volcanic vents branches off towards the east, ranging along the northern coast of Papua or New Guinea, where more than one volcano has been seen in active eruption, as well upon the mainland (of which but little is known) as in the adjacent islands. It is continued through New Britain, the *Solomon* group, *Queen Charlotte's Isles,* the *New Hebrides,* and *New Caledonia,* and, though with a considerable interval, still further southwards, into *New Zealand.* Its course through this extended curve holds a remarkable parallelism to the coast-line of Australia at some distance to the west, which indeed it seems to embrace— very much in the same way as the wreath of volcanic islands last described embraces that of Borneo. The islands composing this prolonged chain are believed to be almost exclusively of volcanic origin; many of them certainly contain active vents. In the Solomon Isles, one, called *Semoya,* has been seen in activity. New Caledonia is said to show some plutonic rocks, together with carboniferous and other sedimentary strata, but is surrounded by basaltic islands. The New Hebrides contain at least two active volcanos, in the isles *Tanna* and *Abrim.* Near New Britain there are two more. *Malicolo,* in the Queen Charlotte Isles, shows much pumice. New Caledonia is chiefly, it is believed, granitic; but an isle called *Matthew's Rock,* S.E. of it, 1180 feet high, was seen in eruption in 1828; and in the archipelago of Santa Cruz are two known volcanos, *Tinahoro* and a smaller one.

New Zealand contains several active volcanos, and a considerable area is covered by the products of very recent eruptions. In the Northern Isle, *Mount Egmont* (8960 feet)—a truncated cone, with a smaller ash-cone on its summit—is occasionally active: its mass consists of clinkstone lavas and scoriæ. So likewise are *Tongariro* (6200 feet), in the centre of the widest part of the island, and *Ruapahu* (9000 feet), rather more to the south. The Lake of Taipu, at the foot of Tongariro, is surrounded by hills of pumice and ash; and thence, in a N.E. direction, a line of solfataras and hot-springs extends to the coast of the Bay of Plenty, in the centre of which, *White Island,* a volcano of considerable activity, rises from the sea. Further northward, in the narrowest part of the island, the district of Auckland consists of horizontal strata of tertiary limestone and sands, studded with the products of very recent eruptions, chiefly from numerous independent vents.   Mr. Heaphy (Journ. Geol. Soc. 1859) divides the volcanic formations of this district into—1. Peaked mountain-masses of trachyte and its conglomerate—a "black boulder-rock"—all older than the tertiary strata. 2. The products of subaqueous eruptions through the tertiary beds at the time they were submerged and in course of deposition, as shown by the interbedding of the volcanic and fossiliferous strata. 3. Eruptions that occurred about the time of the upheaval of those strata, the cones and lavas having risen through faults broken in them. 4. Cinder-cones and lava-streams of very recent aspect. These last show many tuff-craters resembling those of the Phlegræan fields, frequently having lakes, or 'maars,' in the interior, and streams of basaltic lava traceable to them.   The borders of the tuff-craters are often very little above the surrounding plain.   There are also many regular cones, often with breached craters.   One large insular volcano, *Rangitoto,* not at present in activity, is evidently the product of repeated eruptions from the same vent.   It has a central cone and crater, 920 feet high, rising from within an external crater-ring, which is surrounded, in turn, by the ruins of an earlier outer crater, of which the rim reaches about 600 feet above the sea.

It is not known whether the Southern Island of New Zealand contains any active or extinct volcano; but *Chatham Island,* a little to the eastward, is certainly volcanic.

A few scattered islets, the Auckland group, Macquarie, Campbell, and Emerald Isles, appear as if they formed a prolongation of the meridional train of volcanos which has occupied us last, in the direction of Victoria Land, within the Antarctic circle, on the coast of which Sir John Ross observed two lofty fire-emitting volcanic mountains, appropriately named by him (after his vessels) *Mounts Erebus* and *Terror.* He describes them as covered with perpetual snow, except where the hot lava and ashes had melted it, and as seemingly composed of beds of basalt alternating with others of ice.   It is easily understood how a thick coating of scoriæ and ash may protect the surface of a glacier from being thawed even by a stream of glowing lava poured upon it, and the lava in turn may, by its weight, keep the glacier in its place, until a whole mountain has been built up of these strange elements.

The line of the coasts of Victoria, if prolonged past the South Pole, will be found to coincide with that of South Shetland, already noticed, immediately S. of Cape Horn. If we suppose the volcanic fissure through the earth's crust, which we have followed so far, to be continuous along that space, it will literally have made the complete circuit of the great basin of the Pacific Ocean, nearly bisecting the globe. (See pp. 12 & 277.)

The great continental area of *Australia* itself shows marks of volcanic action on several points of its circuit. At its southern extremity, in the *province of Victoria,* some hundreds of recent cinder-cones are to be seen, most of which have produced streams of basaltic lava that deluged the surrounding plains. These eruptions broke through surface-rocks of granite and auriferous palæozoic schists, covered by beds of tertiary basalt, which alternate with sandstones containing Miocene marine shells. Dykes both of the earlier and recent basalts penetrate all these rocks, and connect themselves with the overlying masses. The horizontality of the sedimentary strata has not been disturbed. Mr. Brough Smyth (Journ. Geol. Soc. 1857) estimates the area occupied by the recent lava-fields as at least 3500 square miles. It constitutes the garden of what is called Australia Felix, from the richness of the soil produced from the decomposed basalt. He thinks this was erupted when the surface was beneath a shallow sea, as it is overlain by tertiary drift. Some of the volcanos were, however, undoubtedly subaërial, since, in digging a well through volcanic ash arranged in thin strata, at a depth of 63 feet a layer of turf was met with, formed of the common coarse grass of the country—not scorched, but like dry hay. This locality is only remarkable for the great number and wide dispersion of vents productive of single eruptions, and the absence of great volcanic mountains.

Between Adelaide and the River Murray, *Mount Gambier* is

described as rising 600 feet from its base, and as having three distinct craters, each occupied by a lake of fresh water. *Mount Schenk* is circular, and has one large and two small lateral craters. Along the boundary-line of the two colonies, South Australia and Victoria, trap-rocks occur abundantly, associated with horizontal strata of tertiary limestones, pierced by basaltic dykes, and showing some cones and craters on the surface. The volcanic districts that have been observed in Australia appear to bear much resemblance to those of Central France, of Corsica, the Katakekaumene in Asia Minor, Lancerote, New Zealand, and many others we have already noticed, in which eruptions have from time to time occurred from an early part of the tertiary period down almost to the present time.

In Western Australia also recent volcanic formations have been recognized, and again at Brisbane and other spots along the east coast. The Blue-Mountain-ranges behind Sydney are capped by basalt, at heights above 2000 feet from the sea. But no volcanos in positive activity have yet been observed in any part of this continent. The same may be said of Van Diemen's Land, where numerous old platforms of basalt and its conglomerates occur, but few, if any, traces of very recent development of volcanic energy.

Eastward of the insular chain last described, several more or less detached groups of islands, chiefly volcanic, occur upon a belt which crosses the entire breadth of the Pacific from east to west. Among them are the *Feejee Islands*, full of both basaltic and trachytic lavas, with many hot-springs, and containing many tuff-cones with craters in linear ranges.

The *Navigators' Isles* to the north, the *Friendly* group to the south of the Feejees, are equally volcanic : in the latter, the peak of *Tafua*, 2138 feet high, is always burning ; it has a large crater encircling a central cone of scoriæ. Two other volcanos, *Apia*, 2576 feet, and *Upala*, 3197 feet, are surrounded by extensive fields of scoriform and cavernous lava. Further to the east, *Tahiti*, the largest of the *Society Isles*, is composed of trachytic mountains of an early age, whose foci seem at present extinct. Of these, one (*Orobena*) is said to be 10,000 feet high, and to have a crater on its summit. The group of the *Marquesas*, to the north-east of Tahiti, is likewise, it is believed, for the most part, if not entirely, volcanic.

East of the Society Isles follows the archipelago of the *Low Islands*, which consist almost wholly of flat coralline reefs, as their title indicates, with the exception of the small basaltic group of *Gambier* and *Pitcairn Islands*. It is, however, probable that all rest on a base of volcanic rock. The volcanic range, indeed, is continued further eastward, along the same parallel, in *Easter Islands*, on which Captain Beechey observed a conical mount, 1000 feet high, and a range of craters. *Juan Fernandez*, still nearer the continent of South America, is basaltic.

North of the equator, the Western Pacific is studded with groups of very small islands—the *Carolines, Ladrones* or Marianne Isles (already mentioned), *Marshall's, Gilbert's*, and, further northwards, about the 20th parallel, the group of the *Sandwich Islands*.

The Ladrones contain a train of three or four active volcanos, in a north and south direction. Those to the east and west are volcanic also, but are not believed to be now active. The other islands of these archipelagos are, it is supposed, chiefly coralline, and may or may not rest upon a foundation of volcanic rocks.

The group of the Sandwich Isles seems, from its position, to belong to a more northern trans-Pacific train of fissure, stretching away eastwards from the Loochoo and Bonin Isles. It is wholly volcanic; and having been well observed, especially the greater island of *Hawaii*, it is found to present many features of considerable interest.

According to Mr. Weld (Journ. Geol. Soc. 1856, p. 164), there are three great mountains in Hawaii, all volcanic; indeed, no other rock exists in the island. *Mauna Kea* (13,950 feet), the most northerly, is the highest, and at present inactive, though bearing evident traces of eruption at no very remote period. *Hualalei*, on the west coast, was in eruption a few years since. The third, *Mauna Loa*, is an immense dome-shaped mountain, whose summit, 13,370 feet above the sea, presents a barren area, perhaps forty miles in diameter, covered with lavas and scoriæ, and in frequent eruption from one point or another. (See pp. 219–222, *supra*).

In 1843 an eruption occurred from the summit of Mauna Loa : the lava-flood was thirty miles long. In 1852 another outburst occurred from the same point, a fiery fount of liquid lava-drops being thrown up 500 feet high. In two days this ceased ; and in two more another vent opened, fifteen miles from the first, on the side of the mountain, while a column of fire, scoriæ, lava-filaments, and vapour, 1000 feet in diameter, rose 500 feet high in the air, casting around an intense light by night. This lasted twenty days. The noise was tremendous. The atmosphere was filled with scoriæ, filaments, fine dust, and acid vapours. A vast stream of lava flowed as before some twenty-five miles down the side of the mountain. In August 1855 a yet more fearful eruption burst forth, commencing with a brilliant jet of fiery drops from the summit of the dome, and immediately followed by the emission from an opening about 2000 feet lower, and 11,500 above the sea, on the northern flank of the mountain, of an enormous flow of lava, not accompanied by any proportionate projection of scoriæ or other fragmentary matters. The lava ran with great rapidity into the valley which separates Mauna Loa from Mauna Kea, the main branch being three miles wide. As it reached a more level country, it spread to a width of double that space. It continued to flow for ten months. Before it ceased, it had reached to a distance of seventy miles from the source. In its course it overwhelmed forests, and accumulated on some points to a great thickness.

During the flowing of this lava-stream, the Rev. Mr. Coan ascended the mountain, skirting and occasionally crossing over its hardened surface, while it still flowed on underneath, " like water under ice in a river." " The superficial crust was crackling, and emitting mineralized vapours at innumerable points. Along the margin numerous trees lay crushed, half-charred, and smouldering upon the hardened lava. . . . . We passed opening after opening, through which we looked down upon the igneous river as it rushed through its vitrified duct at the rate of many miles an hour. It was incandescent, and from 25 feet to 100 beneath the surface, the openings (clefts) in which were from one to forty fathoms across. Into these we cast large stones, which, as soon as they struck the surface of the hurrying flood, passed down the stream in an indistinct and instantaneous blaze. We could also see subterranean cataracts of molten rock tumbling down precipices of 25 to 60 feet." On reaching the summit of the mountain they found no regular crater, but "yawning fissures, on each bank of which immense masses of scoriæ, lava, pumice, and cinders were piled in the form of elongated cones, rent longitudinally, their inner walls being hung with stalactites, or festooned with the filamentous lava called *Pele's hair*. These elongated masses overhanging the yawning fissures were often precipitated in avalanches into the gulf below, choking it up for a time. The lava, which flowed off by a lateral subterranean duct probably 1000 feet below the surface, could not be seen at this point. It first makes its appearance through openings some miles lower down the slope of the mount; but the fearful rush of white smoke and sulphureous gases from these summit-fissures is awful and perilous to the spectator." The higher regions of the mountain are described as covered with the recent fragmentary ejections from the volcano, scattered widely on every side. "The smoke (clouds of vapour mixed with ashes) enveloped the summit, darkening the sun, and obscuring every object a few rods distant." From the surface of the lava-current also "clouds of steam rolled up in fleecy wreaths towards heaven *."

It is noticeable, that during the whole of these violent eruptions, the great and permanently active crater of Kilauea, on the other side of the same mountain, at a distance of about fifteen miles, remained in its normal state, showing no sympathy with the extraordinary development of energy which was taking place in the axial centre of the same volcano. This remarkable circumstance has been already noticed. (See p. 262, *supra*.)

It is at a comparatively low point on the flank of Mauna Loa, only 4000 feet above the sea, that a flattened dome-shaped

* Journ. Geol. Soc. xiii. p. 170, 1856.

swelling is seen, of a truncated outline, which contains the wonderful crater of Kilauea. Its upper rim is about seven miles round, and of an irregularly elliptical figure. Within is a vast shifting pool or lake of lava, frequently changing its level, crusted over more or less with consolidated lava-rock, but always boiling up in several places. At times the level of the lava-crust sinks within this crater, leaving a terrace or shelf of solid rock around the deep pit so formed. The lava in this case has escaped by a lateral eruption at a lower level, through a fissure rent in the flank of the mountain. In 1823, according to Mr. Ellis, such an underground tapping lowered the level of the lava in the central pit by about 400 feet. In 1834 Mr. Douglas describes it as more than a thousand feet below the 'black ledge.' (Roy. Geogr. Soc. Trans. vol. iv.) In 1838, according to Captains Chase and Parker (Amer. Journ. xl. p. 117), the lava had risen again up to the level of the ledge, so as to obliterate the lower pit. In 1839 the whole *outer* crater was filled to the brim with boiling lava, more or less crusted over. Suddenly a vent was opened about six miles from Kilauea on the lower slope; the next day another still lower down, and afterwards several more, one below the other upon a line of fissure. From all, torrents of lava flowed with great rapidity a distance of thirty miles into the sea, where it formed two or three islands, killing immense numbers of fish. By this 'tapping' process the original pit-crater was re-formed, the lava-surface having sunk about 1500 feet; but by the year 1844 the whole of this vast chasm had been re-filled to the brim by the renewed up-welling of the lava. In 1849, when visited by Dana, the surface of the inner pit had sunk again about 350 feet below the 'black ledge,' which itself was 650 feet lower than the rim of the outer cliff-range. Since that date the crater has been once more nearly filled up, and again emptied by the last great eruption of 1855.

The floor of the pit is described as usually presenting a crust over a vast pool of lava, which is from time to time broken through by a fresh up-boiling of the incandescent mass beneath. It cools and hardens so rapidly on exposure that it may be walked upon within a few hours after its coagulating.

Sometimes upwards of fifty small cones and craters, more or less in activity, *i. e.* throwing out scoriæ and lava, have been counted on the floor of this great pit. They were from 50 to 100 feet high. (Rev. S. Stewart.) The lava has also been seen to flow from openings in the perpendicular walls of the crater several hundred feet above the bottom, where other lava was at the same time welling up quietly. (See p. 262.) The lavas that issue within the crater are scoriaceous and cellular on the surface; those that escape from vents low down the side of the mountain are more compact and vitreous. The first have been compared to the frothy surface of a liquid in fermentation, the latter to the same liquid issuing clear from a tap below.

Dana describes two kinds of lava in Hawaii; one smooth and solid, taking the surface-forms of concentric cable-like folds, or rounded hillocks and domes, of which the top has often fallen into an oven-shaped cavity beneath (*hornitos*), or of long subterranean glazed passages—all marks of a considerable viscosity and nearly vitreous fusion. (See p. 73, *supra*.) The other kind of lava consists on its surface of scoriform masses, piled in terrific confusion; they are called clinker-fields, and look like "a mountain shattered into a chaos of ruins." The blocks vary in bulk from 1000 to 10,000 cubic feet, and are of all shapes—cubical, or slab-like, with jagged angles or ridges, bristling up in horrible roughness to heights of 20 or 30 feet. The two kinds are often associated in the same eruptive region. In mineral character the lavas of Hawaii are dark-coloured, and generally augitic and ferruginous, containing much olivine and specular iron. Some may properly be classed as basalt; but the greater number are considered by Dana as greystone. In some parts they are highly vitreous and obsidian-like, especially those which, when tossed up in a liquid state, produce the fine filamentous scoriæ called 'Pele's hair.' True felspathic pumice does not appear to occur in the island. Augite is probably the principal ingredient of all its lavas.

The other islands of the Sandwich group are likewise exclusively of volcanic origin. They lie, according to Dana, on two parallel lines, one of which intersects each of the lofty twin mountains of Hawaii (the largest island), Mauna Loa and Mauna Kea. The island of *Mani* has one summit 10,217 feet high; *Eeha*, 6130; *Kaui*, 8000; *Oahu*, two ranges 4000 feet high. Mani exhibits two large craters and vast fields of recent-looking refrigerated lava. One of these craters is 2000 feet deep, of an irregular forked figure, *nine* miles in its greatest diameter, with a floor of lava, on which rise some sixteen or more distinct cinder-cones. The flanks of the mountain are cleft by gorges (barancos) 2000 feet deep and one or two miles wide, opening to the sea. Eruptions took place about two centuries back from a string of lateral openings marked by as many parasitic cones. Oahu shows many beds of greystone lavas containing felspar crystals and olivine, alternating with massive layers of conglomerate and tuff, of which great part of this island and several of its cones are composed : these have usually craters, and are breached on one side : the hollows of some are occupied by lakes. Coral-reefs, elevated 20 to 30 feet above the sea-level, almost surround this island, whose tuff-craters were probably formed below the water-level on a shallow shore, like those of Auckland in New Zealand, &c.

There are " areas of the Pacific, as well as of the Indian Ocean, occupying many hundred thousand square miles, where all the islands consist exclusively of coral, generally in the form of Atolls or circular reefs, none of them rising to a greater height than may be accounted for by the action of the winds and waves on broken and triturated coral." (Lyell, Manual, ed. 1855, p. 789.) These extensive areas are, according to the ingenious theory of Mr. Darwin, which is supported by Sir Charles Lyell, with great plausibility supposed to have been for ages continually, but slowly, subsiding to a lower level. Such subsiding districts are found generally at some distance on one side of the great trains of volcanic islands we have been considering; and their depression may be looked upon as corresponding to, and coeval with, the progressive elevation of the continental or great insular areas on the opposite side. (See p. 282, and p. 308, Conclusion 8.)

It is probable likewise that there may be another con-

necting cause between the development of volcanic action and
the formation of coralline reefs on a large scale, in the abundance of calcareous matter which, as has been seen throughout this work, is habitually discharged by springs issuing
from volcanic districts. The travertins and marls of subaërial volcanic areas are represented, it would seem, in tropical
climates suited to the growth of the zoophytes, by the coralline atolls and fringing reefs of seas through which subaqueous
volcanos are plentifully distributed. In such seas volcanic
rocks will offer very frequently—but, of course, not exclusively—the foundation needed for the work.

In closing this Catalogue of the known Volcanos and volcanic formations of the earth's surface, I would call the attention of those of my readers who have followed me through it, to
the remarkable evidence it affords of the general uniformity
and simplicity of the phenomena of which they have been the
theatre—thus confirming the view expressed in the first
chapter of this volume (p. 4), in opposition to that held by
Humboldt as to their "isolated, variable, and obscure" character.

In every quarter of the globe, and under every degree of latitude, we find the eruptions that have taken place characterized
by the same repeated splitting of the earth's crust in fissures,
generally parallel, but occasionally transverse, accompanied
by earthquake-shocks and other indications of the swelling
and heaving of some subterranean effervescent matter;—
the same explosive outbursts of steam and acid vapours,
throwing up liquid drops and cellular fragments of the wholly
or partially fused mineral substance which we call lava,
accompanied by its expulsion in jets or streams, which either
flow and spread over wide areas—often to considerable distances—or accumulate in bulky masses about the eruptive
vent, according as its greater or less liquidity and specific
gravity may determine. An examination of this mineral matter, when consolidated into a rock, discloses everywhere the
same basalts, greystones, or trachytes, composed for the most
part of the same minerals, though in varying proportions,—

these varieties sometimes occupying distinct localities, but far more generally alternating with one another,—and showing everywhere the same varieties of texture, from glassy obsidian (the result of complete fusion) to the coarsest crystalline and granitoidal rock.   Moreover we find the same composition and structure on the large scale everywhere in volcanic formations, from the smallest cinder-cone with a single stream of lava, to the greatest and loftiest mountain-masses, such as Teneriffe, Etna, Cotopaxi, Ararat, or those of Kamtschatka and Polynesia, each the accumulated result of a long series of successive eruptions; the same general quaquaversal dip of their component beds of lava and conglomerate from the central heights—their angles of inclination just keeping within the limits of those which would be naturally assumed by a talus composed of materials partly loose, partly coagulating as they flow from a higher to a lower level, partly swept downwards by aqueous torrents, sometimes distributed still more widely by marine currents—circumstances to which we are led, by the observed phenomena of active volcanos, to ascribe their production.   Again, we see everywhere the same circular or elliptical hollows, drilled here and there through the axes of these mountain masses, evidently by the force of exploding volumes of steam, but varying greatly in dimensions, and frequently formed, concentrically, one within or by the side of the other, along the line of some fissure.   Finally, we remark the general parallelism over the entire surface of the globe, of the chief trains of volcanic vents, active or extinct, with the outlines of the neighbouring elevated supra-marine or mountainous areas, leading to the supposition that the fissures through which eruptions find their way outwardly are owing to the lateral drag occasioned by the upheaval of some contiguous superficial portion of the earth's crust, overlying a stratum of intensely-heated and highly elastic matter, the tension of which, through increase of temperature, has more or less overcome the resistances opposed to its expansion.   In short, the diagnosis of a volcano, or of a volcanic rock, or volcanic district in one part of the globe is as frequently identical with another at its antipodes, as if they had been produced side by

side. It is the same, as is well known, with the plutonic granites, syenites, gneiss, schists, and the traps or early volcanic
rocks, whose mineral composition, structure, and relations to
the sedimentary strata among which they have been introduced—in a word, their general characters—are the same at
all points of the earth's surface. The sedimentary strata are
far less uniform in mineral nature and disposition, owing to the
greater influence on them of variable conditions of climate and
meteoric agencies, as well as of metamorphism, than are the
plutonic and volcanic rocks. Nor is there, I maintain, any
greater obscurity in the laws by the action of which one of
these classes of formations is produced, than in those of the
other*, if they are studied impartially, and without prepossession, by the light of observation and inductive reasoning.

* A habit has grown up recently among geologists of calling all Elevatory action 'volcanic.' The word would convey a more definite idea
if it were confined (as I have confined it in this work) to true eruptive
action, and that of 'Plutonic' force applied to those upheavals and injections of subterranean heated matter among dislocated rocks, which
there is reason to suppose unaccompanied by outward explosions of
vapour or eruptions of lava. It is true that the boundary-line of the two
kinds of action cannot be clearly defined, as in the case of *dykes*, which
attest both Plutonic upheaval and the eruptive effort—more or less imperfectly accomplished—of volcanic matter. But this is only a defect
common to nearly all the terms employed in geology. Limestones graduate into sandstones, granite into gneiss, trachyte into basalt, conglomerates into lithoidal rocks igneous into aqueous formations, and so on.
Volcanic action I would define as external or superficial; Plutonic, as
internal or subterranean. But, of course, there will be everywhere an intermediate debateable district, the phenomena of which may have as good
a right to the one term as to the other.

INDEX.

ABICH, M., on a dyke in the Val del Bove, 76 ; on Ararat, 398.
Active volcanos, number of, 12 ; activity of, irregular, 16.
Ægean Archipelago, volcanos of, 391.
Africa, North, volcanic formations of, 315.
Aleutian Isles, volcanos of, 455.
Alluvial deposits of volcanic matters, 171–178.
Alteration of sedimentary strata by contact with lavas, 88, 141.
—— of lava-rocks by steam and acid vapours, 142.
—— of lavas before emission, 123.
Alternate eruption of lavas of different mineral characters, 126.
Altitude of volcanic mountains, 169.
Alumine. *See* Solfataras.
America, South, volcanic range of, 434–444 ; plutonic ranges of, 281, 303.
——, North, volcanic range of, 444–454 ; plutonic ranges of, 281, 303.
——, Central, volcanos of, 444.
Amygdaloidal lavas, 144, 245.
Andes, volcanic range of, and sedimentary strata, 176, 248, 277.
——, trachytic domes of, their probable origin, 132, 135.
Angle of inclination of lava-beds in volcanic cones on mountains, 60, 62, 157, 167.
Anticlinal ridges and synclinal troughs, 288, 291.
Apennines, elevated transverse ranges of, 280, 281.
Aqueous lavas, description of, 172.
Ararat, 398.
Ascension Island, laminated trachytes of, 425.
Ashes, volcanic, ejected in vast abundance by eruptions, 22, 57, 185, 195, 204, 207.
Asia Minor, volcanos of, 393.
——, Central, volcanos of, 405.
Astroni, origin of its crater and lava, 134, 322.
Atlantic Ocean, volcanic islands of, 279, 405–430.
Augitic rocks, 111, 112. *See* Basalt.
Auvergne, 58, 361.
Axes of elevated mountain-ranges, 282, 286 ; of dislocation, 48.
Axial upthrust of plutonic rocks, 285, 291.
Azores, 415.

Babbage, on the cause of changes in superficial levels, 271, 274.
Barancos, origin of, 164, 211.
Barren Island, 199.
Basalt, mineral characters of, 111 ; columnar structure of, 94–104.
Beaumont, M. E. de, his Upheaval theory, 168.
Beudant, M., on Hungarian trachytes, 387 ; on the exclusively early production of trachytes, 128.
Bischoff, Prof., on the formation of crystals in lavas, 120, 121, 122, 125, 152.
Blisters in lavas, 80.

Bohemia, volcanic rocks of, 387.
Bombs, volcanic, 56, 116, 410.
Bory de St. Vincent, on Bourbon, 133, 428; on volcanic mountains, 211.
Bourbon, volcano of, 34, 196, 229, 428.
Breaching of cinder-cones, 63, 65.
Breccias, volcanic. *See* Conglomerates.
Brecciated lavas, 140.
British Islands, volcanic outbursts in, 279, 413.
Bromo, large crater of, in Java, 199, 465.
Bubbles of steam explode from exposed lava, 36, 53, 55; vast size of, 203.
Bubble-shaped cavities in lava, 79, 80.

Calcareous sandstone of recent origin on surface of volcanic islands, 254.
—— springs and their deposits, 148.
Calcareo-volcanic strata, 88, 177, 242.
California, volcanos of, 453.
Canary Islands, 418.
Cantal. *See* France, Central.
Cape de Verde Isles, 423.
Caraccas, earthquake of, 432.
Carbonic acid gas exhaled by volcanos, 151.
Caribbean Isles, volcanic range of, 257, 277, 430.
Catalonia, extinct volcanos of, 359.
Catania. *See* Etna.
Caucasus, 399.
Caves in lava, 80.
Cellular lavas, 138, 245.
Chemical theory of volcanos, 308.
Chili, volcanos of, 276 (*note*), 436.
Cimini, Monti, 352.
Cinder-cones. *See* Cones.
Cirques, or encircling cliffs, their origin, 186, 195, 200, 466.
Clinkstone affects pyramidal forms, 135, 136; its lamination, 108.
Columnar structure, how formed, 94–103.
Conclusions, general, on Telluric phenomena, 305.
Cone formed within cone, 186.
Cones, parasitic, 161, 169.
Cones of scoriæ, &c., how formed, 57; structure of, 61.
Conglomerates, volcanic, how formed, &c., 174. *See* Tuffs.
Consolidation of lavas, 93.
Cooling of lavas, 152.
Coral islands based on lava-rocks, 256, 479.
Cordilleras. *See* Andes.
Corrugations of strata, 288–292.
Coseguina, great eruption of, 170, 204.
Crater-lakes, 214–223.
Craters of simple cones, how formed, 58.
—— of volcanic mountains, how formed, 17, 184–195, 202–208.
——, concentric, 186–192.
——, cause of their circular form, 54, 208.
——, their occasionally vast area, 202–208, 345, 353, 466.
—— of Moon, 250.
Crushing action accompanying upheaval, 289.
Crust of the globe, probable formation of, 265; matter underlying the, 267, 270, 305.
Crystalline texture of lavas, 114.

Crystallization of lavas, 120.
Crystals in lavas mostly formed before their eruption, 116 *et seq.*
—— in tuffs, their origin, 178.
Currents of lava, their rate of movement, 68, 84, 131; dimensions, 82, 132.
Curvature of dislocations and axial elevations, 280.
Cutch, its changes of level, 404.

Dana, Prof., on Sandwich Isles, 86; on " volcanos no safety-valves," 262.
Darwin, Mr. C., on the ribboned trachytes of Ascension, 139; on the specific gravity of lavas, 110, 125, 128; on South American volcanos, 248, 297; on volcanic islands, 209, 260; on pyramids of clinkstone, 137; on gneiss, 300; on craters, 206; on coral islands, 256, 257.
Daubeny, Dr., his List of Volcanos, 311, 316, 387; on columnar structure, 102; on craters of elevation, 201; on submarine eruptions, 239.
Daubrée, M., 38, 124, 145, 218.
Davy, Sir H., his theory of volcanic action given up by him, 308.
Degradation of volcanic islands, 209.
Delesse, M., 118, 126, 130, 300.
Denudation of volcanic mountains, 212.
——, craters enlarged by, 208.
Deville, M., 14, 126, 261.
Disposition of erupted matters fragmentary, 55–64, 241; of lavas, 65–88, 244.
Disruption of overlying rocks by subterranean intumescent matter, 49, 259, 263–271.
Distension, inward, of cones, 167; of volcanic mountains, 167, 343.
Divisionary structure of lavas, 94, 95, 109; of tuffs, 178.
Dolomieu, M., on the cause of the liquefaction of lavas, 117.
Dolomitization of limestone by contact with lavas, 89, 355, 357.
Dôme, Puy de, its origin, 132.
Domes of trachyte, how formed, 132, 135.
Dufrénoy, M., his theory on the origin of Monte Nuovo, 324.
Dykes, how formed, &c., 52, 90, 164; their prismatic structure, 98–100; rarely disturb the strata they pierce, 166.

Earth, cause of dislocation of its crust, 302. *See* Crust of the globe.
Earthquakes, theory of, 163, 288, 294; generally precede an eruption, 6, 163; coincidence of, with eruptions, 7–9, 275, 276.
Eifel, volcanos of, 59, 214, 369–384.
Ejections of fragmentary matter, 21, 23, 55–64.
Electricity, frictional, evolved during eruptions, 57.
Elevation-crater theory, 4, 158, 168, 344.
Elevations of land, how produced, 42, 272–275, 283 *et seq.*
—— —— —— alternate with subsidence, 274–276.
—— in mass of volcanic rocks, 226, 249–257.
Eluvial torrents, effect of, on volcanic matters, 171, 373.
Engulfment not the origin of craters, 206, 465.
Eruption, limitation of the term, 282 (*note*).
Eruptions, paroxysmal, described, 19; ordinary, 159; eluvial, 173; generally take place from linear fissures, 159, 259, 276.
Etna, 17, 155, 159, 169, 399.
Expansion of subterranean matter by heat, 30, 41, 54, 270–275.
Explosions, aëriform (of steam), usually characterize an eruption, 21, 30, 35, 40.
Explosive origin of all craters, 54, 185–225.
Faults, how formed, 50, 287, 293.

Felspar the chief element in lavas of every age, 110; its varieties, 113.
Ferroe Islands, 253, 412.
Fissures in rocks overlying intumescent subterranean matter, theory of their formation, 46–50, 259, 272, 273; up-filled by lava form dykes, 53, 164, 259; rent through a volcanic mountain, 158, 160, 162, 221, 228; formed in lavas by shrinkage, 70; transverse, 50, 280, 448, 456.
Flames, glow of red-hot scoriæ mistaken for, 18, 33.
Floods caused by bursting of lakes or melting of snows on volcanic mountains, 171–4, 409.
Fluidity of lavas varies greatly, 68, 131; cause of variation in, 130–137. See Liquidity.
Foci, or subterranean reservoirs of volcanic matter, their probable characters, 46, 266, 272.
Foliation of gneiss not stratification, 298.
Forbes, Mr., on the trachytes of South America, 119, 140.
Fragmentary ejecta of volcanos, 55 et seq., 170, 241.
France, South and Central, volcanos of, 360–368.
Fusion of lavas generally imperfect, 115 et seq.

Galapagos Isles, 442.
Gases, permanent, emanating from volcanos, 151.
Geikie, Mr., on the gneiss of Skye, 286.
Gemmellaro, Signor, 160.
Germany, volcanic formations of, 369, 385.
Geysers of Iceland, theory of, 149, 411.
Glassy lavas, 115, 338.
Globular structure of lava, how formed, 104, 106.
Gneiss, cause of its lamination, 283, 299.
Graham Isle, 61, 345.
Granite, its probable condition at great depths, 282.
Greystone, definition of, 110.
Grotta del Cane, 325.

Hamilton, Mr. W. G., on the volcanos of Asia Minor, 137, 393–395.
Hamilton, Sir W., on Vesuvius, 186; on the flowing of lava, 69.
Hartung, M., on Lancerote, 422.
Heaphy, Mr., on New Zealand volcanos, 215, 226, 247, 273.
Heat, subterranean, its movements the cause of volcanic and plutonic action, 41, 263, 271, 298.
——, ——, origin of, still doubtful, 305, 308.
Hecla, 407.
Helena, St., 226.
Herschel, Sir J. F. W., on changes in superficial levels, 274; on lunar volcanos, 231; on parallelism of volcanic trains with neighbouring coast-lines, 276; on crumpling of strata, 294.
Hindostan, vast lava-field of, 404.
Hopkins, Mr., on the condition of the nucleus of the globe, 264; his theory of elevation and earthquakes, 46, 50, 290; his experiments on pressure and high temperature, 264.
Hornitos of Jorullo, 81, 450.
Humboldt, M., 4, 127, 135, 216, 276; his theory of Jorullo disputed, 81, 449; on the formation of craters, 206; on submarine eruptions, 239.
Hungary, volcanos of, 387.

Ice interbedded with lava, 172, 434.
Iceland, volcanos of, 171, 406.

Independence of proximate vents, 261.
India, volcanos of, 404.
Iron, titaniferous, in lavas, 110, 112; specular and magnetic, 143, 144.
Irregular direction of some axial elevations, 281, 287.
Ischia, 260, 326.
Islands, volcanic, often in part upheaved, 251 *et seq.*
Italy, Central and Northern, volcanos of, 350–358.
——, Southern, volcanos of, 315–332.

Java, volcanos of, &c., 174, 464.
Joints in lava-rocks, how formed, 97.
Jorullo, Humboldt's mistaken view of its upheaval, 81.
Jukes, Prof., on the volcanos of Java, 199, 465: on subterranean changes in lavas, 125.
Junghuhn, M., on Java, 191, 199, 229.

Kamtschatka, volcanos in, 456.
Kilauea, its crater, 35, 219, 476.
Kurile Isles, volcanos of, 458.

Lakes in craters, 214-225.
——, bursting of, the origin of, trass, &c., 172 *et seq.*
Lamination of trachytes, &c., 107, 119, 139.
Lancerote, 163, 421.
Lapillo, how formed, 57, 207.
Lateral or parasitic cones on a volcanic mountain, 58, 155, 161, 169.
—— fissures formed on either side of an upheaved axis, 50.
Lava, its outflow, 23, 68; its disposition, 65–93.
—— in ebullition during an eruption, 30–36.
—— contains water or steam, 37, 124.
——, its consolidation and internal configuration, 93–109.
——, its crystalline texture, when formed, 118–122.
——, its mineral composition, 110, 125.
——, its outflow beneath water, 92.
——, slow cooling of its interior, 84.
——, subaqueous, 244.
——, vast volume of, emitted sometimes, 82, 132, 135.
Leucitic lava, 112, 118; large crystals of, 123.
Lightnings, volcanic, to what owing, 57.
Limestone, dolomitization of, 141.
Lindsay, Dr. L., on Icelandic eruptions, 213.
Linear arrangement of eruptive vents, 161, 259, 277.
Lipari Isles, their volcanos, &c., 332.
Liquidity of lavas variable, 45, 131; to what owing, 116 *et seq.*
Loa, Mount, in Sandwich Isles, its crater, 219, 220, 474.
Lunar craters, 250.
Lyell, Sir C., 166, 172, 192, 201, 206, 213, 339, 416.

Madeira, 416.
Magnesia introduced into limestones from lava, 89, 141.
Mallet, Mr. R., on earthquakes, 3, 295, 301.
Masaya, volcano of, 33.
Mediterranean, line of volcanic eruptions traversing its length, 278, 314.
Mephitic exhalations, 151.
Metamorphism, 88, 124, 140, 298.
Mexico, volcanos of, 448.

Mineral composition of lava-rocks, 110, 129.
Minerals, new, formed by or in lavas, 140–144, 245.
Mont Dore. *See* France.
Monte Bolca, 357.
Monte Nuovo, 179, 323.
Moon, craters of, 230.
Mountains, volcanic, how built up, 155–160, 180.
Mountain-chains, parallelism of, with volcanic fissures, 276.
Moya, how formed, 173, 440.
Mud-eruptions. *See* Moya.
Mud-volcanos, 401, 444.
Murchison, Sir R. I., on the volcanic districts of Italy, 351; on the Ural range, 403.

Naples, volcanic district of, 315.
Naumann, Prof., on the lamination of gneiss, 301 (*note*).
New Zealand volcanos, 470.
Nucleus of globe, question whether fluid or solid, 264.

Obsidian, 113.
Ocean, its saline ingredients probably derived from volcanic submarine emanations, 241.
Olivine, large nodules of, in some lavas, 119.
Oscillations of superficial levels, 304.   *See* Elevation and Subsidence.

Pacific Ocean, chain of volcanos encircling it, 277.
Palma, its crater and baranco, how formed, 201, 212, 251.
Papandayang, truncation of its cone by eruption, 465.
Parallelism of coast-lines to volcanic trains, 276 *et seq.*
—— of volcanic fissures, 49, 260.
Parasitic cones, 58, 161, 169.
Paroxysmal eruptions, description of, 19–25.
Pearlstone, 107.
Peperino, 176, 350–352.
Perrey, M., on earthquakes, 301.
Peru, volcanos of, 439.
Phases, varying, of activity in a volcano, 16.
Phillips, Prof., on the cause of changes of surface-levels, 275.
Phlegræan fields, volcanos of, 179, 247, 249, 315.
Pichincha, 199, 442.
Pitchstone, 113.
Pit-craters, how formed, 216, 223.
Plutonic, its relation to volcanic action, 2, 271, 275.
Ponza Isles, their laminated trachytes, 139, 328.
Prismatic divisionary structure of lava, 94 *et seq.*
Progression of plutonic action, 296, 304.
Puzzolana, how produced, 57.

Quartz in lavas, 110.
Quito, volcanos of, 173, 261.

Radial fissures cleft through a volcanic mountain, 158, 159.
Ramsay, Prof., on the volcanic rocks of Wales, &c., 178, 248.
Rapilli. *See* Lapillo.
Red-Sea volcanos, 396.
Rhine, volcanic district of, 369, 385.

Rifts in lava-currents, how formed, 77.
Rocca Monfina, 134, 380.
Rocks, volcanic. *See* Lava, Tuffs, &c.
——, upheaval and dislocation of, how brought about, 46 *et seq.*, 271.
Rogers, Prof., on gneiss of North America, 284 ; on foldings of strata, 288, 290.
Rome, neighbourhood of, 350.

St. Helena, 426.
St. Jago, 424.
Sandwich Isles, volcanos of, 474.
Santorini, 200, 237, 391.
Sarcouy, Puy de, its origin, 133.
Sardinia, 346.
Scheerer, M., on granite, 282.
Scoriæ, how formed, 23, 35.
Serapis, Temple of, 227, 325.
Serpentine, 112.
Shifting of volcanic vents, 227, 258.
Sicily, its volcanic formations, 338.
Siebengebirge, 384.
Silex held in solution by hot water or steam, 117.
Siliceous trachytes, 111.
Solfataras, 27, 146, 230, 320, 412.
Somma, its eruptive origin, 319.
Spain, volcanos of, 358.
Specific gravity of lavas, influence of, on their fluidity, 131–138.
Spheroidal concretionary structure of lavas, 106.
Springs, thermal and mineral, 147.
Stromboli, 31, 333.
Subaqueous volcanos, 235–257.
Sublimations, metallic, 143.
Subsidence, 225, 273–276.
Subterrestrial changes, 267.
Sugars, analogy of lavas to, 121.
Sulphur in lavas, 143 *et seq.*, 412. *See* Solfataras.
Sumatra, volcanos of, 467.
Sumbawa, volcanos of, 463.
Symonds, Mr., on the elevation of the Malverns, 281.
Sympathy of neighbouring vents, 260.
Syria, volcanic districts of, 395.
Systems of volcanos, 258.

Tabular structure of lavas, how formed, 107.
Tapping of the lava within a volcano by lateral vents, 219, 222.
Telluric phenomena, general conclusions on, 305.
Temperature, subterranean, varying upward flow of, 270 *et seq.*
Temple of Serapis, oscillations of level shown in, 227, 325.
Teneriffe, 197, 229, 251, 418.
Texture of lavas, 113 *et seq.*
Thermal springs, 147 ; act as safety-valves, 148.
Thringvalla, its lava-stream, 77.
Tidal effect on subterranean lava, 301.
Tierra del Fuego, 434.
Tomboro, great eruption of, 171, 463.
Trachyte, definition of, 110.

Trap-dykes, 53, 90.
Trap-rocks, or early lavas, their mineral characters, &c., 112, 129, 253, 245; often of subaqueous origin, 245–247.
Trass of the Rhine, its origin, 174, 371.
Travertin, 354.
Truncation of cones, 191.
Tuff-cones, their origin, 179.
Tuffs, how formed, 171 *et seq.*, 241, 320, 352; varieties of, 177, 243, 352.

Uniformity of laws of volcanic action, 4, 480.
Upheaval theory as to volcanic cones and craters. *See* Elevation-crater theory.
—— of volcanic islands, 249, 252.
Ural, chain of, 403.

Val del Bove, origin of, 211, 228, 339.
Vapour. *See* Water.
——, acid, 26, 141.
Ventotiene, island of, 209.
Vents, shifting of, 192.
Vesicular structure of lavas, 138.
Vesuvius, its history, &c., 17, 196, 315.
Vicentine, rocks of, 356.
Viscous lavas, 72.
Volcanic action related to, yet distinguished from, plutonic, 2, 270, 482.
—— islands, how formed, 235; occasional elevation of, in mass, 249; partly elevated, partly erupted, 250–257.
—— mountains, how formed, 155; their inward distension, 167; how truncated, 191; changes of form in, 191, 193, 227 (*see* Cones); gutted or blown up by paroxysmal eruptions, 264–268; denuded by aqueous erosion, 209, 211–213.
Volcano, island of, 193, 337.
Von Buch, 119, 139.
Vultur, Mount, 329.

Waltershausen, von, his views and map of Etna, 343.
Water in lavas, 37–39, 130.
——, vapour of, the motive force in volcanic action, 39–44.
——, heated, takes both quartz and felspar into solution, 117, 126.
——, ——, deposits other minerals, 145, 148.
Westerwald, the, 385.

THE END.

PRINTED BY TAYLOR AND FRANCIS,
RED LION COURT, FLEET STREET.

Printed in the United States
By Bookmasters